Practical Stress Analysis with Finite Elements

Bryan J. Mac Donald
BEng MSc PhD CEng MIEI

GLASNEVIN
PUBLISHING

www.jion.ie/glasnevin

First Published in 2007 by
Glasnevin Publishing
(A division of Jion Consulting Ltd.)
16 Griffith Parade
Glasnevin
Dublin 11
Ireland
www.jion.ie/glasnevin

The author and publisher of this book have used their best efforts in preparing this book. These efforts include the development, research and testing of the theories and computational models given in the book. The author and publisher make no warranty of any kind, expressed or implied, with regard to any text and the models contained in this book. The author and publisher shall not be liable in any event for incidental or consequential damages in connection with, or arising out of, the furnishing, performance, or use of this text and these models.

British Library Cataloguing-in-Publication Data
A Catalogue record for this book can be obtained from the British Library

ISBN: 978-0-9555781-0-6

1. Finite Element Method. 2. Stress Analysis
620.001
620'001'51535

Printed in Ireland by Gemini International Ltd.

To Helen and Palomi

Contents

Preface

While there are many excellent texts available that cover the theory of finite element analysis (FEA) and the application of that theory to programming the finite element method (FEM), texts that deal with the practical issues of performing a finite element analysis using a commercial FEA package are rare. It has been my experience that even texts that claim to be a practical guide to FEA are typically overloaded with the theory of FEA and contain little information that would be of practical help to the reader who wants to learn how to <u>do</u> a FEA rather than an in depth theoretical discussion of how the method actually works. I feel that it is pointless repeating the theory of FEA in a text that claims to be a practical guide when there are so many widely available texts that cover the matter in great detail. While I agree that it is highly important that anyone who uses FEA seriously must have a good understanding of its underlying theory it is important to also realise that there is much demand for a text that shows the reader how to perform a FEA simply and clearly.

From my experience of teaching undergraduate and postgraduate formal courses on FEA it has become very apparent that the texts available on FEA and the help systems in most FEA commercial solvers do not adequately serve the needs of the novice FEA user. Practical FEA, like most CAD disciplines, is largely competency based. It is not something that can be satisfactorily learned from a theoretical description of the method and it requires a great deal of user experience to have confidence in the results obtained. For these reasons FEA is preferably taught in a scenario where mentors are present to guide the novice. In the case of university environment these mentors are commonly the lecturer and his/her assistants and in the industrial scenario the mentor will generally be a team leader or manager. In many cases, however, a mentor may not be available and the novice is required to teach themselves. This is where I see this book being of great benefit. It is my hope that this book can act as a mentor to the reader, to guide them through those first hazardous analyses by pointing out common mistakes and quickly bringing the reader to a satisfactory level of competence.

It is with this in mind that I decided to write this book. The intended audience is anyone who needs to learn how to use FEA and how to avoid making the common mistakes that newcomers to the method generally make. The text is written in a manner such that it is not specific to any particular FEA software package. Examples are shown using several different commonly available packages and the description of the analysis in each case is kept generic. The intended audience of this text is the undergraduate or postgraduate student who is taking a first course in FEA, however it is my earnest hope that it may also be valuable to researchers and practising engineers who are new to FEA and/or require guidance in its application.

Bryan J. Mac Donald
Dublin, Ireland
June 2007

About the Author

Dr. Bryan Mac Donald B.Eng. M.Sc. Ph.D. C.Eng. MIEI
Chartered Engineer

Dr. Mac Donald is senior lecturer in computer aided design (CAD) at the school of mechanical and manufacturing engineering at Dublin City University, Ireland. He is also the course manager of the B.Eng. in Computer Aided Mechanical and Manufacturing Engineering which is run by the school. He has been responsible for teaching finite element analysis to undergraduate and postgraduate students in mechanical and bioengineering courses for over ten years. In this time he has developed a unique approach to teaching the subject which is competency based rather than purely theoretical.

His research interests include the use of FEA to analyse and design medical devices, optimisation of complex metal forming processes using FE techniques, FEA of impact to aircraft structures and integration of artificial intelligence techniques with FEA. To date, he has published over 80 journal articles and conference papers in these fields. He has acted as an external reviewer for many important journals in these fields such as: Finite Elements in Analysis and Design, The Journal of Biomechanics, Biomechanics in Modelling and Mechanbiology, Clinical Biomechanics, Acta Biomaterialia, The Journal of Materials Processing Technology, The International Journal of Machine Tools and Manufacture and the Proceedings of the IMechE Journal of Engineering Manufacture.

Acknowledgements

I am grateful to my graduate students Damien Comiskey, Emmet Galvin and David Hunt who reviewed parts of the manuscript and gave some constructive suggestions which helped to improve the presentation of some of the topics in this book. I would also like to thank the many undergraduate and postgraduate students I have taught in Dublin City University over the past ten years, whom this material was tested out on, in various forms. Thanks for your patience and the constructive feedback. I am very grateful to my beautiful wife Helen and my daughter Palomi for all their love, care, support and for putting up with long nights of typing. Special thanks to Helen for completing the onerous task of proof reading the text.

Nomenclature

Symbol	Definition	Units
A	Cross sectional area	m^2
A_0	Original cross sectional area	m^2
A_{avg}	Average cross sectional area	m^2
b	Beam width	m
$[B]$	Strain-displacement matrix	-
C	Damping constant	-
$[C]$	Global Damping Matrix	
$[D]$	Material behaviour matrix	-
E	Young's modulus	Pa
E_{Tan}	Tangent modulus	Pa
f	Force	N
f^B	Body force	N
f^S	Surface load	N/m^2
$\{f^B\}$	Nodal body force vector	N
$\{f^S\}$	Nodal surface load vector	N
F^i	Nodal point load on node i	N
$\{F\}$	Global force vector	N
$\{F\}^B$	Global body force vector	N
$\{F\}^S$	Global surface load vector	N
$\{F\}^C$	Global point load vector	N
$[F]$	Flexibility Matrix (for pipe elements)	-
$\{g\}$	Nodal acceleration vector	m/s^2
G	Modulus of rigidity	Pa
h	Beam height	m
I	Second moment of area	m^4
$[J]$	Jacobian matrix	-
k	Stiffness (of a spring)	Pa
k_{eqv}	Equivalent stiffness (of a spring element)	Pa
K_N	Normal stiffness (of contact elements)	Pa
$[K]$	Global stiffness matrix	-
l	Length	m
L	Length	m
$[L]$	Transformation matrix	
M	Bending moment	Nm
$[M]$	Global mass matrix	-
P	Applied load	N
P_N	Normal component of applied load	N
P_S	Shear component of applied load	N
r	Radius	m
R	Radius of curvature	m
$\{R\}$	Reaction force vector	N
S	Surface	m^2
$[S]$	Shape function matrix	-
t	Thickness	m
u	Displacement in the x-direction	m
$\{U\}$	Nodal displacement vector	m
U_i	Displacement of node i	m
$\{\dot{u}\}$	Nodal velocity vector	m/s
$\{\ddot{u}\}$	Nodal acceleration vector	m/s^2
v	Displacement in the y-direction	m
V	Volume	m^3
w	Displacement in the z-direction	m
W	Applied load or Width	N or m

x	Deflection (of a spring)	m

Greek

δ	Deflection	m
ε	Strain	-
ε_1	Maximum principal strain	-
ε_2	Minimum(2D)/Median(3D) principal strain	-
ε_3	Minimum principal strain (3D)	-
ε_f	Failure strain	-
γ	Shear strain	-
γ_{max}	Maximum shear strain	-
η	Vertical natural coordinate	-
Λ	Strain energy	J
μ_s	Coefficient of static friction	-
μ_d	Coefficient of dynamic friction	-
ν	Poisson's ratio	-
θ	Angle	°
ρ	Density	Kg/m^3
σ	Stress	Pa
σ_1	Maximum principal stress	Pa
σ_2	Minimum(2D)/Median(3D) principal stress	Pa
σ_3	Minimum principal stress (3D)	Pa
σ_{pl}	Stress corresponding to the proportional limit	Pa
σ_N	Normal stress	Pa
σ_R	Resultant stress	Pa
σ_{uts}	Ultimate tensile stress	Pa
σ_y	Uniaxial yield stress	Pa
τ	Shear stress	Pa
τ_{max}	Maximum shear stress	Pa
υ	Beam element deflection	m
ξ	Horizontal natural coordinate	-
ζ	Natural coordinate in thickness direction	-

Acronyms

BC	Boundary conditions	-
BQ	Bilinear Quadrilateral (4 node 2D planar element) or Q4	-
CFRP	Carbon fibre reinforced plastic	-
CSR	Cohesion Sliding Resistance	-
CST	Constant Strain Triangle (3 node 2D triangle element)	-
DOF	Degree of Freedom	-
FE	Finite element	-
FEA	Finite element analysis	-
FEM	Finite element method	-
FOS	Factor of safety	-
GFRP	Glass fibre reinforced plastic	-
MPC	Multipoint Constraints	-
LST	Linear Strain Triangles (6 node 2D triangle element)	-
Q4	Four Node Quadrilateral (2D planar element) or BQ	-
Q8	Eight Node Quadrilateral (2D planar element)	-
UTS	Ultimate tensile stress	Pa

1

Overview

1.1.   Why do we need Finite Element Analysis?

This chapter will show you why FEA is used and what exactly it is. Rather than giving a one line definition of what the term "finite element analysis" actually is, we will look at an example which will clearly illustrate the main points of FEA. At this point I would like to encourage the reader to "bear with me" for the next few pages as clearly understanding the points made in this section will make understanding the rest of the text much easier!

Suppose we are asked to solve an engineering problem: A simple cantilever beam of rectangular cross section is used to support a load, W, at it's free end. We are asked, in particular, to find out if the beam will support this load without failing. The beam is made of steel so we can assume relevant material properties. An overview of the problem is shown in figure 1.01.

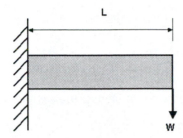

L

W

Figure 1.01: A Simple Engineering Problem

We have a simple theory available that allows us to answer this problem relatively easily:

$$\sigma = \frac{WL}{I} y \tag{1.01}$$

where I is the second moment of area of the beam cross section, and y is the distance from the beams neutral axis during bending. Using this formula we can estimate what the maximum stress in the beam will be. As the beam height is defined as h then we can conclude that the maximum stress will occur at y=h/2. The I value for this beam cross section is given by $I = bh^3/12$. If the distance L is measured from the point of load application then L will be a maximum and hence stress will be maximised at the built in end of the beam. Putting all these values into equation 1.01 gives us an equation for the maximum stress in the beam:

$$\sigma_{max} = \frac{6WL_{max}h}{bh^3} \tag{1.2}$$

This equation will now give us the maximum stress in the beam due to the applied load, however, we are asked to make a judgement as to whether the beam will fail under the applied load. In order to do this we must compare the value for maximum stress obtained to the yield and ultimate tensile stress values for steel. These values can conveniently be obtained from a material property database. So, if σ_{max} is greater than the material yield stress then we can say that permanent plastic deformation will take place, and if σ_{max} is greater than the material's ultimate tensile stress (UTS) then we can say that complete failure of the beam will occur. This is summarised in the figure below.

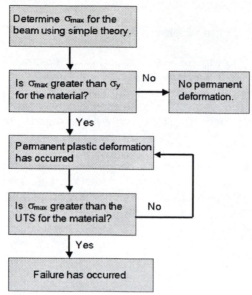

Figure 1.02: Determining if the beam will fail under the applied load

Now, suppose we are asked to solve a more complex version of the same problem, as illustrated in figure 1.03.

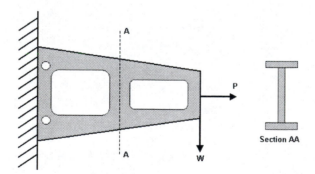

Figure 1.03: A More Complex Engineering Problem

We are now faced with a number of problems:
- There is a taper in the beam which our theory doesn't take account of
- There are a number of holes in the beam which our theory doesn't take account of
- The cross section is more complex so the I value in our theory must be changed.
- There is a more complex loading system present which our theory doesn't allow for.

In order to overcome these problems we can make some fairly large assumptions, such as ignoring the effect of the holes or ignoring the effect of the taper, and still use our simple theory as illustrated above. If we decide to take this course of action then we must understand that these assumptions will affect the results we will obtain and it is quite likely that we could underestimate the load required to cause failure. If this is the case and the beam is put into service then it could result in catastrophic failure and, depending on the application of the beam, could even result in death or serious injury to users. In order to avoid these unpleasant outcomes we must introduce a "factor of safety" (FOS) into our calculations which takes account of our assumptions. In this case the FOS will typically reduce the calculated maximum load the beam can endure before failure by a value of between 1.5 and 2. This process is summarised in the chart below.

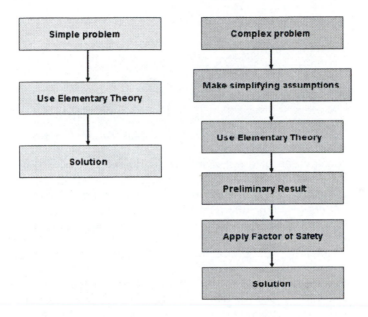

Figure 1.04: Overview of the Solution of Simple and Complex Problems

It should be obvious to the reader that this method of solution is far from ideal. The introduction of the FOS into the solution process will result in over-designing. Over-designing will mean that far more material than is needed will be used in the product thus making it far more expensive, much bigger and much heavier than it needs to be. Depending upon the application, over-designing can lead to very serious problems. If we take the example of a component for use on an aircraft, the fact that is has been over-designed means it is much heavier than it needs to be and thus will result in higher operational costs during service. This is because anything on an aircraft that takes up valuable weight is increasing fuel use and reducing the amount of passengers or cargo that can be carried. The cost of manufacturing the over-designed product will also be greater as it will require more material usage. This can become a serious problem if a very specific and expensive material (e.g. titanium, nickel based alloys etc.) is required.

So, it should be clear that if over-designing is such problem then we cannot use our elementary theories and a FOS to design complex products. Then what can we do? The most obvious thing we can do is to design using data obtained from experimental investigations. We could make a preliminary design using the FOS method and then build a prototype and test it. Based on the results from this experiment we could then change the design in order to make it more optimal and then build a new prototype of this new design and test that. This process can go on and on until we are fairly confident that we have arrived at the best possible design. This process is called a "build and break" cycle and is summarised in the flow chart below.

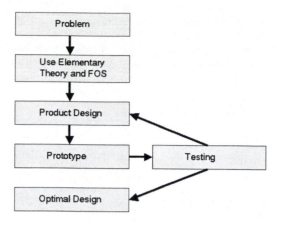

Figure 1.05: Overview of a "build and break" Design Process

This method of design can give excellent results however there are a few inherent disadvantages in its use. Firstly it is very expensive as it requires the manufacture of a product and it's testing during each loop of the "build-test-build" cycle. Not only are the costs of raw materials an issue, but personnel costs required for the manufacture and testing processes will be high. The second major disadvantage of this method is that it takes a long time to get an answer. Generally speaking the designer will not know much about the behaviour of the product until near the end of the process.

Clearly, it would be very useful if we had some method of learning more about the behaviour of the design before we committed to manufacturing a prototype and this method also avoided or reduced over-designing. The good news is that we do have such a method available to us! It is called "The Finite Element Method" and its application to the analysis of engineering problems is called "Finite Element Analysis".

1.2. What is Finite Element Analysis?

The finite element method (FEM) is a mathematical technique used to obtain approximate solutions to complex problems that cannot be solved using basic theories.

> The fundamental concept of the FEM is that it splits up a complex problem into a greater number of simpler problems and uses complex mathematics to "glue" together the answers to all the simple problems to give an approximate solution to the complex problem.

The above statement is a simple explanation of how the FEM works. A more complex and more accurate way of stating the above is that "the FEM divides the domain of interest into a finite number of simple sub-domains and uses variational concepts to construct an approximation of the solution over the collection of sub-domains". There are a number of important points in this statement: there are a finite (i.e. known) number of sub-domains, these "sub-domains" are more usually called "elements" and this leads to the name "finite element method".

Most practical problems in engineering and the applied sciences can be represented by mathematical models of the actual physical problem. These problems are generally governed by differential or integral equations and we can easily express the problems in terms of these equations. The difficulty we have is that because of complexities in the geometry, complex boundary conditions or other complexities which are found in most real-world problems we cannot solve these differential or integral equations. The FEM is a very valuable numerical procedure that is used to approximately solve these equations and thus give us an approximate solution to the problem. Let's look at an example to illustrate the above points:

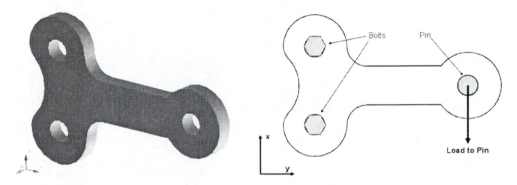

Figure 1.06: A Stress Analysis Problem

Figure 1.06 shows a bracket that is used to support a load via a pin at its free end. We are asked to find the location and magnitude of the maximum stress in the bracket. The bracket is bolted to a wall and so we can assume that it is rigidly fixed at the bolt holes. The bracket has a number of complexities such as filleted corners and holes which make analysis using an elementary beam impossible as it would be unable to predict the location of the maximum stress as required. Since we cannot use beam theory we must find another mathematical model that can be used to find the location and magnitude of the maximum stress.

If we examine the problem description in figure 1.06 we can see that the bracket is flat, it is symmetrical about its mid-plane, its thickness dimension is much smaller than the other dimensions and all the loads are on the x-y plane. This means that some form of two-dimensional analysis will suffice in this case. If we further assume that displacements, stresses and strains are uniform through the thickness and hence stresses in the z-direction are zero: this leads to a suitable mathematical model governed by differential equations namely: a linear elastic two dimensional plane stress theory.

Let's now look at the differential equations that govern this plane stress problem. By considering equilibrium we can establish the equilibrium equations for the bracket to be:

$$\left.\begin{array}{l} \dfrac{\partial \sigma_x}{\partial x} + \dfrac{\partial \tau_{xy}}{\partial y} = 0 \\[2ex] \dfrac{\partial \tau_{yx}}{\partial x} + \dfrac{\partial \sigma_y}{\partial y} = 0 \end{array}\right\} \text{ in the domain of the bracket} \qquad (1.03)$$

We also know that shear stresses will be zero on all surfaces except at the point of load application and at imposed zero displacements (i.e. at the bolted holes).

Since we have decided to carry out a linear elastic plane stress analysis the stress-strain relationship for these conditions is:

$$\begin{bmatrix} \sigma_x \\ \sigma_y \\ \tau_{xy} \end{bmatrix} = \frac{E}{1-\upsilon^2} \begin{bmatrix} 1 & \upsilon & 0 \\ \upsilon & 1 & 0 \\ 0 & 0 & (1-\upsilon)/2 \end{bmatrix} \begin{bmatrix} \varepsilon_x \\ \varepsilon_y \\ \gamma_{zz} \end{bmatrix} \tag{1.04}$$

Strain-displacement relationships are given by:

$$\varepsilon_x = \frac{\partial u}{\partial x}; \quad \varepsilon_y = \frac{\partial v}{\partial y}; \qquad \gamma_{xy} = \frac{\partial u}{\partial y} + \frac{\partial v}{\partial x} \tag{1.05}$$

Equations 1.03, 1.04 and 1.05 represent the plane stress mathematical model which represents the geometry of the bracket more accurately than a simple beam model and assumes a two-dimensional stress situation in the bracket. Once we solve the above equations we will have the stress distribution in the bracket and can thus determine the magnitude and location of the maximum stress.

The problem is an analytical solution of this model is not possible (i.e. it is impossible to solve this problem "by hand") so a numerical solution (i.e. FEA) must be used. Figure 1.07 shows a finite element "mesh" of elements which was used in the solution of the plane stress mathematical model. It can be seen from the figure that the bolts that fasten the bracket to the wall are not modelled instead the lines around the bolt holes are constrained in all directions (represented by the triangles) to simulate the clamping effect of the bolt. Similarly the pin through which the load is applied is not modelled; instead a concentrated point force (represented by the arrow) is applied to a node to simulate the transfer of load from the pin to the bracket.

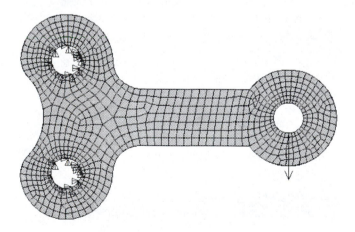

Figure 1.07: Finite Element Model Used for Plane Stress Analysis

The figures below show the results obtained from the finite element solution of the plane stress problem. Figure 1.08 shows the predicted deformed shape of the model due to the applied load and figure 1.09 shows the distribution of maximum principal stress throughout the bracket due to the applied load.

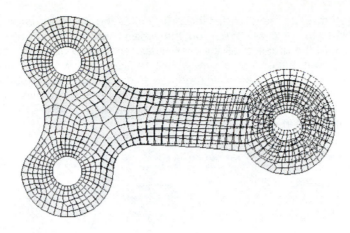

Figure 1.08: Deformed Shape Predicted by FEA Solution of Plane Stress Problem.
Deflections are exaggerated and shown over the original mesh.

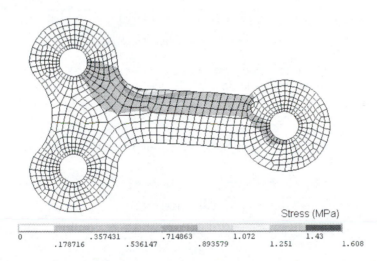

Figure 1.09: Contour Plot of the Distribution of Maximum Principal Stress in the Bracket as
Predicted by FEA Solution of Plane Stress Problem.

So the FEM allowed us to solve the plane stress problem which we otherwise were unable to solve. Now that we understand what FEA is capable of doing, the next obvious is question is "how exactly does it work?"

1.3.   How does Finite Element Analysis Work?

As we already know, the finite element analysis process begins by dividing up the structure into small pieces, called elements that are easier to analyse. All of the elements make up a "mesh" which is an approximation of the problem. Each of these elements can easily be analysed using simple equations for stress and strain. As the number of elements increases (increasing mesh size), their size decreases and the approximate solution will, theoretically, become more accurate.

In order to illustrate these points neatly we can compare the finite element analysis process to trying to find the area under a simple curved line on a graph. We know that we can get an exact solution for the area under the curve by integration. If we didn't know this then we could break up the area under the curve into a series of rectangles and add the areas of all the rectangles together to obtain an approximate solution for the area under the curve. While the solution will not give an exact solution, it can be simpler than integrating to find the area particularly when the function describing the curve is not known. In order to increase solution accuracy the width of the rectangles can be decreased to better approximate the curve. This process is summarised in figure 1.10.

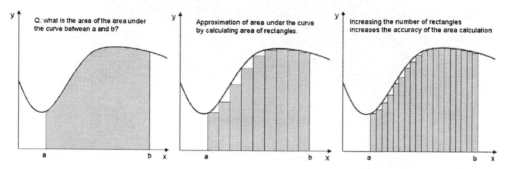

Figure 1.10: Approximate Method of Determining the Area under a Simple Curve

The following chapters in this book will discuss in detail how FEA works and how it should be used. Here we will try to describe how FEA works by considering a simple example and detailing the entire FEA process for modelling and obtaining an approximate solution to this problem.

All finite element analyses share common procedural steps no matter how simple or how complex the problem:

Step1: Evaluate the problem and make assumptions

Step 2: Describe how the finite elements will behave

Step 3: Build the finite element model

Step 4: Form element equations

Step 5: Assemble each element equation into a global problem equation

Step 6: Specify loads and boundary conditions

Step 7: Solve the global problem

Step 8: Evaluate the results.

In the following chapters we will discuss each of these steps in great detail, but for now let's just accept that these are the steps we have to follow. Now let's look at an example of finite element analysis in action.

Example 1.1

A tapered steel beam of rectangular cross section is used to support a horizontal load F = 10,000 N as shown in the figure below, we are required to use FEA to determine the deflection at the free end of the beam and the maximum stress in the beam.

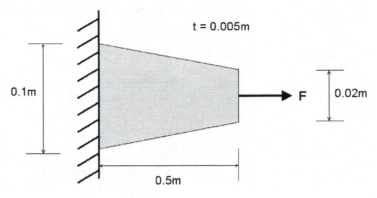

Figure 1.11: Example 1.1 Problem Definition

Step1: Evaluate the problem and make assumptions

This is a relatively simple problem that has an analytical solution and probably doesn't require FEA to obtain an approximate solution, however, for the purposes of example let's assume that an analytical solution does not exist. We are told that the beam is made from steel so let's assume that the load isn't great enough to cause permanent plastic deformation which means we can assume a linear elastic material model governed by Hooke's law with usual material properties for steel.

Assumption 1:   Linear elastic material model with E = 210 x 10^9 Pa

We know that the beam is embedded in a concrete wall so it is reasonable to assume that the left hand edge of the beam cannot move.

Assumption 2:   The left hand edge of the beam is fixed

The only applied load is on the right hand side of the beam and is acting in a horizontal direction. The beam's own weight will, however, cause it to deflect slightly downwards. If we consider the effect of the beams own weight to be negligible then we can assume that this is a one-dimensional problem and the beam can only deflect in the horizontal direction.

Assumption 3:   A one-dimensional analysis is sufficient to solve this problem

We are not told much about the manner that the load is applied to the right hand edge of the beam. Let's assume that the load is applied very slowly. This means that dynamic (vibration) effects can be ignored and that only a linear static analysis, obeying Hooke's law, is required.

Assumption 4:   A linear static analysis is sufficient to solve this problem

We are also not told about the environment in which the beam is placed. Is the immediate environment hot/cold, damp/dry, subject to radiation etc? All of these conditions can affect the behaviour of the beam (e.g. thermal stressing). As a first approximation let's assume that these factors are not significant and hence only a structural analysis is required.

Assumption 5:   A structural analysis is sufficient to solve this problem

These five assumptions will now determine how we approach the remainder of the steps in the finite element analysis process.

Step 2: Describe how the finite elements will behave

From the above, assumptions number 1 and number 4 essentially dictate how the elements will behave. From these assumptions we know that have a material which obeys Hooke's law.

Hooke's law (in general): $\sigma = E\varepsilon$ where: σ = stress
ε = strain
E = Elastic modulus

Hooke's law (for a spring):   F = kx   where: F = force
k = stiffness
x = deflection

so we can model the beam using one-dimensional line elements that behave similar to springs governed by Hooke's law. When a load is applied to the bar it will extend proportionally to the applied load just like a linear elastic spring.

The average stress σ in the beam is given by: $\sigma = F/A$
(applied force / cross sectional area of the beam)

The average strain in the beam is given by: $\varepsilon = \Delta l/l$
(change in length / original length)

In the elastic region Hooke's law governs the expansion: $E = \sigma/\varepsilon$
(Elastic modulus = stress/strain)

Putting these three equations together gives us:

$$F = \left(\frac{AE}{l}\right)\Delta l \qquad\qquad (1.06)$$

If we compare this equation to the equation for a linear spring, F = kx, we can see that a centrally loaded member of uniform cross section may be modelled as a spring with an equivalent stiffness of:

$$k_{eqv} = \frac{AE}{l} \qquad\qquad (1.07)$$

This allows us to model our bar as a series of elastic springs (elements) and the elastic behaviour of each element is modelled by an equivalent linear spring according to the equation:

$$f = k_{eqv}x \qquad\qquad (1.08)$$

Step 3: Build the finite element model

Now we know what type of element we are going to use and how they will behave we can decide how we will split our problem up into finite elements. In this case we will split the tapered beam into three elements as shown in the figure below:

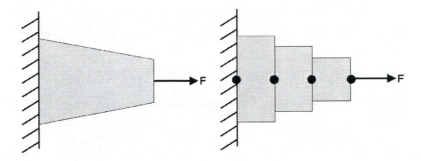

Figure 1.12: Idealisation of Tapered Beam into Three Elements

The bar is modelled using three elements, with each element having a uniform cross section. The cross sectional area of each element is represented by an average area of the cross sections at the nodes that make up that particular element. The points at each end of each element are called "nodes" and are indicated in figure 1.12. Figure1.13 shows the finalised model where the beam is modelled as a series of three elastic springs with equivalent stiffness to the three uniform cross section elements shown in figure 1.12.

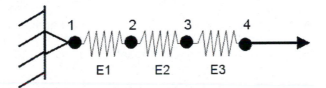

Figure 1.13: Finite Element Model of Tapered Beam

Step 4: Form element equations

From figure 1.13 it is obvious that the deflection or change in length of each element will be given by the difference in the displacement of the two nodes that make up the particular element. So we can say that for any element defined by nodes i and i+1, the change in element length is given by:

$$\Delta_I = U_{i+1} - U_i \qquad (1.09)$$

Where U_i is the displacement of node i. Now, putting equation 1.09 into equation 1.06 gives us:

$$F = \left(\frac{A_{avg}E}{I}\right)(U_{i+1} - U_i) \qquad (1.10)$$

Where, A_{avg} is the average cross sectional area of the part of the beam that makes up that particular element and, in this case, is given by:

Practical Stress Analysis with Finite Elements

$$A_{avg} = \frac{(W_{i+1} + W_i)t}{2} \qquad (1.11)$$

Where, W_i is the width of the tapered beam at node i, and, provided each element is the same length, is given by:

$$W_i = W_{i+1} + \frac{(W_{i-1} + W_{i+1})}{N_E} \qquad (1.12)$$

Where, N_E is the number of elements in the finite element model.

This allows us to calculate the governing equation for each element in the finite element model:

$$f_1 = k_1(U_1 - U_2) = \frac{(0.0004333)(210x10^9)}{0.166}(U_1 - U_2) = 5.48x10^8(U_1 - U_2) \qquad (1.13)$$

$$f_2 = k_2(U_2 - U_3) = \frac{(0.0002999)(210x10^9)}{0.166}(U_2 - U_3) = 3.795x10^8(U_2 - U_3) \qquad (1.14)$$

$$f_3 = k_3(U_4 - U_3) = \frac{(0.0001665)(210x10^9)}{0.166}(U_4 - U_3) = 2.106x10^8(U_4 - U_3) \qquad (1.15)$$

Where f_1 is the force in element 1, k_1 is the stiffness of element 1 and U_1 is the displacement of node 1 etc.

Step 5: Assemble each element equation into a global problem equation

Creating the global problem essentially involves combining equations 1.13, 1.14 and 1.15 into a single equation which describes the entire problem. This is conveniently done using matrices. The global problem statement for any structural problem is:

$$[M]\{\ddot{u}\} + [C]\{\dot{u}\} + [K]\{u\} = \{F\} \qquad (1.16)$$

Where, $[M]$ is the global mass matrix, $\{\ddot{u}\}$ is the nodal acceleration vector, $[C]$ is the global damping matrix, $\{\dot{u}\}$ is the nodal velocity vector, $[K]$ is the global stiffness matrix, $\{u\}$ is the nodal displacement vector and $\{F\}$ is the vector of nodal point forces.

In this case since we have assumed that a linear static analysis is sufficient we can assume that acceleration and velocity are zero and thus equation 1.16 reduces to:

[stiffness matrix] [displacement vector] = [force vector]

Since in this case we have four nodes with one degree of freedom (DOF) at each node (i.e. each node can only move in one direction) then we know that our global stiffness matrix will be a 4x4 square matrix. Each element has two nodes with one DOF so each individual element stiffness matrix will be a 2x2 square matrix.

The stiffness matrix for element 1 is given by:

$$[K]^{(1)} = \begin{bmatrix} k_1 & -k_1 \\ -k_1 & k_1 \end{bmatrix}$$

and it's position in the global stiffness matrix is given by:

$$[K]^{(1G)} = \begin{bmatrix} k_1 & -k_1 & 0 & 0 \\ -k_1 & k_1 & 0 & 0 \\ 0 & 0 & 0 & 0 \\ 0 & 0 & 0 & 0 \end{bmatrix}$$

Similarly for element 2, it's stiffness matrix is given by:

$$[K]^{(2)} = \begin{bmatrix} k_2 & -k_2 \\ -k_2 & k_2 \end{bmatrix}$$

and it's position in the global stiffness matrix is given by:

$$[K]^{(2G)} = \begin{bmatrix} 0 & 0 & 0 & 0 \\ 0 & k_2 & -k_2 & 0 \\ 0 & -k_2 & k_2 & 0 \\ 0 & 0 & 0 & 0 \end{bmatrix}$$

The stiffness matrix for element 3 is given by:

$$[K]^{(3)} = \begin{bmatrix} k_3 & -k_3 \\ -k_3 & k_3 \end{bmatrix}$$

and it's position in the global stiffness matrix is given by:

$$[K]^{(3G)} = \begin{bmatrix} 0 & 0 & 0 & 0 \\ 0 & 0 & 0 & 0 \\ 0 & 0 & k_3 & -k_3 \\ 0 & 0 & -k_3 & k_3 \end{bmatrix}$$

The global stiffness matrix is given by:

$$[K]^{(G)} = [K]^{(1G)} + [K]^{(2G)} + [K]^{(3G)}$$

$$[K]^{(G)} = \begin{bmatrix} k_1 & -k_1 & 0 & 0 \\ -k_1 & k_1+k_2 & -k_2 & 0 \\ 0 & -k_2 & k_2+k_3 & -k_3 \\ 0 & 0 & -k_3 & k_3 \end{bmatrix} = 1x10^8 \begin{bmatrix} 5.48 & -5.48 & 0 & 0 \\ -5.48 & 9.275 & -3.795 & 0 \\ 0 & -3.795 & 5.901 & -2.106 \\ 0 & 0 & -2.106 & 2.106 \end{bmatrix}$$

Notice how the stiffness matrix is banded, there are non-zero terms down the diagonal and zero terms outside the banded diagonal. This is a common feature of most finite element analyses. We can now assemble the global problem equation given by equation 1.16

$$1x10^8 \begin{bmatrix} 5.48 & -5.48 & 0 & 0 \\ -5.48 & 9.275 & -3.795 & 0 \\ 0 & -3.795 & 5.901 & -2.106 \\ 0 & 0 & -2.106 & 2.106 \end{bmatrix} \begin{bmatrix} U_1 \\ U_2 \\ U_3 \\ U_4 \end{bmatrix} = \begin{bmatrix} f_1 \\ f_2 \\ f_3 \\ f_4 \end{bmatrix}$$

This equation in, its current form, cannot be solved as there are too many unknowns. We must know at least one nodal displacement (U_i) and one nodal point force (f_i) before we can solve the problem. In practical terms this means we must apply boundary conditions and loads. This is the reason why all finite element analyses must have boundary conditions and applied loads. There is simply no way of avoiding this fact. Where boundary conditions do not exist in the problem then we must create them, for example by exploiting symmetry, as we will see in later chapters.

Step 6: Specify loads and boundary conditions

As the beam is fixed on the left hand side, at node 1, we can apply the boundary condition $U_1 = 0$. We also know that a load of 10,000 N is applied at the right hand edge of the beam, corresponding to node 4, thus $f_4 = 10,000$. Our global problem equation now becomes:

$$1x10^8 \begin{bmatrix} 5.48 & -5.48 & 0 & 0 \\ -5.48 & 9.275 & -3.795 & 0 \\ 0 & -3.795 & 5.901 & -2.106 \\ 0 & 0 & -2.106 & 2.106 \end{bmatrix} \begin{bmatrix} 0 \\ U_2 \\ U_3 \\ U_4 \end{bmatrix} = \begin{bmatrix} 0 \\ 0 \\ 0 \\ 10,000 \end{bmatrix} \tag{1.17}$$

Step 7: Solve the global problem

We can now solve this system of equations to obtain the nodal unknowns U_1, U_2 and U_3. It should be obvious from the above equation that the first line of the equation is redundant and can thus be eliminated. This leaves us with three equations and three unknowns. There are a number of mathematical methods available for solving the above system of equations such as Gaussian elimination, iterative methods and Newton-Raphson methods. In this, relatively simple, case a simple Gaussian elimination is all that is required. Multiplying the bottom three lines of the matrices in equation 1.17 gives:

$$1x10^8 (9.275U_2 - 3.795U_3) = 0$$
$$1x10^8 (-3.795U_2 + 5.901U_3 - 2.106U_4) = 0$$
$$1x10^8 (-2.106U_3 + 2.106U_4) = 0$$

Solving these equations gives:

$U_1 = 0$, $U_2 = 7.66x10^{-6}$ m, $U_3 = 1.86x10^{-5}$ m, $U_4 = 6.58x10^{-5}$ m

This gives us the displacement of each node in our finite element model and immediately provides us with an answer to one of the questions we were asked: the deflection of the free end of the beam is = U_4 = <u>6.58x10^{-5} m</u>

Step 8: Evaluate the results.

Generally speaking we are interested in more information that the displacement of the nodes when we do a FEA. In order to get this information we must further manipulate the nodal displacements to get the information we need.

Since the displacement of each of the nodes is now known, we can easily obtain the element strains, from:

$$\varepsilon = \frac{U_{i+1} - U_i}{l} \qquad\qquad (1.18)$$

Where, l is the original length of the element.

Also, since we know the value of Young's modulus for the beam material we can obtain element stresses via element strains using Hooke's law:

$$\sigma = E\varepsilon = E\left(\frac{U_{i+1} - U_i}{l}\right) \qquad\qquad (1.19)$$

In this case the element strains and stresses were calculated using equations 1.18 and 1.19 and are presented in the table below:

Element No	Strain	Stress (Pa)
1	4.61×10^{-5}	9.69×10^6
2	6.59×10^{-4}	13.8×10^6
3	2.84×10^{-4}	59.7×10^6

From the above table we can answer the second question we were asked: the maximum stress in the beam occurs at the region near the point of load application and is 59.7 MPa

The above example illustrates the application of the finite element method to the solution of a simple engineering problem. The application of the method to more complex problems will still be quite similar to the basic procedure used in the example above. If you can understand this example then you are well on your way to understanding finite element analysis and will be prepared for the following chapters.

If you don't understand it then please take the time to work through the example yourself, write out each stage of the procedure on a piece of paper and do all the calculations yourself. Make sure you are happy that you understand where each figure came from before you move to the next stage of the process. I am sure that if you take the time to study the example properly you will quickly understand the procedure.

1.4.   Summary of Chapter 1

After completing this chapter you should:
- Be able to explain why finite element analysis is an important tool for engineering analysis and design.
- Be able to explain what finite element analysis is and how it works.
- Be able to carry out a simple one-dimensional finite element analysis of an engineering problem on a piece of paper, by following the methodology of example 1.1.
- Understand the various steps required for any finite element analysis

1.5.   Problems

P1.1.   List the advantages of a product design process using FEA over a traditional "build and break" style process.

P1.2.   List the essential steps in any finite element analysis.

P1.3.   Using the methodology outlined in example 1.1 determine the stress distribution in the stepped cylindrical shaft shown in the figure below. The left hand edge of the shaft is rigidly clamped and cannot move. You may assume that the shaft is made of steel with a Young's modulus of 210 GPa.

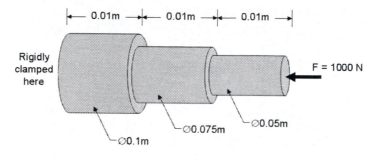

Answer Outline: K_1 = 1.65e10, K_2 = 9.28e9, K_3 = 4.12e9
U_1 = 0, U_2 = -6.06e-8, U_3 = -1.64e-7, U_4 = -4.11e-7
σ_1 = -127,260 Pa, σ_2 = -226,380 Pa, σ_3 = -509,460 Pa

P1.4.   Using the methodology outlined in example 1.1 determine the stress distribution in the stepped cylindrical shaft shown in the figure below. In this case you can assume that, due to the applied loading, the right hand edge of the shaft comes into contact with the wall at the point q as illustrated in the figure. Again, assume that E = 210 GPa for the shaft material.

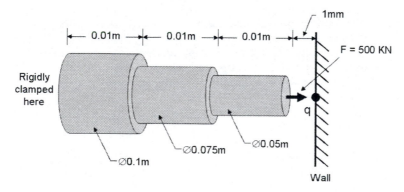

Hint: since the right hand edge touches the wall we know the displacement of any node that will be placed at that point!

Answer Outline: U_1 = 0, U_2 =8.25e-4, U_3 = 8.79e-4, U_4 = 0.001
σ_1 = 1.73 GPa, σ_2 = 0.113 GPa, σ_3 = 0.255 GPa

2

Fundamentals of Stress Analysis

2.1. Preliminary Material

There are a number of engineering concepts which you must be familiar with before we begin to look at Finite Element Analysis in detail in the following chapters. We will be examining advanced concepts such as stress analysis, material modelling, predicting yield and failure analysis in later chapters and the purpose of this chapter is to provide you with a refresher course in the basic engineering concepts that are required to understand the more advanced topics we will cover later. This chapter will also be a handy reference if you encounter problems in understanding the more complex topics later in the book and I will regularly refer you back here at the appropriate point. Please take the time to read this chapter in detail as it will greatly assist your understanding of both stress analysis and finite element analysis.

2.2 Units and Dimensional Analysis

The finite element method is a mathematical method that does not understand or require units. If you want to ensure that your results are correct and you are reading the units correctly you must ensure that you use a consistent unit system at all times.

The basic units required for any engineering analysis are length, weight and time. All other units are derived from these units.

Example 2.1:   The SI Unit System

In the SI unit system, length is in metres (m), weight is in Kilogram's (Kg) and time is in seconds (s). From these will we derive all the other required units:

Force = mass x acceleration = $Kg\ m/s^2$ = Newton (N)
Pressure and Stress = Force/Area = N/m^2 = Pascal (Pa)
Density = mass/volume = Kg/m^3

So, in this system you would enter the dimensions for your geometry in metres, enter the Young's Modulus value for steel as $210x10^9$ Pa, you would specify applied forces in N and you would read the stress results as being expressed in Pa.

It should be clear from the above example that if you alter any of the basic units then you will, by consequence, change all of the derived units. A simple mistake that many new users to FEA make is to decide to use the SI unit system (i.e. use metres as the unit of length) and then enter the dimensions for their model in millimetres rather than meters! Since force, stress, density and many other derived quantities are derived from length this means that all these derived quantities will no longer adhere to the SI unit system. This can be very dangerous, particularly if you think your results are in N/m^2 when in fact they are in another unit, e.g. N/mm^2.

The example below illustrates these points.

Example 2.2: The Modified SI Unit System

Suppose you wanted to specify your model dimensions in mm instead of m. This means that all the units will change accordingly and if you still think that your stress results are in Pa then you are wrong!!!
In this case, length = mm, weight = Kg and time = s and our derived units are:

Force = mass x acceleration = Kg mm/s^2 = N/1000 = mN
Pressure and Stress = Force/Area = N/mm^2 = 1,000,000Pa = MPa
Density = mass/volume = Kg/mm^3

So in this case you would enter your dimensions in mm, Young's Modulus for steel would be: 210,000, you would specify applied forces in N/1000 and read the stress results as being expressed in MPa.

If you don't understand this concept then take a look at example 1.1 on page 9 again. Write out the example on a piece of paper but instead of entering 0.1m or 0.5m for a particular length into a calculation, enter 100 or 500mm. Complete all the calculations in the example and see what happens!

The table below shows systems of consistent units, each system is usually determined by how you want to enter the dimensions of your model (i.e. either m or mm) and, for dynamic analysis, how you want to deal with time (i.e. either seconds or milliseconds).

Name	Mass	Length	Time	Force	Stress	Energy
SI system	Kg	m	sec	N	Pa	Joule
Altered SI (small dims)	Kg	mm	sec	N/1000	MPa	Joule/1000
Altered SI (small loads)	Kg	µm	sec	µN	MPa	1e-12 Joule
Altered SI (fast loads)	Kg	mm	msec	KN	GPa	KN-mm
Imperial	Slug	ft	s	lbf	psf	lbf-ft

Please note that using a inconsistent unit system is the <u>number one reason why new users of FEA get incorrect results!</u> If you are at all unsure about this you should stick to the SI system. In fact my advice to you as a beginner is to <u>always</u> use the SI unit system!

Tip:

Always use the SI unit system to ensure accurate results. Enter your model dimensions in meters (i.e. 100mm is entered as 0.1), enter your material properties in Pa (i.e. a Young's modulus of 130GPa is entered as 130x10^9) and enter applied loads in N (i.e. a load of 10KN is entered as 10,000).

2.3   Material Properties

2.3.1. Homogenous and Non-Homogenous Materials

A homogenous material is a pure material which doesn't contain any impurities or traces of other materials. Most engineering materials are considered homogenous with the obvious exception of composite materials. For example, when we buy a piece of steel from a supplier we can be reasonably confident that the steel will 99.9% pure and will not contain any significant amounts of any other material.

In contrast a non-homogenous material will contain a significant amount of different materials which will affect its performance under load. Good examples of non-homogenous engineering materials are carbon fibre reinforced plastic (CFRP), glass fibre reinforced plastic (GFRP), metal matrix composites and reinforced concrete. CFRP contains carbon fibre reinforcement embedded in a matrix of, usually, an epoxy resin. So we have two distinct constituents that make up the composite material. Each of these constituents affects the mechanical behaviour of the composite material. The carbon fibres will carry most of the load and the epoxy resin will ensure that the fibres are kept in an orderly pattern. When fibres are discontinuous, the epoxy resin matrix also helps to transfer load among the reinforcements.

Reinforcements come in a variety of forms: Long continuous fibres, short discontinuous fibres and particulates. Long fibres are popular in CFRP and some forms of GFRP. Short chopped fibres are most common in GFRP, while particulates are common in metal matrix composites (as added powders) and concrete (as gravel). Metal matrix composites typically consist of a metal matrix that makes up around 60% of the volume (e.g. aluminium) and a particulate that makes up the remaining 40% of the material volume (e.g. silicon carbide). Composites can have more than one type of reinforcement, for example reinforced concrete which has both a particulate (gravel) and long fibres (steel rods).

Identifying a material as being homogenous or non-homogenous is one of the key steps before choosing an appropriate element type and material model in FEA. Most FEA software has a number of composite element types available for modelling non-homogenous materials.

2.3.2. Isotropic and Non-Isotropic Materials

Isotropic Materials

An isotropic material is a material which has the same properties in all directions. This is true of most metals but not of composites or biological materials such as wood or bone.

If we take the example of a linear elastic material which obeys Hooke's law: $\sigma = E\varepsilon$, an isotropic material will have the same material properties in each direction so the relationship: $\sigma = E\varepsilon$ (or $\varepsilon = \sigma/E$) will be valid no matter what direction the load is applied. This will be valid for a one dimensional, two dimensional or three dimensional model.

Orthotropic Materials

Now let's consider a thin sheet of wood or a sheet of carbon fibre material: these will have different material properties in two directions. It will be stronger in the direction along the grain/fibres than in the direction across the grain/fibres. This type of material is known as an orthotropic material and requires us to develop our relationship into two dimensions.

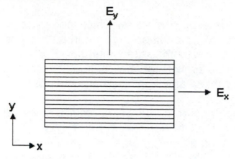

Figure 2.01: An Orthotropic Material with Higher Strength in the X-direction compared to the Y-direction.

Now, let's apply Hooke's law to this material:

$$\varepsilon = \sigma/E \quad => \quad \begin{bmatrix} \varepsilon_x \\ \varepsilon_y \\ \gamma_{xy} \end{bmatrix} = \begin{bmatrix} \dfrac{1}{E_x} & \dfrac{-\nu_{xy}}{E_y} & 0 \\ \dfrac{-\nu_{xy}}{E_y} & \dfrac{1}{E_y} & 0 \\ 0 & 0 & \dfrac{1}{G_{xy}} \end{bmatrix} \begin{bmatrix} \sigma_x \\ \sigma_y \\ \tau_{xy} \end{bmatrix} \qquad (2.01)$$

This equation is only valid for a 2-D model where loads are applied in either the x-direction (along the grain) or the y-direction (across the grain), where: ε is direct strain, γ is shear strain, ν is Poisson's ratio, E is Young's modulus; G is Shear modulus, σ is direct stress and τ is shear stress.

In the above example, as we are going from strain to stress the large matrix is known as the "compliance matrix". If we invert the equation and go from stress to strain, the large matrix is known as the "stiffness matrix".

Anisotropic Materials

For a fully non-isotropic material – which has different material properties in all three directions, we must use the full 3-D strain/stress matrix relationship for Hooke's law:

$$\begin{bmatrix} \varepsilon_x \\ \varepsilon_y \\ \varepsilon_z \\ \gamma_{xy} \\ \gamma_{yz} \\ \gamma_{zx} \end{bmatrix} = \begin{bmatrix} D_{11} & D_{12} & D_{13} & D_{14} & D_{15} & D_{16} \\ D_{21} & D_{22} & D_{23} & D_{24} & D_{25} & D_{26} \\ D_{31} & D_{23} & D_{33} & D_{34} & D_{35} & D_{36} \\ D_{41} & D_{24} & D_{43} & D_{44} & D_{45} & D_{46} \\ D_{51} & D_{25} & D_{53} & D_{54} & D_{55} & D_{56} \\ D_{61} & D_{26} & D_{63} & D_{64} & D_{65} & D_{66} \end{bmatrix} \begin{bmatrix} \sigma_x \\ \sigma_y \\ \sigma_z \\ \tau_{xy} \\ \tau_{yz} \\ \tau_{zx} \end{bmatrix} \qquad (2.02)$$

Where each of the D_{ij} terms is different and must be determined via a programme of extensive materials testing. This equation is valid for any 3-D model where loads are applied in any one of the three directions x, y, z or any combination of loads along these directions.

2.3.3   Linear and Non-Linear Materials

Linear Elastic Materials

The most basic material type used in stress analysis is a linear elastic material. This material extends proportionally to the load applied to it and always returns to its original state when the load is removed (i.e. obeys Hooke's law). Plastic deformation is not possible for this type of material.

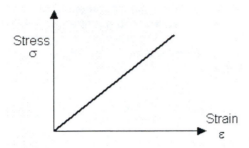

Figure 2.02: A Linear Elastic Material that Obeys Hooke's Law

In reality very few materials are linearly elastic, but most metals behave this way before they yield and experience plastic deformation. Therefore, we can use a linear elastic material model if we are sure that the material won't experience stresses above the yield stress. This is often very useful as a linear elastic analysis is very easy to set up and cheap (in terms of computing resources) to run. It is important to note that if you decide to adopt this approach the linear elastic assumption is only valid for very small strains. Take for example, the stress-strain curve obtained from a tensile test on a sample of annealed copper shown in figure 2.03.

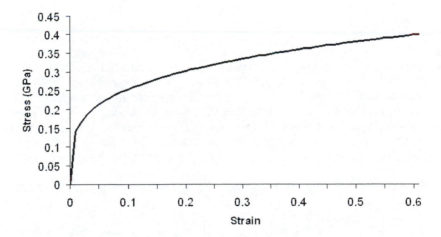

Figure 2.03: A Typical Stress-strain Curve for an Engineering Metal

We can easily see from the plot that the linear portion of the stress-strain curve makes up a very small portion of the entire curve. The material only behaves linearly up to a strain of approximately 0.01%. This means that only loads that produce extremely small deformations can be modelled using a linear elastic material model for this material. This very low level of strain in which a linear elastic material model is valid is usually referred to as "infinitesimal strain".

Nonlinear Elastic Materials

In direct contrast to linear elastic materials are non-linear elastic materials. These materials also do not undergo any permanent deformation, so after the applied loading is removed they will return to their original shape, however, these materials can deform quite significantly under an applied load and this deformation is non-linear. Materials in this category include rubbers, polymer networks and many biological materials (e.g. arterial wall tissue).

The most important type of non-linear elastic materials are hyperelastic materials. A stress-strain curve for a typical hyperelastic material model is shown in figure 2.04. There are several different material models available for modelling hyperelastic behaviour such as: Neo-Hookean, Mooney-Rivlin, Ogden and Blatz-Ko. These will be discussed in detail in chapter 5.

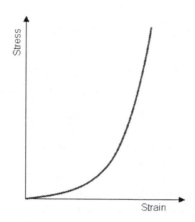

Figure 2.04: A Typical Stress-Strain Curve for a Hyperelastic Material.

Nonlinear Plastic Materials

Most engineering metals are non-linear materials which experience permanent plastic deformation once the yield stress has been exceeded. Even if the load is removed, after this point the material will not return to its original state. This condition is the basis of metal forming and is used to manufacture many products. After the initial linear phase of the stress strain curve the curve becomes highly non-linear after the yield stress, as already shown in figure 2.03. A non-linear plastic material model will be required when permanent deformation is expected in your analysis, for example in: metal forming, impact damage, prediction of ductile failure etc.

There are a number of ways of modelling this highly curved section in engineering analysis:

A bilinear model assumes that the plastic portion of the curve is linear and that this line is a tangent to the actual curve. The slope of this line is known as the Tangent Modulus (E_{Tan}). This method is rather crude but can give reasonable results in some cases

A "multi-linear" or "piecewise linear" model divides the stress strain curve into a number of straight lines and the slope of each line is individually defined. This method can be quite accurate but takes a lot of time to input.

A power law model uses a mathematical relationship to describe the geometry of the curve. This method is very accurate and is easy to use, but it requires a sufficient amount of experimental data to be available. (see also section 2.4.3)

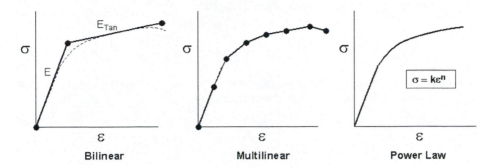

Figure 2.05: Methods for Modelling Non-Linear Plastic Material Behaviour

Each of these methods, and others, will be discussed in detail in chapter 5.

2.3.4. Brittle and Ductile Materials

Materials can generally be classed as either brittle or ductile materials. Brittle materials will fracture at much lower strains than ductile materials, so they will not deform significantly before failing. Brittle materials generally have a large Young's modulus and high failure strength (i.e. UTS). Although brittle materials generally have high strength they fail suddenly and without warning. Brittle materials are particularly poor when subjected to impact loads as they will shatter almost immediately. Good examples of brittle materials are glass, ceramics and cast iron.

In comparison, ductile materials, such as steel, aluminium and copper, will fracture at much higher strains than brittle materials. This allows for large deformation before failure, which is the basis of many metal forming and manufacturing processes. For ductile materials the yielding region of the stress-strain curve is significant, whereas for brittle materials it is practically nonexistent. In contrast to brittle materials, ductile materials will fail slowly and will give considerable warning, via necking, before failure. This characteristic is obviously very desirable when designing load bearing products as a ductile material is capable of absorbing much more energy before failure than a brittle material.

It is worth noting that it is not always easy to classify a material as either ductile or brittle; environmental considerations may have to be taken into account as many ductile materials become brittle at very low temperatures. Many composite materials are combinations of ductile and brittle materials, for example the metal matrix composite Al-Si-C contains brittle silicon carbide particles in an aluminium matrix.

2.4. Stress and Strain

2.4.1 Engineering Stress and Strain

Engineering stress, σ, is defined as applied load, P, divided by the original cross sectional area, A_0, to which this load is applied:

$$Stress, \ \sigma = \frac{Load}{Area} = \frac{P}{A_0} \tag{2.03}$$

The units of engineering stress are Mega-Pascal's (MPa) where MPa = 1×10^6 N/m^2.

Engineering strain, ε is defined as the deformation elongation or change in length, Δl, at some instant, as referenced to the original length, l_0:

$$Strain,\ \varepsilon = \frac{l_i - l_0}{l_0} = \frac{\Delta l}{l_0} \qquad\qquad (2.04)$$

There are three major ways in which load can be applied and thus stress and strain generated: tension, compression and shear (see figures 2.06, 2.07 and 2.08).

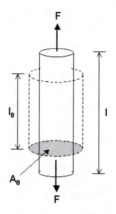

Figure 2.06: A tensile load produces an elongation and positive linear strains. The dashed lines represent the shape before deformation.

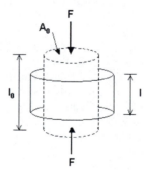

Figure 2.07: A compressive load produces contraction and a negative linear strain. The dashed lines represent the shape before deformation.

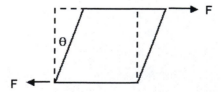

Figure 2.08: A shear load F causes a distortion in which a shear strain γ results where γ=tan θ

Shear strains are caused by shear stresses which tend to change the shape of a material without particular volume change. The shape change is evaluated by measuring the relative change in the angle between initially perpendicular sides of a differential element of material (shear strain). A simple definition of shear stress is the components of stress at a point that act parallel to the plane in which they lie.

In engineering analysis it is difficult to encounter pure shear loading and many loads are torsional rather than pure shear. Shear stresses are important in rotating shafts, and riveted and bolted joints. Many beams experience composite loading consisting of shear, tensile, and compressive stresses.

2.4.2 Poisson's Ratio

When a tensile stress σ_z is imposed on a metal specimen an elastic elongation and accompanying strain ε_z will occur in the direction of the applied stress. As a result of this elongation there will be constrictions in the lateral direction perpendicular to the applied stress. From this constriction the lateral compressive strain ε_x can be determined. A parameter named Poisson's ratio, υ, is defined as the ratio of the lateral and axial strains:

$$\upsilon = -\frac{\varepsilon_x}{\varepsilon_z}$$

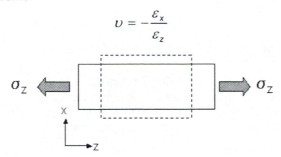

Figure 2.09: Poisson's Ratio – The dashed lines represent the un-deformed shape.

For many metals and other alloys values of Poisson's ratio range between 0.25 and 0.35. The maximum theoretical value of Poisson's ratio is 0.5 with any volume change taking place (i.e. without material being removed). For isotropic materials, shear and elastic modulii are related to each other and to Poisson's ratio according to:

$$E = 2G(1+v) \tag{2.05}$$

2.4.3. True Stress and Strain

If we consider one dimensional tensile deformation, as shown in figure 2.06, we know that engineering stress (sometimes also called "nominal stress") is defined as $\sigma = F/A_0$, where A_0 is the initial cross-sectional area prior to the application of the load. We also know, however, that when any material is stretched, its cross-sectional area reduces by an amount that depends on the Poisson's ratio of the material, as shown in figure 2.09. Engineering stress neglects this change in area. The stress axis on a stress-strain graph is often engineering stress, even though the sample may undergo a substantial change in cross-sectional area during testing.

True stress is an alternative definition of stress in which the initial area is replaced by the current area. For small deformation, the reduction in cross-sectional area is small and the distinction between engineering and true stress is insignificant. This isn't so for the large deformations typical of elastomers and plastic materials when the change in cross-sectional area can be significant as shown in figure 2.10 below.

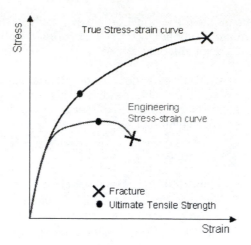

Figure 2.10: Comparison of Engineering and True Stress-Strain Curves

In one dimension, true stress is related to engineering stress via:

$$\sigma_{true} = \sigma(1 + \varepsilon) \tag{2.06}$$

where ε is engineering strain and σ is nominal stress. In uniaxial tension, true stress is then greater than the engineering stress. In compression the opposite is true, with engineering stress being greater than true stress. We can evaluate a similar equation for relating true strain to engineering strain via a natural log:

$$\varepsilon_{true} = \ln(1 + \varepsilon) \tag{2.07}$$

Where ε is engineering strain.

If we plot a true stress-strain curve using a logarithmic scale for both stress and strain we get a plot of the form shown in figure 2.11.

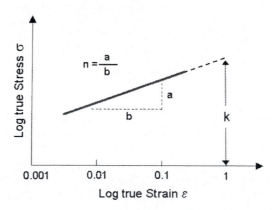

Figure 2.11: Log/log plot of true stress-strain curve

From figure 2.11 it is clear where the power law material model discussed in section 2.3.3 arrives. From figure 2.11 it can be shown that the true stress-strain curve can be described using a power law equation:

$$\sigma = K\varepsilon^n \tag{2.08}$$

Where, K is the strength coefficient which is found by extending the log/log plot of true stress strain to a log strain of 1 and measuring stress at this point, and n is the strain hardening exponent and is defined as the slope of the log/log true stress-strain curve, as shown in figure 2.11.

2.4.4. 3-D Stress at a Point

As we know from equation 2.03, engineering stress, is applied load, P, divided by the cross sectional area, A_0, to which this load is applied. The direction of action of the load P may not necessarily perpendicular to the area A_0, so we resolve P into two components: P_N normal to the plane of A_0 and P_S acting on the plane A_0, as illustrated in figure 2.12.

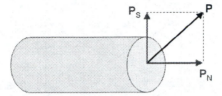

Figure 2.12: Normal and Shear Loads

The stresses associated with these components are:

Normal stress, $$\sigma_N = \frac{P_N}{A_0}$$ (2.09)

Shear stress, $$\tau = \frac{P_S}{A_0}$$ (2.10)

Obviously, the resultant stress is given by: $\sigma_R = \sqrt{\sigma_N^2 - \tau^2}$ (2.11)

Note that stress is not strictly speaking a vector quantity as in addition to magnitude and direction we must also know the plane on which the stress acts. Thus, strain is a tensor.

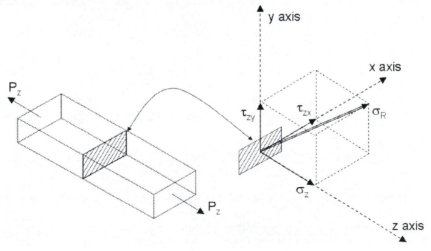

Figure 2.13: Three Orthogonal Planes

Now, let's cut the body under investigation into three orthogonal planes as shown in figure 2.13. The resultant stress, σ_R, acting on each plane can be resolved into a direct stress, in this case σ_z, and two shear stresses, in this case τ_{zx} and τ_{zy}. Shear stresses are given the notation: τ_{AB} where A is the plane on which they act and B is the direction in which they act. So, τ_{zx} acts on the z-plane in the x-direction.

So we can completely describe the state of stress at any point in a body by specifying components of shear and direct stress acting on the faces of an element of sides δ_x, δ_y, δ_z.

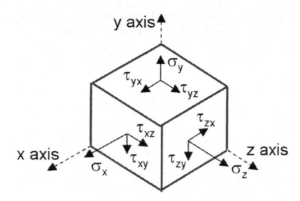

Figure 2.14: Illustration of stresses acting on a differential volume δ_x, δ_y, δ_z. Note that opposite faces will have equal and opposite stresses acting on them (not shown here)

In figure 2.14 normal stresses directed away from a related surface are taken as positive, thus tensile stresses are positive and compressive stresses are negative.

Two types of external force can produce the above stress system:
1) Surface forces (such as hydrostatic pressure) which are distributed over the surfaces of the body and can be resolved into components parallel to the orthogonal x, y, z system of axis and are given the symbols: f_x^S, f_y^S and f_z^S

2) Body forces (such as gravity and inertia) which are distributed over the volume of the body and the components of body force per unit volume are given the symbols: f_x^B, f_y^B and f_z^B

If we consider equilibrium it will become clear that, contrary to what was indicated in figure 2.14 stresses on either side of the element will not be equal but will differ by a small amount because of the fact that the stress quantities will be changed as they travel through the volume of the element. Thus we can say that if the direct stress acting normal to the z-plane is σ_z then the direct stress acting on the z+δz plane will be given by:

$$\sigma_z + \left(\frac{\partial \sigma_z}{\partial z}\right)\delta z \qquad (2.12)$$

So if we consider our differential element again and just look at one direction, figure 2.15 shows the actual stresses acting on either face of the differential volume. Obviously this method can be extended to all three directions to obtain the full set of stresses acting on each face of the differential volume of sides δ_x, δ_y, δ_z.

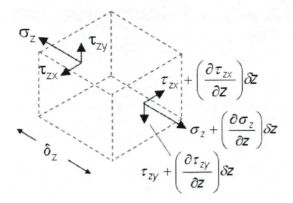

Figure 2.15: Equilibrium in the z-direction for a differential element of sides δ_x, δ_y, δ_z.

From this we can develop the equilibrium equations for three dimensional stress analysis, which are:

$$\frac{\partial \sigma_x}{\partial x} + \frac{\partial \tau_{xy}}{\partial y} + \frac{\partial \tau_{xz}}{\partial z} + f_x^B = 0$$

$$\frac{\partial \tau_{xy}}{\partial x} + \frac{\partial \sigma_y}{\partial y} + \frac{\partial \tau_{yz}}{\partial z} + f_y^B = 0 \qquad (2.13)$$

$$\frac{\partial \tau_{xz}}{\partial x} + \frac{\partial \tau_{xy}}{\partial y} + \frac{\partial \sigma_z}{\partial z} + f_z^B = 0$$

Many thin sheet structures have negligible stresses across the thickness of the sheet and thus "plane stress" analysis method can be used. Let's assume that the z-direction is the thickness direction of the thin sheet, thus: $\sigma_z = 0$, $\tau_{zx} = 0$ and $\tau_{zy} = 0$. This simplifies the equilibrium equations in equation 2.13 to the plane stress equilibrium equations shown in equation 2.14.

$$\frac{\partial \sigma_x}{\partial x} + \frac{\partial \tau_{xy}}{\partial y} + f_x^B = 0$$

$$\frac{\partial \tau_{yx}}{\partial x} + \frac{\partial \sigma_y}{\partial y} + f_y^B = 0 \qquad (2.14)$$

By satisfying equilibrium at the boundary of the body under investigation we can obtain the boundary conditions:

$$f_x^S = \sigma_x l + \tau_{xy} m + \tau_{xz} n$$

$$f_y^S = \tau_{yx} l + \sigma_y m + \tau_{yz} n \qquad (2.15)$$

$$f_z^S = \tau_{zx} l + \tau_{zy} m + \sigma_z n$$

Where f_x^S, f_y^S and f_z^S are surface loads in the x, y and z directions and l, m and n are the direction cosines which define the angle of the normal to the plane under consideration. The direction cosines are defined by: l = cos α, where α is the angle between the normal and the x-axis, m = cos β, where β is the angle between the normal and the y-axis, and n = cos χ, where χ is the angle between the normal and the z-axis.

2.4.5. Stress on an Inclined Plane

More than often we will be required to determine stress on a plane that is not one of our three orthogonal planes. In such cases it is important to be able to deal with stresses on planes that are inclined to the orthogonal planes. In order to explore these concepts let's examine the 2-D plane stress problem shown in figure 2.16.

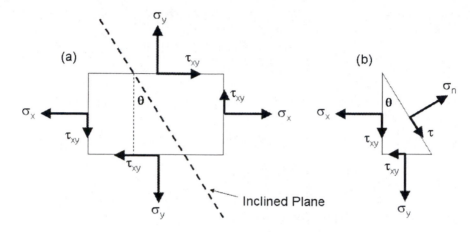

Figure 2.16: Two dimensional stress on an inclined plane

Figure 2.16 (a) shows the full two dimensional stress system, while figure 2.16 (b) focuses on the stresses acting on the inclined plane. It can be seen that there are two stresses acting on the inclined plane: σ_n which is the stress normal to the inclined plane and τ which is the shear stress on the inclined plane. Using trigonometry it can be shown that:

$$\sigma_n = \sigma_x \cos^2 \theta + \sigma_y \sin^2 \theta + \tau_{xy} \sin 2\theta \qquad (2.16)$$

and

$$\tau = \frac{(\sigma_x + \sigma_y)}{2} \sin 2\theta - \tau_{xy} \cos 2\theta \qquad (2.17)$$

In three dimensions, it can be shown that the normal stress on an inclined plane is given by:

$$\sigma_n = \sigma_x l^2 + \sigma_y m^2 + \sigma_z n^2 + 2(\tau_{xy} lm + \tau_{yz} mn + \tau_{zx} nl) \qquad (2.18)$$

Where l, m and n are the direction cosines of the plane under investigation.

2.4.6. Principal Stresses

By examining equations 2.16 and 2.17 it can easily be seen that for certain values of θ, σ_n and τ will be at a maximum or a minimum. Thus, we can find certain inclined planes in any stress problem whereby σ_n is at a maximum and a minimum and $\tau = 0$. The stresses on these inclined planes are known as "principal stresses" and the planes on which they act are known as "principal planes". The principal stresses are notated as follows:

σ_1 is the maximum (or major) principal stress
σ_2 is the minimum (or minor) principal stress

where:

$$\sigma_1 = \frac{\left(\sigma_x + \sigma_y\right)}{2} + \frac{1}{2}\sqrt{\left(\sigma_x - \sigma_y\right)^2 + 4\tau_{xy}^2} \qquad (2.19)$$

$$\sigma_1 = \frac{\left(\sigma_x + \sigma_y\right)}{2} - \frac{1}{2}\sqrt{\left(\sigma_x - \sigma_y\right)^2 + 4\tau_{xy}^2} \qquad (2.20)$$

Similarly, we can find a plane of maximum shear stress which will be midway between the principal planes (i.e. inclined at 45° to the principal planes as shown in figure 2.17). The maximum shear stress on this plane, τ_{max} is given by:

$$\tau_{max} = \frac{1}{2}\sqrt{\left(\sigma_x - \sigma_y\right)^2 + 4\tau_{xy}^2} = \frac{\sigma_1 - \sigma_2}{2} \qquad (2.21)$$

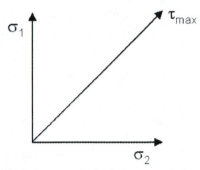

Figure 2.17: Relationship between principal planes and plane of max shear stress.

In three dimensions in can be shown that the principal stresses are found by solving for σ_p the characteristic equation:

$$\sigma_p^3 - I_1\sigma_p^2 + I_2\sigma_p - I_3 = 0 \qquad (2.22)$$

Where,

$I_1 = \sigma_x + \sigma_y + \sigma_z$

$I_2 = \sigma_x\sigma_y + \sigma_y\sigma_z + \sigma_z\sigma_x - \tau_{xy}^2 - \tau_{yz}^2 - \tau_{zx}^2$

$I_3 = \sigma_x\sigma_y\sigma_z + 2\tau_{xy}\tau_{yz}\tau_{zx} - \sigma_x\tau_{xy}^2 - \sigma_y\tau_{yz}^2 - \sigma_z\tau_{zx}^2$

Solving equation 2.22 will give three values for σ_p, i.e. the three principal stresses σ_1, σ_2 and σ_3.

2.4.7. 3-D Strain at a Point

We already know from section 2.4.1 that strain in one dimension is given by $\varepsilon = \Delta l/l_o$. For a three dimensional body, strain is given by:

$$\varepsilon_x = \frac{\partial u}{\partial x} \qquad \varepsilon_y = \frac{\partial v}{\partial y} \qquad \varepsilon_z = \frac{\partial w}{\partial z} \qquad (2.23)$$

Where, u, v and w are displacements in the x, y and z directions, respectively. In this case, shear strains are given by:

$$\gamma_{xy} = \frac{\partial v}{\partial x} + \frac{\partial u}{\partial y}$$

$$\gamma_{yz} = \frac{\partial w}{\partial y} + \frac{\partial v}{\partial z} \qquad (2.24)$$

$$\gamma_{zx} = \frac{\partial w}{\partial x} + \frac{\partial u}{\partial z}$$

Using a similar manner to that used for strains we can determine the principal strains for any given 3-D strain field as being:

$$\varepsilon_1 = \frac{(\varepsilon_x + \varepsilon_y)}{2} + \frac{1}{2}\sqrt{(\varepsilon_x - \varepsilon_y)^2 + 4\gamma_{xy}^2} \qquad (2.25)$$

And $\qquad \varepsilon_2 = \frac{(\varepsilon_x + \varepsilon_y)}{2} - \frac{1}{2}\sqrt{(\varepsilon_x - \varepsilon_y)^2 + 4\gamma_{xy}^2} \qquad (2.26)$

With the maximum shear strain given by:

$$\frac{\gamma_{max}}{2} = \frac{\varepsilon_1 - \varepsilon_2}{2} \qquad (2.27)$$

2.4.8. Experimental Measurement of Surface Strains

Strain gauges are often used to experimentally measure strains on an object under investigation. These gauges measure strain via a change in electrical resistance. If ε_1 and ε_2 are principal strains at a point and ε_a, ε_b and ε_c are the experimentally measured strains in the direction θ, $\theta+\alpha$ and $\theta+\alpha+\beta$ respectively, then we can show that:

$$\varepsilon_a = \frac{(\varepsilon_1 + \varepsilon_2)}{2} + \frac{(\varepsilon_1 - \varepsilon_2)}{2}\cos\theta \qquad (2.28)$$

$$\varepsilon_b = \frac{(\varepsilon_1 + \varepsilon_2)}{2} + \frac{(\varepsilon_1 - \varepsilon_2)}{2}\cos 2(\theta + \alpha) \qquad (2.29)$$

$$\varepsilon_c = \frac{(\varepsilon_1 + \varepsilon_2)}{2} + \frac{(\varepsilon_1 - \varepsilon_2)}{2}\cos 2(\theta + \alpha + \beta) \qquad (2.30)$$

Let's take the example where $\alpha = \beta = 45°$, we can rearrange the above equations to give:

$$\varepsilon_1 = \frac{(\varepsilon_a + \varepsilon_b)}{2} + \frac{1}{\sqrt{2}}\sqrt{(\varepsilon_a + \varepsilon_b)^2 + (\varepsilon_a - \varepsilon_b)^2} \qquad (2.31)$$

$$\varepsilon_2 = \frac{(\varepsilon_a + \varepsilon_b)}{2} - \frac{1}{\sqrt{2}}\sqrt{(\varepsilon_a + \varepsilon_b)^2 + (\varepsilon_a - \varepsilon_b)^2} \qquad (2.32)$$

So if we can experimentally measure ε_a, ε_b and ε_c then we can determine the principal strains ε_1 and ε_2.

Similarly, we can determine what direction the principal planes are from the strain gauge rosette from:

$$\tan 2\theta = \frac{2\varepsilon_b - \varepsilon_a - \varepsilon_c}{\varepsilon_a - \varepsilon_c} \qquad (2.33)$$

2.4.9   Plane Stress and Plane Strain

<u>Plane Stress</u>

In plane stress analysis we assume that all stresses act on the one plane – normally the x-y plane. Due to Poisson's effect there will be a strain in the z-direction, but we assume that there is no stress in the z-direction as shown in figure 2.18. In this case, stresses σ_z, τ_{xz} and τ_{yz} will all be zero.

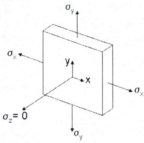

Figure 2.18: Plane Stress

The governing stress-strain relationship for plane stress in shown in equation 2.34:

$$\begin{bmatrix} \sigma_x \\ \sigma_y \\ \tau_{xy} \end{bmatrix} = \frac{E}{1-v^2} \begin{bmatrix} 1 & v & 0 \\ v & 1 & 0 \\ 0 & 0 & (1-v)/2 \end{bmatrix} \begin{bmatrix} \varepsilon_x \\ \varepsilon_y \\ \gamma_{zz} \end{bmatrix} \tag{2.34}$$

<u>Plane Strain</u>

In plain strain analysis we assume that all strains act on the x-y plane and hence, there is no strain in the z-direction. In this case σ_z will not equal zero as we may require a stress to prevent displacement from occurring in the z-direction. In this case, strains ε_z, χ_{xz} and χ_{yz} will all be zero.

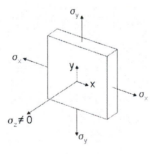

Figure 2.19: Plane Strain

The governing stress-strain relationship for plane strain in shown in equation 2.35:

$$\begin{bmatrix} \sigma_x \\ \sigma_y \\ \tau_{xy} \end{bmatrix} = \frac{E}{(1+v)(1-2v)} \begin{bmatrix} 1-v & v & 0 \\ v & 1-v & 0 \\ 0 & 0 & \frac{1}{2}-v \end{bmatrix} \begin{bmatrix} \varepsilon_x \\ \varepsilon_y \\ \gamma_{zz} \end{bmatrix} \qquad \text{Equation 2.35}$$

2.5.   Yielding and Plastic Deformation

2.5.1. The Stress-Strain Curve

If we examine a typical stress-strain curve for a steel material, as shown in figure 2.20 we can easily see that the portion of the curve describing elastic deformation is relatively small compared to the portion describing plastic deformation. It is important to note that a stress-strain curve is most often plotted using engineering stress and strain measures, because the reference length and initial cross-sectional area of the sample are easily measured. Stress-strain curves generated from tensile test results can help us model the constitutive relationship between stress and strain for the tested material.

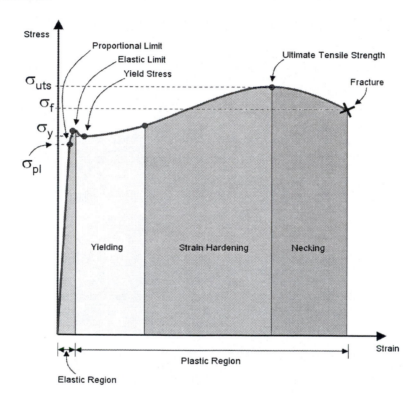

Figure 2.20: Details of a Typical Stress-Strain Curve

There are several distinct regions in the stress-strain curve:
- The elastic region
- Yielding
- Strain hardening
- Necking leading to failure

In the elastic region, as we already know, stress is linearly proportional to strain. When the applied load exceeds a value corresponding to the yield strength, σ_y, the specimen undergoes gross plastic deformation and will be permanently deformed even if the load is subsequently reduced to zero. "Strain hardening" is a term used to describe the fact that after yielding the stress required to produce continued plastic deformation increases with increasing plastic strain. During this phase of the deformation the volume of the specimen will remain constant, and as the specimen elongates its cross sectional area will decrease uniformly along the gauge length.

Initially strain hardening will more than compensate for this decrease in area and engineering stress will continue to rise with increasing strain. A point will eventually be reached, however, where the decrease in the cross sectional area of the specimen will be greater than the increase in load arising from strain hardening. This will happen at some point in the specimen that is very slightly weaker than the surrounding material and so all further plastic deformation will be concentrated at this point and the specimen begins to "neck" or locally deform. During necking the cross sectional area will reduce far more rapidly than before, so the load required to deform the specimen reduces and stress will decrease until fracture occurs.

Let's now look at each of these regions in detail:

2.5.1.1.   The Elastic Region

This part of the curve is linear and stress is related to strain via Young's modulus according to Hooke's law. For most metallic materials elastic deformation only occurs to strains of about 0.005. As the material is deformed beyond this point the stress is no longer proportional to strain and Hooke's law is no longer valid. After this point permanent non-recoverable plastic deformation occurs. Measuring this transition from elastic to non-elastic deformation is not easy and a number of methods of determining when Hooke's law breaks down have been developed.

The proportional limit, σ_{pl} is defined as the initial departure from linearity of the stress-strain curve. It should be noted that it is possible for some metals to behave elastically but not linearly elastically. It is clear from figure 2.20 that a small portion of the elastic region is above the proportional limit and in this region strain is increasing non-linearly with increasing stress. After passing through this non-linear elastic region of the curve the material will reach the elastic limit. Beyond the elastic limit any additional loading will cause non-elastic (i.e. permanent) deformation. Some metals such as soft copper and cast iron will have a mostly non-linear elastic region but for many materials the elastic limit and the proportional limit will be so close that it is difficult to distinguish between them.

2.5.1.2.   Yielding

Most structures are designed to ensure that only elastic deformation will result when a stress is applied. It is therefore vital to know the stress level at which plastic deformation begins – i.e. the point where yielding occurs.

If stress is increased beyond the yield stress, σ_y then permanent deformation of the material will occur. In the yielding region of the curve there will be large increases in strain for almost negligible increases in stress. This phenomenon where there is a near zero slope to the stress-strain curve is known as "perfect plasticity". The stress at which yielding is observed to begin depends on the sensitivity of the strain measurements. Many materials will exhibit a gradual transition from elastic to plastic behaviour, and the point at which yield occurs is hard to define with precision.

In most cases it is very difficult to determine this position exactly so a convention has been established where a straight line is drawn parallel to the elastic portion of the stress-strain curve at some specified strain offset, usually 0.002 (i.e. 0.2% strain). The stress corresponding to the intersection of this line and the stress-strain curve as it bends over in the plastic region is defined as the yield strength σ_y. This method of determining the yield strength is known as the *Offset Yield Strength* or *Proof Stress* method.

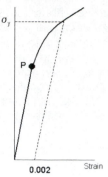

Figure 2.21: The Proof Stress / Offset Method of Determining Yield

A good way of looking at offset yield strength is that after a specimen has been loaded to its 0.2 percent offset yield strength and then unloaded it will be 0.2 percent longer than before the test. The yield strength obtained by this method is commonly used for design and specification purposes because it avoids the practical difficulties of measuring the elastic limit or proportional limit. As previously mentioned, some materials have essentially no linear portion to their stress-strain curve and hence the offset method cannot be used. In such cases the usual practice is to define the yield strength as the stress to produce some total strain, for example, $\varepsilon = 0.005$.

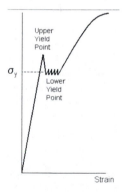

Figure 2.22: Determination of the Yield Stress for a Ductile Material

Some steels and other materials have a very well defined yield point which is then followed by a reduction in stress. Continued deformation fluctuates slightly about some constant stress value, known as the lower yield point and then stress again rises with increased strain. For materials that exhibit this effect, as shown in figure 2.22, the yield strength is taken as the average stress associated with the lower yield point.

In order to determine our stress-strain curve we perform a tensile test where the strain in the specimen is gradually increased and no unloading occurs. If, however, the yield stress was exceeded during loading and the specimen was then unloaded, at the end of the unloading there would be a permanent strain in the material, as illustrated in figure 2.23.

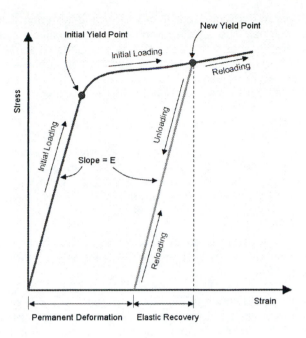

Figure 2.23: Stress-strain curve for a Process of Loading past the yield point, unloading and subsequent reloading.

The unloading process follows the slope of the elastic portion of the curve, so there will be some non-permanent strain "lost" in the unloading process. This phenomenon is known as *elastic recovery*. Subsequent reloading will proceed up this new loading curve, with the slope again being the Young's Modulus of the material. In this region the material will again behave linear elastically until it reaches the new yield stress, as indicated in figure 2.23.

2.5.1.3.   Strain Hardening and UTS

After yielding the stress necessary to continue plastic deformation in metals increases to a maximum point (the Ultimate Tensile Stress – UTS) and then decreases to the eventual fracture point. The UTS is the stress at the highest point on the stress-strain curve and corresponds to the maximum stress that a material can withstand in tension – if this stress is applied and maintained then failure will occur.

The tensile strength (UTS) is given by the maximum load divided by the original cross-sectional area of the specimen. The tensile strength is the value most often quoted from the results of a tension test; yet in reality it is a value of little fundamental significance with regard to the strength of a metal. For ductile metals the tensile strength should be regarded as a measure of the maximum load, which a metal can withstand under the very restrictive conditions of uniaxial loading. Later on we will see that this value bears little relation to the useful strength of the metal under the more complex conditions of stress, which are usually encountered.

For many years it was customary to base the strength of load carrying structural members on the tensile strength, suitably reduced by a factor of safety as described in chapter 1. In more modern times the more rational approach of basing the static design of ductile metals on the material yield strength has been used. However, because of the long practice of using the tensile strength to determine the strength of materials, it has become a very familiar property, and as such it is a very useful identification of a material. The tensile strength is also relatively easy to determine and is quite reproducible, thus it is useful for the purposes of specifications and for quality control of a product. Extensive empirical correlations between tensile strength and properties such as hardness and fatigue strength are often quite useful. For brittle materials, the tensile strength is a valid criterion for design.

2.5.1.4. Necking

When the loading is continued beyond the ultimate stress, the cross-sectional area decreases rapidly in a localized region of the test specimen. Since the cross-sectional area decreases, the load carrying capacity of this region also decreases rapidly. The load (and stress) keeps dropping until the specimen reaches the fracture point.

All deformation up to this point is uniform, however, at this point necking will occur and all further deformation will be localised to the point at which necking occurs: this is indicated in figure 2.24.

The fracture strength corresponds to the stress at fracture and is generally not used in engineering analysis as, for metals, it is usually less than the UTS. Generally speaking when the strength of a metal is listed in a reference it is the yield strength that is used. This is because by the time the UTS has been exceeded the structure will have undergone significant plastic deformation and will generally be useless. However, if we are analysing a metal forming operation both the yield stress and the UTS will be needed as we will want to deform the structure but ensure that it does not approach the UTS.

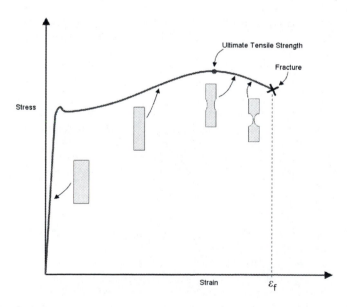

Figure 2.24: Stress-strain curve illustrating reduction in cross sectional area of test specimen during uniaxial tensile test.

2.5.2 Ductility

Ductility is a measure of how far a material can be deformed before it fractures. In general measures of ductility are qualitative and subjective. In other words, we can say that one material is "more ductile" than another but measuring how *much* more ductile it is, is quite difficult to quantify. Measures of ductility are very important in metal forming as they indicate to what a metal can be deformed without fracture in metalworking operations such as drawing, rolling and extrusion. Measures of ductility can also be useful to the designer as they indicate the ability of the metal to flow plastically before fracture. A high ductility indicates that the material is "forgiving" and likely to deform locally without fracture should the designer make an error in stress calculations or the prediction of severe loads. Ductility measurements may also be used, in certain cases, to assess material quality even though no direct relationship exists between the ductility measurement and performance in service.

The conventional measure of ductility obtained from a tensile test is the engineering strain at fracture ε_f. This is also known as the *elongation* or *elongation at break*. Another measure sometimes used is the reduction of area at fracture. When determining ε_f it is clear that a significant amount of the plastic deformation will be concentrated in the necked region of the tensile specimen, so the value of ε_f will depend greatly on the gauge length L_0 over which the measurement was taken. The smaller the gauge length the greater the contribution to the overall elongation from the necked region and this will lead to higher values of ε_f. Therefore, when reporting values of ε_f the gauge length L_0 is always given.

2.5.3. Work Hardening

In the majority of structural designs the structural members are designed to be in service below the yield stress as once the load exceeds the yield stress, the members will exhibit large deformations that are undesirable (e.g. a bridge sagging). Thus, in general, materials with larger yield strength are preferable.

If we take the example of steel: mild steel has a uniaxial yield strength of between 250 and 400 MPa. Work hardening, which is the process of loading the material beyond it's yield stress and then unloading it again, can be used to increase the yield strength of mild steel so that it approaches that of a high strength steel. This process is summarised in figure 2.25.

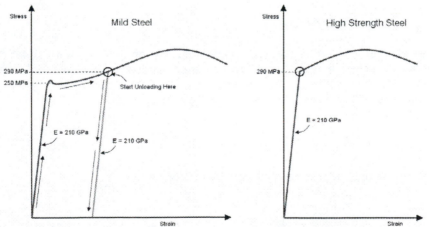

Figure 2.25: After work hardening, the stress-strain curve of mild steel (left) resembles that of high-strength steel (right)

The obvious negative aspect of work hardening is that we have "used up" some of the available strain in the material so there will be some loss in ductility of the material. It is worth noting that some materials like mild steel are often recycled and because of this the yield strength may be a little higher than expected.

2.6. Yield Criteria and Yield Analysis

A major problem in engineering stress analysis is how we determine yield stress from experimental measurements or finite element results. For a simple one dimensional problem we can simply compare the measured stress against the yield stress for the material as measured from a uniaxial tensile test or as published in a reference book. This reference yield strength is commonly denoted as σ_y.

For a two and three dimensional problem we are placed in a predicament. We will have stresses in two or three directions and corresponding shear stresses. Which stress should we use to compare with σ_y? Yield does not always occur just because one of the directional stresses exceeds the yield stress, so assuming this would be too conservative, also, yield will definitely occur some time before all of the directional stresses exceed the yield stress. Thus we need some intelligent predictor of when yield will occur based on the measured directional stresses – such a device is called a *yield criteria*. There are a number of yield criteria used in general engineering analysis all of which are based on principal stresses, we will now examine each of them in turn.

2.6.1. The Rankine Yield Criterion

The Rankine criterion is also known as the "maximum normal stress" criterion as it assumes that yield will occur when the maximum normal stress at any point reaches the uniaxial yield stress, σ_y, for the material. In a three dimensional analysis this criteria assumes that yield in tension will occur when $\sigma_1 = \sigma_y$, and yield in compression will occur when $\sigma_3 = \sigma_y$.

The Rankine criterion is obviously quite conservative and is generally suitable for brittle materials (e.g. concrete, cast iron) but is unsuitable for ductile materials as it takes no account of shear stresses which heavily influence the plastic flow of ductile materials.

2.6.2. The Tresca Yield Criterion

The Tresca criterion predicts that yielding will occur when the maximum shear stress at any point reaches a maximum allowable shear stress for that material. For this reason the Tresca criterion is often referred to as the *maximum shear stress criteria*. In one dimension the maximum shear stress is given by:

$$\tau_{max_{1-D}} = \frac{\sigma_y}{2} \qquad (2.36)$$

In three dimensions the maximum shear stress is given by:

$$\tau_{max_{3-D}} = \frac{\sigma_1 - \sigma_3}{2} \qquad (2.37)$$

Putting equations 2.36 and 2.37 together gives the Tresca yield criterion in three dimensions:

Yield will occur when $\sigma_1 - \sigma_3 = \sigma_y$ $\qquad (2.38)$

2.6.3. The Maximum Normal Strain Criterion

This criterion predicts that yielding will occur when the largest of the three principal strains becomes equal to the strain corresponding to the yield strength, i.e. when $\varepsilon_1 = \varepsilon_y$. In one dimension the maximum normal strain is given by:

$$\frac{\sigma_1}{E} = \varepsilon_y \tag{2.39}$$

In three dimensions the maximum shear stress is given by:

$$\frac{1}{E}\left[\sigma_1 - \upsilon(\sigma_2 + \sigma_3)\right] = \varepsilon_y \tag{2.40}$$

or:

$$\left[\sigma_1 - \upsilon(\sigma_2 + \sigma_3)\right] = \sigma_y$$

The maximum normal strain theory is not useful for ductile materials but can be appropriate for some non-metals such as Perspex type materials.

2.6.4. The Von-Mises Yield Criterion

The von-Mises criterion is based on the assumption that failure occurs when the energy of distortion reaches the same energy for yield/failure in uniaxial tension. It is much less conservative than other yield criteria (e.g. Rankine, Tresca etc.) and thus can help to eliminate over designing when used with ductile materials. The use of the von-Mises criterion is summarised in the table below.

Problem Type	Stresses	Method of determining yield
1-dimensional		Yield occurs when: σ_1 is greater than σ_y.
2-dimensional		Von-Mises stress in 2-D Yield occurs when: $\sigma_1^2 - \sigma_1\sigma_2 + \sigma_2^2 = \sigma_y \qquad (2.41)$
3-dimensional		Von-Mises stress in 3-D Yield occurs when: $\frac{1}{\sqrt{2}}\sqrt{(\sigma_1 - \sigma_2)^2 + (\sigma_2 - \sigma_3)^2 + (\sigma_3 - \sigma_1)^2} = \sigma_y \qquad (2.42)$

Table 2.1: Summary of Use of the Von-Mises Yield Criterion

When using the von-Mises criterion to analyse stress results generated from a finite element analysis we generally get the FE software postprocessor to generate a contour plot of von-Mises stress throughout the problem and when this stress is seen to exceed σ_y then we may assume that yield has occurred at those locations. We shall discuss this further in chapter 9 when we deal with post-processing of results.

2.7.   Failure Analysis

In some cases it can be appropriate to extend the yield criteria discussed in section 2.6. and use the same equations to predict when failure will occur based on the ultimate tensile strength (UTS) obtain from a tensile test performed on a sample of the material. There are, however, a number of approaches available and their use will be briefly discussed here.

2.7.1   Modes of Failure

In structural applications the most common types of failure encountered are fracture, yielding, buckling, fatigue and creep. Fracture occurs when new cracks are initiated or existing cracks are extended. There are two types of fracture, brittle and ductile, which will be discussed below. We have already discussed yielding in the sections above. Buckling is a sudden loss of stability or stiffness under an applied load. Fatigue occurs in structures that are subjected to variable or cyclical loading due to the fact that the structure loses strength over time under the influence of the varying load. Creep occurs as a slow deformation in bodies that are loaded over a significant length of time.

2.7.2.   Failure of Brittle Materials

The most basic brittle fracture criterion corresponds to the Rankine yield criteria and states that fracture is initiated when the greatest tensile principal stress in the solid reaches the UTS, i.e. $\sigma_f = \sigma_1 = \sigma_{UTS}$

To apply this criterion, we must measure σ_{UTS} for the material by conducting a large number of tensile tests on specimens – it is important to test a large number of specimens because the failure stress is likely to show a great deal of statistical scatter. The tensile strength can also be measured using beam bending tests. The failure stress measured in a bending test is referred to as the *modulus of rupture*, σ_r, for the material. It is nominally equivalent to σ_{UTS} but in practice usually turns out to be slightly higher. This criterion is appropriate for use with materials such as glass or polycrystalline ceramics.

2.7.3.   Failure of Ductile Materials

A ductile failure criterion (also known as a *ductile fracture criterion*) attempts to predict how far a ductile material (normally a metal) can be deformed before cracks form. This allows for the prediction of where and when a crack will form during ductile deformation.

The hypothesis of ductile failure criteria is that ductile failure will occur when the maximum damage of the material exceeds a critical damage value, CDV. The various ductile fracture criteria available typically have an integral form, as shown in equation 2.43, which represent the effect of the deformation history of the process parameters:

$$\int_0^{\bar{\varepsilon}_f} F_{PP} \, d\bar{\varepsilon} = CDV \tag{2.43}$$

Where $\bar{\varepsilon}_f$ is the equivalent strain at fracture, $\bar{\varepsilon}$ is equivalent strain and F_{PP} is a function of the process parameters that is causing the ductile deformation under consideration.

A study conducted by Cockroft and Latham concluded that ductile failures generally occur in the region of the largest tensile stress and as such they proposed the Cockroft and Latham ductile failure criterion given by equation 2.44:

$$\int_0^{\bar{\varepsilon}_f} \sigma_{max} \, d\bar{\varepsilon} = C_1 \tag{2.44}$$

Where σ_{max} is the largest tensile stress and C_1 is a material constant obtained from material tests.

Brozzo modified the Cockroft and Latham criterion by including a hydrostatic stress term, as shown in equation 2.45.

$$\int_0^{\bar{\varepsilon}_f} \frac{2}{3}\left(1 - \frac{\sigma_h}{\sigma_{max}}\right)^{-1} d\bar{\varepsilon} = C_2 \tag{2.45}$$

Where σ_h is the hydrostatic stress and C_2 is a material constant.

Oyane derived a ductile failure criterion from the equations of plasticity theory for porous materials, this criterion is given by:

$$\int_0^{\bar{\varepsilon}_f} \left(\frac{\sigma_h}{\bar{\sigma}} + a\right) d\bar{\varepsilon} = b \tag{2.46}$$

Where $\bar{\sigma}$ is equivalent stress and, a and b are material constants.

In order to use these ductile failure criteria to estimate when failure will occur in a finite element model, they are rewritten in terms of an integral value as follows:

$$I_{C-L} = \frac{1}{C_1} \int_0^{\bar{\varepsilon}_f} \sigma_{max} \, d\bar{\varepsilon} \tag{2.47}$$

$$I_{Brozzo} = \frac{1}{C_2} \int_0^{\bar{\varepsilon}_f} \frac{2}{3}\left(1 - \frac{\sigma_h}{\sigma_{max}}\right)^{-1} d\bar{\varepsilon} \tag{2.48}$$

$$I_{Oyane} = \frac{1}{b} \int_0^{\bar{\varepsilon}_f} \left(\frac{\sigma_h}{\bar{\sigma}} + a\right) d\bar{\varepsilon} \tag{2.49}$$

Using stresses and strains obtained from a finite element solution and the relevant material constants C_1, C_2, a and b, the integrals above can be calculated for each element in the finite element mesh. Ductile failure is assumed to have occurred when and where the integral equals 1.

These criteria are very useful for analysing ductile deformation processes such as metal forming and are valuable for their ability to predict when and where failure will occur, thus allowing for appropriate design to avoid failure during the metal forming process.

2.8.   Summary of Chapter 2

After completing this chapter you should:
- Understand the importance of using a consistent unit system in stress analysis problems.
- Be able to classify materials as either: isotropic or non-isotropic, homogenous or non-homogenous and linear or non-linear.
- Be able to explain the difference between engineering stress/strain and true stress/strain and appreciate the importance of this difference.
- Be able to calculate the principal stresses and strains for a given stress analysis problem.
- Be able to identify the different regions and important points on a stress-strain curve.
- Understand the principles of strain hardening and work hardening.
- Understand and be able to apply various yield and failure criteria.

2.9.   Problems

P2.1. Test your understanding of unit systems and dimensional analysis by answering the following questions:

   (a)   Using the SI unit system, express the following quantities in terms of the basic units of length, mass and time.
   (i)   Force
   (ii)   Pressure
   (iii)   Work (in 1-D, force × distance)
   (iv)   energy (e.g. potential energy = mass*gravity*height)
   (b)   If E = aL sin(bt), where E is energy, L is length and t is time:
   (i) What are the dimensions and SI units of b?
   (ii) What are the dimensions and SI units of a?
   (c)   Given the following equation:

   $$F = -2\pi r L \frac{v}{R}\eta$$

   Where: F = force, r = radius, L = length, v = speed, R = distance
   Determine the SI units of viscosity, η.

   Answers: (a) (i) Kgm/s^2 (ii) Kg/ms^2 (iii) Kgm^2/s^2 (iv) Kgm^2/s^2
   (b) (i) $b = s^{-1}$ (ii) a = N
   (c) $kgm^{-1}s^{-1}$

P2.2. The figure below shows the engineering stress-strain curve obtained from a uniaxial tensile test of a sample of material. From the figure determine:
(a) the materials yield stress
(b) the materials failure strain
(c) the materials ultimate tensile stress

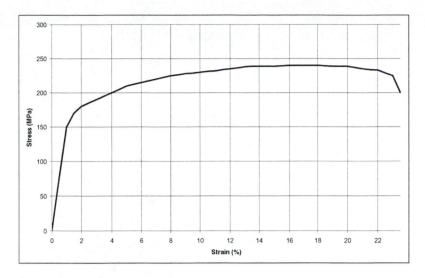

P2.3. The state of stress at a point in a thin plate is given by: σ_x = 14 MPa, σ_y = 35 MPa and τ_{xy} = -7 MPa.
(a) Determine the normal and shear stresses on a plane aligned at θ = 45° to the y-axis.
(b) Determine the principal stresses σ_1 and σ_2.

Answers: (a) $\sigma_{n@45}$ = 17.5 MPa, $\tau_{@45}$ = 24.5 MPa
(b) σ_1 = 49.74 MPa, σ_2 = 0.74 MPa

P2.4. The state of stress at a point in a steel block under complex loading is such that, σ_x = 55.6 MPa, σ_y = 0 MPa, σ_z = 444 MPa, τ_{xy} = 94.3 MPa, τ_{yz} = 333 MPa and τ_{zx} = -157 MPa.
(a) Determine the normal stress acting on a plane whose normal is defined by l = 0.5345, m = -0.2672 and n = 0.8018.
(b) Determine the principal stresses σ_1, σ_2 and σ_3.

Answers: (a) σ_n = -2.848 MPa
(b) σ_1 = 1000 MPa, σ_2 = 500 MPa and σ_3 = -1000 MPa

P2.5. A strain gauge rosette mounted on the surface of a metal plate under stress gave the following readings:
Gauge A (at 0°) +0.000592
Gauge B (at 45°) +0.000308
Gauge C (at 90°) -0.000432
The angles were measured clockwise from gauge A.
Determine the magnitudes of the principal stresses and their direction relative to gauge A. You may assume that E = 200 GPa and ν = 0.3

Answers: σ_1 = 190.1 MPa (tension) at an angle 12° anticlockwise
σ_2 = -63.4 MPa (compressive) at an angle 102° anticlockwise

P2.6. The loads on a bolted joint can be resolved into an axial force along the bolt of 12 KN and a transverse shear force of 4 KN, as shown in the figure below. Estimate the minimum bolt diameter required to support these loads without yielding using both the von-Mises and maximum shear stress criteria.

Assume that the yield stress in tension for the bolt, σ_y = 240 MPa and use a factor of safety of 2 in your calculations.

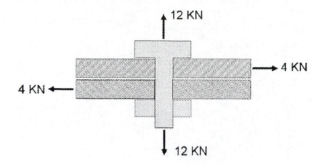

Answer outline: σ_x = axial load/cross sectional area of bolt
τ_{xy} = shear load/cross sectional area of bolt
Calculate principal stresses based on these loads
$d_{\text{von-Mises}}$ = 12.4mm
$d_{\text{max shear}}$ = 12.12 mm

P2.8. A finite element analysis of an aircraft structure predicted a 3D stress regime given by: σ_1 = 200 MPa, σ_2 = 100 MPa, σ_3 = -50 MPa. If the structure is made from Aluminium alloy AL7075-T6 with σ_y = 500 MPa, will the structure experience plastic deformation during service?

3

Finite Element Procedure

3.1. Introduction

In this chapter we will examine the FEA process in detail from both a practical and a theoretical point of view. The major emphasis of this book is on practical FEA and the first half of this chapter will show you how to do a finite element analysis, where to start, what questions to ask yourself and where you can go to find more help in later chapters. The second half of the chapter examines the theoretical FEA method and relates theory to practical decisions we make when performing a FEA. Even though this book is not concerned with the detail of FE theory, it is important to understand how the process is implemented in commercial FE software, as often error messages are produced that from a practical point of view are meaningless but when related to the theory can provide a method for fixing the analysis and obtaining a solution. From this point of view, a general knowledge of the theory of FEA is required and is therefore provided here.

This chapter discusses FEA process as a whole and examines each part of the process generally; each subsequent chapter will discuss each part of the process in great detail.

3.2. Overview of the FEA Process

A basic overview of the procedure used in any finite element analysis is shown in figure 3.01.

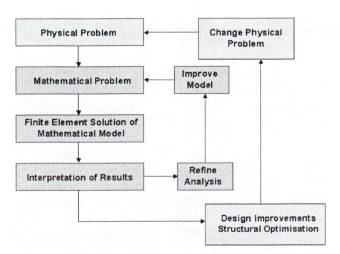

Figure 3.01: Overview of the FEA Process

The inner loop in figure 3.01 is the process of building the finite element model, obtaining a solution for the nodal unknowns and post-processing the results. In recent years many computer aids such as graphic interfaces and CAD modelling have helped to make this part of process much easier for the user to master. The engineering decision making process, represented by the outer loop in figure 3.01, is now generally recognised as the source of major difficulty and the involvement of the engineering expert in the process represents the major time required to perform the analysis.

Before a FE analysis can be generated and run (inner loop) an experienced user must make major decisions on how the physical problem under investigation can be transformed into a mathematical model for FE analysis (outer loop). In order to turn the physical problem into a mathematical problem (i.e. go from the outer loop to the inner loop) we make certain assumptions that lead to a system of differential equations which govern the mathematical problem. These assumptions are generally related to geometry, material laws, loading, boundary conditions etc. We have already looked at some of these types of assumptions in example 1.1 in chapter 1.

Obviously, the FE solution will only solve the selected mathematical model and all the assumptions we made in order to generate this mathematical model will be reflected in the predicted response/results. We can't expect any more information than the information contained in the mathematical model. We can't, for example, assume that a structural analysis is sufficient and then later look for results relating to temperature effects! Therefore the mathematical model is crucial and the key step in engineering analysis is choosing an appropriate mathematical model and creating a representative finite element model based on this.

A more detailed overview of the stages of the FE process is shown in figure 3.02. The process begins with identification of the physical problem and design objectives and ends with a model of a product satisfying specific design criteria. As the design process proceeds, knowledge about the model is increased and this allows for modifications of the model based on new knowledge.

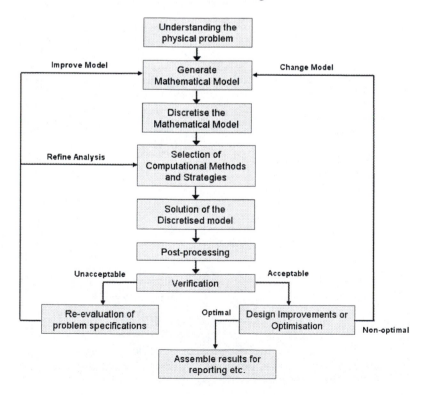

Figure 3.02: Overview of Consecutive Stages of an FE Analysis

Let's now discuss each stage of this process, with reference to figure 3.02, in detail:

3.2.1   Understanding the Physical Problem

The first step in the procedure outlined in figure 3.02 is: understanding the physical problem. It has been the author's experience that this is the step that causes the greatest amount of difficulty to novice FEA users. When novice users are presented with a problem the usual first course of action is to go directly to the FE software and begin building a model or looking for answers in the online help system. It is a curious phenomenon that users with several years of engineering knowledge behind them tend to totally ignore it when presented with a FEA problem. This phenomenon may be partly explained by the fact that FEA is sold as the optimal answer to all problems, with the implication that one doesn't really need an engineering background if a powerful piece of software is available. Another possible explanation may be that new users don't have confidence in other analysis methods or consider them "too simple" to be effective in today's high tech work environment.

What the user should actually be doing at this stage is identifying all distinctive features of the structure and principle design objectives. The type of loading inherent in the problem should be identified and the importance of any other environmental influences should be determined. Traditionally this step is performed by the designer based on their engineering expertise, by asking themselves a series of questions which will eventually lead to an explicit statement of the physical problem. This stage of the process is very much a "pen and paper" or "blackboard" type of analysis.

3.2.2.   Generating the Mathematical Model

Once the essential features of the physical problem have been identified it is necessary to translate these features into a mathematical representation of the problem. This stage involves two major tasks: firstly, definition of the problem domain (i.e. physical shape of the problem) and secondly, selection of a mathematical formulation which best represents the physical behaviour of the structure. As a practical example a problem may be defined as requiring a plane strain analysis, thus a 2-D model is required and the presence of symmetry in the model may mean that only a ¼ of the actual structure is required as the problem domain. Identifying, at this point, that a particular element type is required (e.g. plane stress, axisymmetric etc.) is crucial. These decisions are made before the discretisation process is begun and hence before the FE software is even started up. Novice users commonly make the mistake of attempting to make these decisions while using the software rather than in the planning stage, before the software is used.

3.2.3.   Discretisation of the Model

Once the element type and formulation has been decided from the previous step, the next step is to actually split up the model into these elements. Most commercial FE software can now automatically perform discretisation of the problem domain using automatic mesh generators. Automatic meshing, however, cannot predict phenomena such as the likely location of stress concentrations, and hence, will not know where a finer mesh is required. Thus, the user will generally be required to specify some mesh refinements based on knowledge gained from the "understanding the physical problem" phase of the process.

Some FE software contain advanced adaptive mesh refinement algorithms and, with these, there is less need for the user to anticipate the nature of the final solution and to, thus, refine the mesh in certain regions. These capabilities should be used with caution and cannot substitute for lack of experience or lack of engineering know how.

3.2.4. Selection of Computational Methods and Strategies

Most commercial FE software offer a wide selection of strategies for the solutions of various problems and the selection of which strategy to use is largely based on the previous experience of the user. For a specific problem the user must select a particular solution method as well as time step or load step size.

3.2.5. Numerical Analysis of the FE Model

At this stage of the process, when working with most commercial FE codes, it is assumed that all decisions have already been made and the software follows a prescribed procedure to produce results or give error messages and stop the solution. In other words, this is the point where you tell the computer to go and solve the problem that you have specified.

When an error occurs there is usually no suggestion how to fix it and the interpretation of the error is reliant on the experience of the user. A further problem occurs if the solution of a complex analysis diverges as there is usually no reason given for the divergence or suggestions as to how to modify the FEA in order to achieve convergence. Even if results are obtained there is usually no aid available for the user to estimate the reliability of the results and in many cases a full cycle of solutions need to be obtained just to learn that the results are not acceptable and the model needs modification!

3.2.6. Post-Processing, Model Verification and Validation

Post processing provides essential information required for the acceptance or rejection of the solution and for modification of the input data in order to obtain a satisfactory solution. Post processing via a graphical user interface is perhaps the most straightforward part of the FE process and new users, provided that they know what they are looking for, rarely have much difficulty in actually obtaining and manipulating results. The interpretation of the results by novice users is, however, a source of much difficulty. Asking pertinent questions, such as those suggested later in this chapter, can aid in ensuring that the error of solution is within specified tolerances, that the maximum stress predicted does not exceed specified limits, that displacements are not too large etc. It is precisely at this stage that the novice user should be asking these types of questions of their analysis.

3.3. Detailed Practical FEA Procedure

3.3.1 Step One: Understanding the Physical Problem

This is perhaps the most important step in the FE procedure as it will determine all steps that follow. In order to successfully complete this stage of the process, I suggest that you follow the procedure outlined below:

1. Sit down at a desk with a piece of paper. Ask yourself the question "what exactly do I want to do?" You can answer this question with drawings of the problem and/or a number of statements. Some examples of answers to this question maybe "I want to design a pressure vessel", "I want to determine the maximum stress in a plate due to an applied pressure load", "I want to optimise the shape of a steel beam so it can carry a load of 250 KN".

Basically what you want to have on your piece of paper is a statement of the physical problem to be analysed. This statement can be one sentence, a paragraph, a set of sketches or a combination of all of these.

2. Identify all distinctive features of the structure to be analysed. Write down everything you know about the problem. Try to answer as many of the following questions as possible:
   - What are the dimensions of the structure?
   - Which dimensions are particularly important?
   - Can the structure be simplified into a number of simpler structures for analysis?
   - Can the structure be simplified by removing any geometric entities that will not effect the analysis results? (i.e. defeaturing)
   - Does the structure exhibit any symmetry?
   - What material is the structure made from?
   - What type of loading is present?
   - Are there any locations where stress concentrations are to be expected (e.g. holes, fillets etc.)?
   - Are the loads expected to be large enough to induce yielding, permanent deformation or failure?
   - Are the loads constant or do they vary with time?
   - Are the loads symmetrical or applied all in one plane?
   - Is the structure expected to come in contact with other structures as a result of the loads or as a result of an impact event etc?
   - Are there any significant environmental factors which will effect the structures response?
   - For a design project - what are the design objectives?

   After answering these questions, and any others you can think of, you should have a great deal of information about the physical problem. These questions should also have helped you to identify some characteristics of the problem that will help you in the next stage of the process.

3.3.2   Step 2: Determination of the Mathematical Model

The selection of an appropriate mathematical model requires a great deal of engineering theoretical background and experience of going through the FEA process. The factors influencing the selection of the mathematical model are:
1. The geometry of the structure
2. The type of material used in the structure
3. The loads to which the structure is subjected
4. The manner in which the structure is supported or fixed
5. The manner in which the structure interacts with other structures and the immediate environment
6. The facilities available to the user in the FE software and the computational resources available to solve the problem

The geometry of the structure will often dictate that it can be modelled as a 1-D structure (as in example 1.1) or a 2-D structure. A structure may also exhibit symmetry which will mean that only a portion of the structure may need to be modelled. For example if a structure is recognised as only requiring a 2-D analysis, as there is no variation in cross section in one particular direction, then a plane stress, plane strain or axisymmetric mathematical model may be appropriate for the analysis.

For any structural problem there is generally a hierarchy of mathematical models that can be used to represent the mechanical behaviour of the structure. Generally speaking more comprehensive models (which include more effects or are more "realistic") require more complex mathematics and are computationally more expensive to solve. It is difficult to describe a hierarchy of models that covers all aspects of structural analysis, however, if we make some assumptions and deal only with the more common mathematical models we can specify a hierarchy of geometric models for structural analysis as shown in figure 3.03.

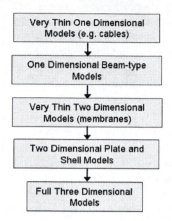

Figure 3.03: Geometric Hierarchy of Mathematical Models for Structural Analysis

Similarly we can construct a general hierarchy of kinematic mathematical models structural analysis as shown in figure 3.04 and a general hierarchy of constitutive mathematical models for modelling material behaviour as shown in figure 3.05.

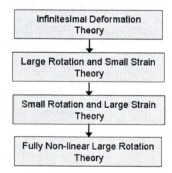

Figure 3.04: An Example of a Hierarchy of Kinematic Models for Structural Analysis

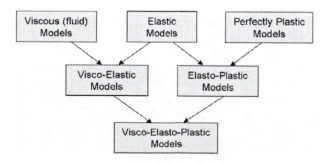

Figure 3.05: A General Hierarchy of Constitutive Models for Structural Analysis

Each box in figure 3.05 represents a number of different types of constitutive models, for example, Elastic models can be linearly elastic, non-linearly elastic, hyperelastic etc. Elasto-plastic models can be any of the myriad of such models available which are all based on different principles and calculate yield in different manners. A detailed description of available material models is undertaken in chapter 5.

The general hierarchies presented in figures 3.03, 3.04 and 3.05 are useful in terms of helping to identify a possible mathematical model by forcing us to think about the essential features of the problem when we consider which "box" our problem fits into. In order to further aid this process, Table 3.01 presents a general hierarchy of the key assumptions required in any FEA and lists the mathematical models available under the four key headings of geometric models, material models, loading and boundary conditions.

Level	Geometry	Material Model	Loading	Boundary Conditions
1	1-D Spar or Truss Uniaxial compression/tension	Linear Elastic Isotropic	Static Point Force	Regions fixed in all degrees of freedom
2	2-D Spar or Truss Uniaxial compression/tension	Linear Elastic orthotropic	Static Distributed Force	Regions fixed in specific DOF but free in others
3	3-D Spar or Truss Uniaxial compression/tension	Linear Elastic anisotropic	Static Pressure Load	Prescribed displacement on specific DOF
4	2-D Beam Tension, compression and bending	Non-linear Elastic	Static Inertia Load	2-D Contact Surfaces
5	2-D Plane Stress or Plane Strain Membrane effect	Visco-Elastic	Time varying point force	3-D Contact Surfaces
6	2-D Axisymmetric Membrane effect	Elasto-Plastic Rate Independent	Time varying distributed force	Time dependant boundary conditions
7	2-D Axisymmetric with Torsion Membrane effect and Torsion	Elasto-Plastic Rate Dependant	Time varying pressure load	
8	3-D Thin Shell Membrane effect and bending	Non-metal and specialized Plasticity Models	Applied Velocity	
9	3-D Beam Tension, compression, torsion and bending	Models incorporating damage	Applied Acceleration	
10	3-D Solid Full 3-D stress regime	Specialized Material Models		

Table 3.01: General Hierarchy of Mathematical Models for Use in Structural FEA

So, the next stage in our practical FEA procedure involves using table 3.01 and identifying which level in the table our problem fits into under each heading. Let's briefly examine each heading in turn:

3.3.2.1.   Geometry

Level 1:   1-D Spar or Truss

This type of geometry is characterised by the fact that there is no variation in cross sectional area along the length of the structure and that the structure can only change in length along its greatest dimension. An example of this type of structure would be a cable in uniaxial tension. A "spar" or "truss" is a term used for structure which can only extend or compress in one direction. In practice the applications of such a structure are very limited, can generally be described using elementary theory and are rarely, if ever, used for a finite element analysis. If you think your structure falls into this category then you would be better off doing a paper based analysis and will in all probability not need to perform a finite element analysis

Level 2:   2-D Spar or Truss

2-D truss structures are framework type structures in which every member within the structure is assumed to carry only axial loads and has no resistance to bending. The members are assumed to be connected via frictional pin joints. Typical examples of structures that are treated as 2-D Truss structures are bridges and roof trusses used in building construction, as shown in figure 3.06.

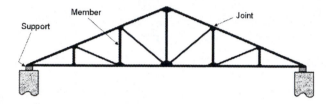

Figure 3.06: Typical Example of a 2-D Truss Structure

The assumption of frictionless pin joints between the members is obviously not representative of the way the structure would actually behave. Typically such structures will have members joined via gusset plates which will be fixed to the members via rivets, nails, welding etc. and because of this the members will carry some bending loads. From experience, however, it is known that this assumption is valid and provides satisfactory results for design purposes. Truss analyses do not consider the shape of the truss cross section to be important (as bending is not considered) and thus only require the definition of cross sectional area.

Level 3:   3-D Spar or Truss

This type of structure is essentially a 3-D version of the 2-D Truss structure described above. It is a more complex truss structure and is generally seen in bridges, towers and space frames. An example of a space frame, 3-D truss structure is shown in figure 3.07.

Figure 3.07: Typical Example of a Space Frame (3-D Truss Structure)

Level 4:   2-D Beam

A beam is a structural element whose length is large compared to its width and height. Two dimensional beam analyses take account of axial loading and bending of the beam. Figures 1.01 and 1.03 in chapter 1 describe 2-D beam problems. It is important to note that 2D beam problems do not have to consist of just a single beam. For example, the Truss structure in figure 3.06 could be modelled as a 2-D beam mathematical model if bending of the members during loading was thought to be significant. The essential difference between a 2-D Truss and a 2-D beam is that bending loads are considered and the properties of the beam cross section are considered. Thus, a 2-D beam analysis will differentiate between the use of a "C" section and a "T" section beam.

Level 5:   2-D Plane Stress

As discussed in chapter 2, a plane stress mathematical model assumes that there are zero stresses (and non-zero strains) through the thickness of the structure. This type of model is only valid for plate like structures where the thickness dimension (z direction) is much smaller than the other two dimensions (x and y directions). All loads must act on the x-y plane and all predicted responses will be on the x-y plane. This model is generally used for flat plates subjected to in-plane loading as shown in figure 3.08.

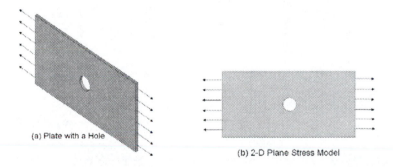

(a) Plate with a Hole

(b) 2-D Plane Stress Model

Figure 3.08: Typical Example of Appropriate Use of a Plane Stress Model

Level 6:   2-D Plane Strain

As discussed in chapter 2, a plane strain mathematical model assumes that there are zero strains (and non-zero stresses) through the thickness of the structure (normally the z-direction). This model is used for long, constant section structures, where the thickness dimension is much larger than the other two dimensions.

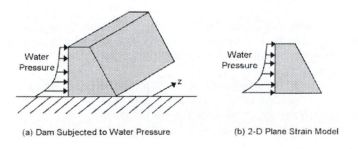

(a) Dam Subjected to Water Pressure   (b) 2-D Plane Strain Model

Figure 3.09: Typical Example of Appropriate Use of a Plain Strain Model

All loads must act on the x-y plane and all predicted responses will be on the x-y plane. This model is generally used for long structures such as beams and splined shafts, which experience a constant load along their length. Plain strain analysis is also very useful for certain metal forming processes such as rolling, drawing and forging.

Level 7:   2-D Axisymmetric

An axisymmetric mathematical model assumes that the geometry, loading and boundary conditions are symmetric around a central axis and thus can be described by rotating a 2-D section 360° around the central axis. This mathematical model is often used in the analysis of pressure vessels, straight pipes and shafts. It is also very useful in the analysis of certain metal forming processes such as deep drawing and sheet bulging. There will always be a constant distribution of displacement in the circumferential direction and loading and deformations will be confined to the radial and axial directions.

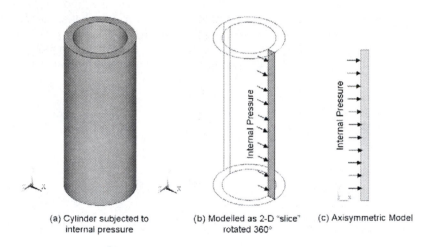

(a) Cylinder subjected to internal pressure   (b) Modelled as 2-D "slice" rotated 360°   (c) Axisymmetric Model

Figure 3.10: Use of an Axisymmetric Mathematical Model in the Analysis of a Pressure Vessel

Level 8:   2-D Axisymmetric with Torsion

This mathematical model is a more complex version of the 2-D Axisymmetric model and is essentially the same except in this case a torsion load may also be applied to the structure. This model is useful for modelling non-axisymmetric loads on an axisymmetric structure such as torque on a shaft.

Level 9:   3-D Thin Shell

This mathematical model is useful for thin panels or curved surfaces. A thin shell model includes both bending and membrane or stretching effects and assumes that stresses through the thickness of the material are not significant.

Typical applications of this type of mathematical model include analysis of thin walled pressure vessels, analysis of sheet metal forming processes, analysis of aircraft and spacecraft structures, analysis of ship and submarine structures, automotive vehicle body panels, thin concrete structures such as reinforced concrete domes, etc.

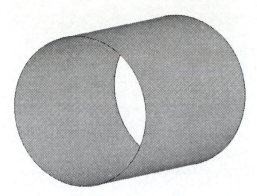

Figure 3.11: Thin Shell Mathematical Model of an Aircraft Fuselage Section

Level 10: <u>3-D Beam</u>

This mathematical model is a more complex version of the 2-D beam model described in level 4 above. Here the beam is defined in 3-D space, the shape of the cross section is considered and axial, bending and torsional loadings are all considered. This mathematical model is highly suited to the analysis of 3-D beam structures such as space frames where bending and torsion loads are considered important. Figure 3.13 shows a typical example of a 3-D beam structure.

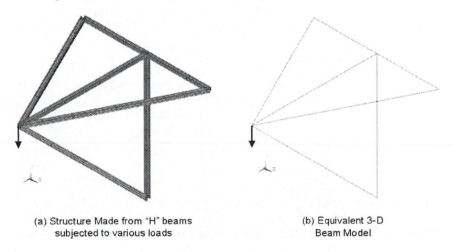

(a) Structure Made from "H" beams   (b) Equivalent 3-D
subjected to various loads   Beam Model

Figure 3.12: Typical Example of a 3-D Beam Structure

Level 11: <u>3-D Solid</u>

A 3-D solid representational mathematical model is used to model structures which because of their geometries, materials, loading or the detail of the required results cannot be satisfactorily modelled with any of the simpler mathematical models. This model is the last resort, when none of the lower level models will suffice, and is highly complex and difficult to formulate.

You should really think carefully before deciding if a 3D solid model is required. As humans we have a natural tendency to prefer something that "looks like" the structure we are trying to analyse. While this is understandable, in many cases using 3D solid elements incorrectly can result in inaccuracies.

3.3.2.2.   Material Model

Level 1:   <u>Linear Elastic Isotropic</u>

The simplest material model available is the linear elastic isotropic model which assumes that Hooke's law holds in all directions. This model assumes that stress is always linearly proportional to strain and the direction of loading does not affect this relationship. This model is useful for modelling the behaviour of metals that are not expected to experience yielding.

Level 2:   <u>Linear Elastic Orthotropic</u>

The linear elastic orthotropic model again assumes that Hooke's law holds, but allows different values for Young's modulus in two different directions. This model is useful for modelling the behaviour of wood, bone and reinforced composite materials, where yielding or failure is not expected.

Level 3:   <u>Linear Elastic Anisotropic</u>

This model is a further development of the previous model and assumes again that Hooke's law holds but, in this case, there is a different value for Young's modulus in each direction. This model is useful for representing materials that are highly anisotropic such as bone and certain composite materials. Again, this model assumes that loads will not be great enough to cause yielding or failure of the material.

Level 4:   <u>Non-Linear Elastic</u>

Non-linear elastic material models are useful for representing the behaviour of materials such as rubber, foams, biological tissues and certain polymers that behave in this manner. There are many types of non-linear elastic models available such as: Blatz-Ko, Mooney-Rivlin, Neo-Hookean and Ogden. These will be discussed in detail in chapter 5.

Level 5:   <u>Visco-Elastic Models</u>

A visco-elastic material exhibits both elastic and viscous behaviour. The viscosity of a visco-elastic material results in a strain rate dependent on time. Elastic materials do not dissipate energy when a load is applied, then removed. A visco-elastic material, however, will lose some energy when a load is applied, then removed. Consequently, hysteresis is observed in the stress-strain curve, with the area of the loop being equal to the energy lost during the loading cycle.

If we take the example of a polymeric material which exhibits visco-elastic behaviour: when a stress is applied to the material portions of the long polymer chain will change position. This rearrangement is called Creep. The material will remain solid even when these parts of their chains are rearranging in order to accompany the stress, and as this occurs, it creates a back stress in the material. When the back stress is the same magnitude as the applied stress, the material no longer creeps. When the original stress is taken away, the accumulated back stresses will cause the material to return to its original form, thus exhibiting elasticity.

There are a number of material models available for describing this type of behaviour such as the Maxwell, Kelvin-Voight and standard linear solid models. These will be discussed in detail in chapter 5.

Level 6:   Elasto-Plastic Rate Independent

Material models under this heading consist of models that allow for yielding and subsequent plastic deformation of the material due to the applied loading. In this case the plastic deformation is assumed to be independent of strain rate, so the speed at which the deformation takes place will not affect the resultant response of the structure. There are many different types of model available under this heading such as: bilinear isotropic models, multi-linear isotopic models, bilinear kinematic models, multi-linear kinematic models, power law models, anisotropic plasticity models etc. Each of these options will be discussed in detail in chapter 5.

Level 7:   Elasto-Plastic Rate Dependant

In contrast to the previous model, the elasto-plastic rate dependant model considers the speed at which deformation takes place to be very important and understands that it will affect the resultant response of the structure. Models in this category require a definition of strain rate, $\dot{\varepsilon}$, in their formulation, and include: visco-plastic models, Cowper-Symonds models, Ramburgh-Osgood models and creep models, all of which will be further discussed in chapter 5.

Level 8:   Specialized Plasticity Models

Some materials, although behaving in a plastic manner do not conform to standardised plasticity models. Examples of such materials are cast iron, concrete, soils, rock, materials exhibiting high creep rates etc. Specialized plasticity models have been developed to describe the plastic behaviour of these materials.

Level 9:   Damage Models

These models are used to model damage to composite or reinforced concrete materials, which typically occurs due to impact loading. These models are highly complex and typically involve the specification of many constants obtained from materials testing such as stiffness and failure stresses in three directions.

Level 10:   Specialized Material Models

Some materials exhibit a particular type of behaviour which does not neatly fit into any of the above categories and, as such, specialized models have been developed to describe their behaviour. Examples of such materials are shape memory alloys such as Nitinol.

3.3.2.3.   Loading

Level 1:   Static Point Force

The simplest method available of modelling a load is to assume that it is applied at a specific point and only acts through this point. For example a load may be assumed to be applied to the free end of a beam, the midpoint of a beam, the centre of a disc etc. In reality point loads to not occur as it is practically impossible to ensure that an applied load is focused at a certain point. The assumption of a point load, however, is a valid modelling method for certain types of problems, e.g. a cantilever beam subjected to a tip load. A static load is a load which is applied so slowly that it does not introduce any dynamic effects into the system. By assuming that a load is *static* we can assume that velocity and acceleration terms in the analysis may be ignored, thus considerably simplifying the analysis process.

Level 2:   Static Distributed Force

A distributed load is one of the most common types of loading encountered in structural analysis. Distributed loads on beams are found in many civil engineering applications in building and structure design. Distributed forces are also important in stress analysis of aircraft components, where aerodynamic loads are distributed over wings etc. This type of loading assumes that a force is applied and distributed over a certain length of the loaded geometry. The form of the distribution can vary from simple linear distributions to more complex non-linear distributions. Again, assuming that this load is static will considerably simplify the analysis.

Level 3:   Static Pressure Load

A pressure load assumes that an applied force is distributed over a certain area of the loaded geometry. The form of distribution can vary from a constant pressure to a highly non-linear distribution across the loaded area. Pressure loads are very common in engineering analysis and are found, for example, in: analysis of pressure vessels, analysis of metal forming techniques, analysis of aircraft structures and analysis of underwater structures.

Level 4:   Static Inertia Load

This category of loading refers to applied inertia loads that will not result in motion of the structure under investigation but will effect predicted stresses/strains. The main type of loading that fits into this category is self weight due to gravity. It is important to realise that, particularly if other types of load are not significant and a large structure is being analysed, self weight may cause significant deformation of the structure.

Level 5:   Time Varying Point Force

The static loading described in level 1 was applied slowly in order to avoid introducing dynamic effects into the system. Here this is not the case, a point load can be applied at any speed and can vary with time. For example, a load could increase and decrease according to a sine wave. Since time is now a factor, velocity and acceleration in the system will now have to be considered.

Level 6:   Time Varying Distributed Force

This level describes a distributed force which varies with time. The analysis of such a loading can be highly complex as one function may describe the distribution of load on the structure and a second function may describe how this variation changes with time.

Level 7:   Time Varying Pressure Load

This type of loading is an even more complex version of the previous level, where time varying forces which are distributed over specific areas of the model are used. As with the previous level, multiple functions may be needed to describe the loading regime and a dynamic analysis will be required. This type of loading is commonly used in non-linear analysis of metal forming operations where pressure loads are applied to work-pieces or tools.

Level 8:   Velocity

A velocity load can be applied to a whole model or to a certain part of a model. Where a velocity is applied to a whole model, analyses such as that of rotating machinery, turbine blades, and propellers can be carried out. Examples of where a velocity load is applied to only part of a model include impact analyses where a velocity is applied to a projectile.

Level 9:   Acceleration

Similar to velocity, acceleration loads can also be applied to a whole model or to a portion of the model. Acceleration loads are very useful in the analysis of space vehicles and aerospace components.

3.2.3.4   Boundary Conditions

Level 1:   Regions Fixed in All Degrees of Freedom

The simplest type of boundary condition is to fix a certain region of the problem in all degrees of freedom. This means that a particular part of a problem is held in space. A good example of such an approach is to assume that a beam acts as a cantilever, thus assuming that the built-in end of the beam is held in all degrees of freedom and hence cannot move.

Level 2:   Regions Fixed in Specific Degrees of Freedom

It is common to have regions in the problem that cannot move in one direction but are free to move in others. For example, consider a block sandwiched between two larger blocks, the block in the middle will not be able to move up or down because the larger blocks will constrain it from doing so; however it may still be able to move in the horizontal plane.

Level 3:   Prescribed Displacement of Specific Degrees of Freedom

This type of loading is useful when we know that a particular part of the structure moves a particular distance in a particular direction. For example in a metal forming process we may know that during the process the die moves a distance of 20mm while deforming the blank, thus we can prescribe a displacement on the die to model this loading. This type of loading has many applications and is applicable anywhere when we know exactly how far a point or points on the model move due to an applied load.

Level 4:   2-D Contact Surfaces

Two dimensional contact surfaces are used when two separate parts of the model come into contact during a 2-D analysis. A good example of this is in the analysis of many metal forming processes, such as the one shown in figure 3.13, where the work-piece is deformed via coming into contact with various forming tools. In this case 2-D contact surfaces will exist between the die and the work-piece, the clamp and the work-piece and the punch and the work-piece. Friction is obviously very important in such a process so the definition of such contact models will always include friction parameters.

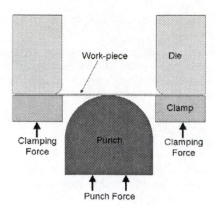

Figure 3.13: Example of 2-D Contact Surfaces in a 2-D Stretch Forming Model

Level 5:   <u>3-D Contact Surfaces</u>

Problems that require a three-dimensional mathematical model will often involve contact between 3-D bodies and will thus require 3-D contact surface models. Typical examples of problems requiring this type of contact are: metal forming processes which involve large amounts of relative sliding between work-pieces and forming tools (which cannot be satisfactorily described in 2-D), impact analysis problems, analysis of fastened joints etc. The main difference between this level and the previous level, is that here the contact surface will be an area whereas in the previous level the contact surface was a line.

Level 6:   <u>Time Dependant Boundary Conditions</u>

In certain cases it will be necessary to model boundary conditions which may be active at a certain stage of the analysis but will not be part of the analysis at a later stage, or vice versa. Normally the definition of such boundary conditions is dependant on time. Let's take the example of a square plate supported on two opposite edges, at the beginning of the analysis the plate is supported on both edges, however if we assume that part of the supporting structure will fail after a certain time or once a certain load is reached then the support at one of the edges will have to be removed at this point in the analysis. This concept is summarised in figure 3.14.

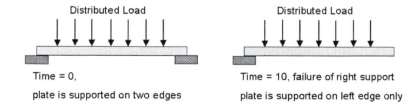

Figure 3.14: Example of Time Dependant Boundary Conditions

3.3.3. Step 3: Generation of the Finite Element Model

Once the mathematical model to be used has been determined from the previous step using the model hierarchy presented in table 3.1 then the actual process of generating the finite element model can begin. Choice of element type and material model should be clear from the previous step. For example, if a plane stress mathematical model was identified from table 3.1 then a 2-D structural solid type element that supports plane stress formulation is required in the finite element model. Similarly, if a particular type of mathematical material model was identified from table 3.1 then the corresponding model should be chosen in the finite element software. In most cases, if you have spent enough time thinking about the problem in the preceding step then these choices should be clear. In some cases there may be a number of modelling options available, for example in the case of a pressure vessel it may be possible to use either an axisymmetric, plane strain or thin shell element. In such cases it is best practice to try the simplest type of analysis first (in this case axisymmetric) as it may give you all the answers you need! We will discuss this further in chapters 4 and 6.

Once an element type and material model have been chosen the next task is to convert the geometric model of the entire problem (most likely available as a CAD file) into the geometry required for the FE analysis of the mathematical model (generally requiring defeaturing, removal of symmetrical features etc.). Again, this should be done with reference to decisions made in the previous step and by referring to your answers to the questions asked on pages 50/51. Particular emphasis should be placed on taking advantage of any symmetry inherent in the problem (figure 3.15) and removing any small geometric features that will not affect the analysis (figure 3.16). This will be discussed in detail in chapter 6.

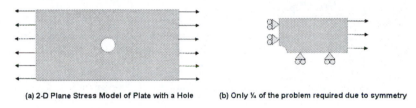

(a) 2-D Plane Stress Model of Plate with a Hole   (b) Only ¼ of the problem required due to symmetry

Figure 3.15: Taking advantage of Symmetry to Reduce Model Size

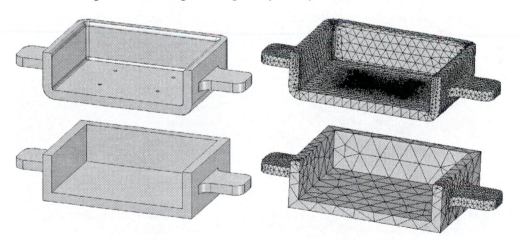

Figure 3.16: Example of Model Defeaturing – Solid Models on the Right Result in Meshes on the Left. Bottom Model has been De-featured to Simplify Mesh.

Once the geometry of the mathematical model is available it must now be "discretised", or split up into finite elements. Most contemporary FE software has automatic mesh generation capabilities with advanced algorithms that significantly aid this process. We will discuss the discretisation process in detail in chapter 6, but the main concern at this stage is to ensure that the mesh is suitable to adequately represent the change in stress throughout the geometry. This brings us back to step 1 where we should have identified possible regions of high stress (where stress will change rapidly from element to element) and should now ensure that we have a fine mesh in these regions with a suitable transition to less refined regions where stresses are not expected to change rapidly.

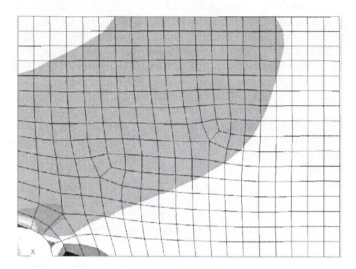

Figure 3.17: Example of an Unsuitable Mesh to Analyse the "Plate with Hole" Problem; Plot of Element Stresses Shows Discontinuities

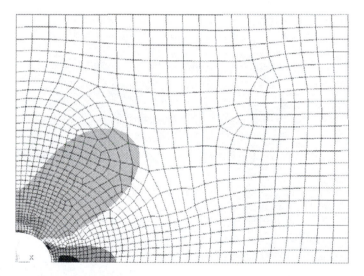

Figure 3.18: Example of a Suitable Mesh to Analyse the "Plate with Hole" Problem; Plot of Element Stresses Shows Smooth Transitions

Once a representative finite element mesh has been obtained the next stage in the FE process is to apply loads and boundary conditions. The types of loads and boundary conditions should have been identified in the previous step, with reference to table 3.1 and so, should just be a matter of implementing the required choices in the FE software. So far as is possible, one should avoid applying point loads to a problem as this will cause artificial stress concentrations to be introduced into the problem. Other types of loads can be applied with relative ease. This stage in the process will be discussed in detail in chapter 7.

3.3.4. Step 4: Selection of Computational Methods and Obtaining a Solution

Most commercial FE software's have a variety of computational strategies available for solution of the problem. Selecting the most appropriate method is a difficult task for an inexperienced user. This is made even more difficult due to the fact that many methods need to be closely monitored during solution and have certain parameters adjusted in order to ensure convergence and stability. The basic structural computational methods offered by most FE software are:

- Static Analysis
  This is the most basic type of analysis available which assumes that there are no dynamic effects in the system and hence that all loads are applied very slowly without introducing dynamic effects. A static analysis can consider steady inertia loads such as gravity or rotational velocity. Static analyses can be both linear and non-linear. Examples of non-linearity allowed in a static analysis include: large deformations, contact, plasticity, creep, stress stiffening, hyper-elasticity etc.

- Modal Analysis
  A modal analysis is used to determine the natural frequencies and associated mode shapes of a structure. A number of different mathematical methods for formulating the problem and obtaining a solution are usually available in order to solve many different types of problem.

- Harmonic Analysis
  A structure that is exposed to a sustained cyclic load will respond with a sustained cyclic response (i.e. a harmonic response). A harmonic response analysis allows for the prediction of the sustained dynamic behaviour of a structure, thus providing information as to whether a design will successfully overcome resonance, fatigue, and other harmful effects of forced vibrations. The structure to be analysed is generally subjected to loads that vary sinusoidally (harmonically) with time. The response of the structure is examined at several frequencies and generally a graph of some response quantity (usually displacements) versus frequency is produced. Peak responses on the graph are then identified and stresses reviewed at those peak frequencies. There are normally a number of different solution methods available for harmonic analysis which are individually applicable to particular problems.

- Transient Dynamic Analysis
  A transient dynamic analysis (sometimes called time-history analysis) is a method used to determine the dynamic response of a structure due to the action of any type of time-dependent loading. This type of analysis is used to determine time-varying displacements, strains, stresses, and forces in a structure as it responds to any combination of static, transient, and harmonic loads. Unlike a static analysis, in this case the time scale of the loading is such that the inertia or damping effects are considered to be important. Again, there are normally a number of different transient solution methods available depending on the type of structure and problem being analysed.

- Explicit Dynamic Analysis
  Explicit dynamic analyses are used to obtain fast solutions for short-time, large deformation dynamic, quasi-static problems with large deformations and multiple nonlinearities, and complex contact/impact problems. The explicit

solution method is different to the more common implicit solution method used in most other types of analysis. Explicit dynamic analysis is essentially used for complex problems that are too difficult to solve using a transient dynamic analysis.

- Other Specialised types of Analysis
    o Spectrum Analysis: the results of a modal analysis are used with a known spectrum to calculate displacements and stresses in the model. Used in place of a time-history analysis to determine the response of structures to random or time-dependent loading conditions such as earthquakes, wind loads, jet engine thrust, rocket motor vibrations etc.
    o Buckling Analysis: used to determine buckling loads (critical loads at which a structure becomes unstable) and buckled mode shapes (the characteristic shape associated with a structure's buckled response).

As mentioned several times above, each of the solution types will normally have a number of solution methods available for use and the choice of which solver to use will be dependant on the particular type of problem being analysed. These solution types will be discussed in detail in chapter 8.

3.3.5. Step 5: Post-Processing, Model Verification and Validation

This is the stage where you query your solved model and attempt to extract information required to allow you to make design decisions, to help determine if the model accurately models the physical problem and to help determine any problems with the modelling or solution process.

A structural finite element analysis will always result in a list of displacements of each node in the finite element mesh. Post-processing converts these displacements into more useful results such as contour plots of stress and strain throughout the model. At this stage it is very important to check your results and ask questions such as:

- Has the yield stress of the material been exceeded?
- Has the UTS of the material been exceeded?
- Are displacements excessively large?
- Are stresses/strains excessively large?
- Is stress transitioning smoothly through the model in element plots?
- Do error estimation plots show acceptable levels?

If a linear elastic material has been used and the material yield stress has been exceeded due to the correctly applied loading then a new analysis which includes post yielding material behaviour must be begun, otherwise the applied loading should be suitably scaled down to bring predicted stresses to sub yield values. If excessively large displacements/strains or stresses are noticed in the model then there is clearly an error in the model generation or the assumptions made at the beginning of the process. This is normally due to incorrect units being used, incorrect material model parameters being used or incorrect/excessive loading. The model should be checked for errors and adapted accordingly.

If the model is checked and found to be sound based on the results the next stage is to validate the model against experimental results or a higher order model. This may result in required changes to assumptions made at the beginning of the process and require another analysis cycle or a series of analysis cycles to be completed in order to validate the finite element model. This stage of the process will be discussed in detail in chapter 9

3.4. Theoretical Overview of the FEA Procedure

3.4.1. Fundamentals

Let's consider a three-dimensional body of volume V and having an external surface S, with points on the body defined via an x, y, z coordinate system. A portion of the surface of the body, S_u, is constrained as shown in figure 3.19. A distributed force per unit area, f^S, (i.e. surface load) acts on a portion of the boundary surface, S_{fs}.

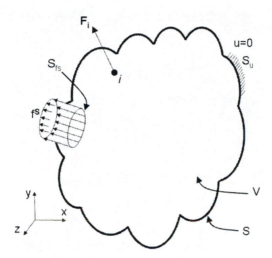

Figure 3.19: A three-dimensional body subjected to various loads

The body will obviously deform due to this surface load and the deformation of any point in the body is given by the three components of its displacement:

$$\{u\} = \begin{bmatrix} u \\ v \\ w \end{bmatrix} = [u, v, w]^T \tag{3.01}$$

where u is displacement in the x-direction, v is displacement in the y-direction and w is displacement in the z-direction. The superscript T at the right hand side of equation 3.01 indicates the transpose of a matrix or vector and indicates that the rows in the matrix should be turned into columns or vice versa. This notation will be used throughout this section to allow for better flow of the text. The reader should remember that wherever the transpose symbol is seen the horizontal matrix should be written as a vertical column etc.

A distributed force per unit volume, f^B, (i.e. body load) may act on the body and is given by:

$$\{f^B\} = \begin{bmatrix} f_x^B & f_y^B & f_z^B \end{bmatrix}^T \tag{3.02}$$

Examples of body loads are gravity effects such as weight per unit volume.

The surface load will be given by its component values at points on the surface:

$$\{f^s\} = \begin{bmatrix} f_x^s & f_y^s & f_z^s \end{bmatrix}^T \tag{3.03}$$

Examples of surface loads are pressure loads or contact force due to contact with another body.

A point load F acting on a point i is given by it's component values:

$$\{F^i\} = \begin{bmatrix} F_x^i & F_y^i & F_z^i \end{bmatrix}^T \tag{3.04}$$

Since we have a three-dimensional body the stresses acting on any elemental volume of the body will be given by:

$$\{\sigma\} = \begin{bmatrix} \sigma_{xx} & \sigma_{yy} & \sigma_{zz} & \tau_{xy} & \tau_{yz} & \tau_{zx} \end{bmatrix}^T \tag{3.05}$$

The strains corresponding to these stresses are:

$$\{\varepsilon\} = \begin{bmatrix} \varepsilon_{xx} & \varepsilon_{yy} & \varepsilon_{zz} & \gamma_{xy} & \gamma_{yz} & \gamma_{zx} \end{bmatrix}^T \tag{3.06}$$

We can relate strains to displacements via:

$$\{\varepsilon\} = \begin{bmatrix} \dfrac{\partial u}{\partial x} & \dfrac{\partial v}{\partial y} & \dfrac{\partial w}{\partial z} & \dfrac{\partial u}{\partial y} + \dfrac{\partial v}{\partial x} & \dfrac{\partial v}{\partial z} + \dfrac{\partial w}{\partial y} & \dfrac{\partial w}{\partial x} + \dfrac{\partial u}{\partial z} \end{bmatrix}^T \tag{3.07}$$

We can relate stress to strain for a linear elastic material via Hooke's law:

$$\{\sigma\} = [D]\{\varepsilon\} \tag{3.08}$$

Where [D] is the material matrix given by:

$$[D] = \frac{E}{(1+v)(1-2v)} \begin{bmatrix} 1-v & v & v & 0 & 0 & 0 \\ v & 1-v & v & 0 & 0 & 0 \\ v & v & 1-v & 0 & 0 & 0 \\ 0 & 0 & 0 & 0.5-v & 0 & 0 \\ 0 & 0 & 0 & 0 & 0.5-v & 0 \\ 0 & 0 & 0 & 0 & 0 & 0.5-v \end{bmatrix} \tag{3.09}$$

3.4.2. The Minimum Potential Energy Method

Our problem here is to determine the displacement of the body shown in figure 3.19 while satisfying the equilibrium equations for three-dimensional stress analysis given in equation 2.13. From the above we know that stresses are related to strains (equation 3.08) which in turn are related to displacements (equation 3.07). This requires the solution of 2nd order partial differential equations. Solving these equations is not possible for complex geometries so we will introduce a potential energy method which is less stringent and allows for an approximate solution to the problem to be obtained.

The total potential energy Π of an elastic body is given by the sum of total strain energy and the sum of the work done by external forces:

$$\Pi = \text{strain energy} + \text{external work} \qquad (3.10)$$

For a linear elastic material the strain energy per unit volume in the body is given by $\frac{1}{2}\{\sigma\}^T\{\varepsilon\}$. For the elastic body shown in figure 3.191 the total strain energy Λ is given by:

$$\Lambda = \frac{1}{2}\int_V \{\sigma\}^T\{\varepsilon\}dV \qquad (3.11)$$

The work done by the external forces; a combination of body forces, surface loads and point loads, is given by:

$$\text{External work} = -\int_V \{u\}^T\{f^B\}dV - \int_S \{u\}^T\{f\}^S dS - \sum_i \{U_i\}^T\{F^i\} \qquad (3.12)$$

Thus, the total potential energy for the general elastic body shown in figure 3.19 is:

$$\Pi = \frac{1}{2}\int_V \{\sigma\}^T\{\varepsilon\}dV - \int_V \{u\}^T\{f^B\}dV - \int_S \{u\}^T\{f^S\}dS - \sum_i \{U_i\}^T\{F^i\} \qquad (3.13)$$

The minimum total potential energy principle simply states that for a stable system, the displacement at the equilibrium position occurs such that the value of the systems total potential energy is a minimum, i.e.:

$$\frac{\partial \Pi}{\partial U} = 0 \qquad (3.14)$$

In order to illustrate the above points lets consider an example:

Example 3.1

Figure 3.20 shows a one dimensional finite element model which consists of a series of springs.

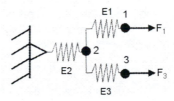

Figure 3.20: A System of Springs

The total potential energy for the system is given by:

$$\Pi = \frac{1}{2} k_1 \delta_1^2 + \frac{1}{2} k_2 \delta_2^2 + \frac{1}{2} k_3 \delta_3^2 - F_1 U_1 - F_3 U_3$$

where δ_i is the extension of spring i. We can express the extension of each spring in terms of the displacement of each node that defines the spring, thus: $\delta_1 = U_1 - U_2$, $\delta_2 = U_2 - 0$, $\delta_3 = U_3 - U_2$, and our equation becomes:

$$\Pi = \frac{1}{2} k_1 (U_2 - U_1)^2 + \frac{1}{2} k_2 (U_2)^2 + \frac{1}{2} k_3 (U_3 - U_2)^2 - F_1 U_1 - F_3 U_3$$

Where u_i is the displacement of node i.

Since this is a three degree of freedom system (i.e. we have three nodes that can move in one-dimension), for equilibrium we must minimize Π with respect to u_1, u_2 and u_3, i.e.

$$\frac{\partial \Pi}{\partial U_i} = 0 \quad \text{for i = 1,2,3}$$

Which is:

$$\frac{\partial \Pi}{\partial U_1} = k_1 (U_1 - U_2) - F_1 = 0$$

$$\frac{\partial \Pi}{\partial U_2} = -k_1 (U_1 - U_2) + k_2 U_2 - k_3 (U_3 - U_2) = 0$$

$$\frac{\partial \Pi}{\partial U_3} = k_3 (U_3 - U_2) - F_3 = 0$$

These equilibrium equations can be put in the matrix form of KU=F as usual:

$$\begin{bmatrix} k_1 & -k_1 & 0 \\ -k_1 & k_1 + k_2 + k_3 & -k_3 \\ 0 & -k_3 & k_3 \end{bmatrix} \begin{bmatrix} U_1 \\ U_2 \\ U_3 \end{bmatrix} = \begin{bmatrix} F_1 \\ 0 \\ F_3 \end{bmatrix}$$

It should be clear that this is the same set of equations that we would have arrived at had we used the direct method of formulating the finite element analysis as detailed in example 1.1. in chapter 1. Thus, the minimum potential energy approach provides a valuable method of arriving at the governing equation for a finite element analysis without any reference to free body diagrams etc. This makes the potential energy method very attractive for formulating large and complex problems, which is why it is implemented in many commercial FE software packages.

Now that we know how the minimum potential energy method works, via example 3.1, we can now explore the method in more depth and examine how it is specifically implemented by most FE codes.

Equation 3.13 gives the general form of the equation for potential energy for a continuum. In a finite element analysis we will have discretised the problem into a discrete problem represented by a number of finite elements which are interconnected at nodes on the element boundaries. In the finite element approach the displacements measured within each element are assumed to be a function of the displacements of the nodes. Thus, the displacement within the element is related to the displacement of the nodes via an interpolation function known as a *shape function.* We shall discuss shape functions in detail in chapter 4, however for now we just need to know that, for any element:

$$\{u\}^{(m)} = [S]^{(m)} \{U\} \tag{3.15}$$

where $[S]^{(m)}$ is the shape function matrix for element m and $\{U\}$ is the nodal point displacement vector of the element. For illustrative purposes we can show that if we are using a two node one-dimensional linear element then equation 3.15 will be given by:

$$\{u\}^{(m)} = [S_i \ S_j] \begin{Bmatrix} U_i \\ U_j \end{Bmatrix}$$

where U_i is the displacement of node i and U_j is the displacement of node j. So, if we know the displacements of all of the nodes that make up an element we can describe how any point in the element displaces.

Using equation 3.15 we can rewrite equation 3.13 in terms of a discretised system as:

$$\Pi = \sum_m \frac{1}{2} \int_{V_m} \{\sigma\}^T_{(m)} \{\varepsilon\}_{(m)} \, dV_{(m)} - \sum_m \int_{V_m} \{u\}^T_{(m)} \{f^B\}_{(m)} \, dV_{(m)} - \sum_m \int_{S_m} \{u\}^T_{(m)} \{f^S\}_{(m)} \, dS_{(m)} - \sum_i U_i^T F^i \tag{3.16}$$

Note that in equation, 3.16 all terms refer to elements, *m*, except for the final term which assumes that point loads are applied to the nodes, *i*.

Let's now concentrate on the first term in equation 3.16 which deals with element stress, $\sigma_{(m)}$ and element strain, $\varepsilon_{(m)}$. From equation 3.15, which relates element displacement to nodal displacement we can obtain:

$$\{\varepsilon\}^{(m)} = [B]^{(m)} \{U\} \tag{3.17}$$

where [B] is the element *strain-displacement matrix* and relates element strains to nodal point displacements. So, from equation 3.17, if we know the displacements of all of the nodes that make up an element we will know the strain in that element.

The stress within a finite element is related to the element strains and initial stresses by

$$\{\sigma\}^{(m)} = [D]^{(m)} \{\varepsilon\}^{(m)} \tag{3.18}$$

where $[D]^{(m)}$ is a matrix which relates stress to strain in element m. $[D]^{(m)}$ is essentially a matrix which contains the material behaviour of the element and may be isotropic or anisotropic and may vary from element to element. Putting Equations 3.18 and 3.17 together gives us an alternative expression for the first term in equation 3.16:

$$\sum_m \frac{1}{2} \int_{V_m} \{\sigma\}^T_{(m)} \{\varepsilon\}_{(m)} \, dV_{(m)} = \sum_m \frac{1}{2} \int_{V_m} [D]_{(m)} [B]^T_{(m)} \{U\}^T [B]_{(m)} \{U\} dV_{(m)} \qquad (3.19)$$

Since, $\{U\}$ is independent of the element being considered the right hand side of equation 3.A19 can be rewritten as:

$$\{U\}^T \left[\sum_m \frac{1}{2} \int_{V_m} [B]^T_{(m)} [D]_{(m)} [B]_{(m)} \, dV_{(m)} \right] \{U\} \qquad (3.20)$$

Replacing the first term in equation 3.16 with equation 3.20 gives:

$$\Pi = \{U\}^T \left[\sum_m \frac{1}{2} \int_{V_m} [B]^T_{(m)} [D]_{(m)} [B]_{(m)} \, dV_{(m)} \right] \{U\} - \sum_m \int_{V_m} \{u\}^T_{(m)} \{f\}^B_{(m)} dV_{(m)} - \sum_m \int_{S_m} \{u\}^T_{(m)} \{f\}^S_{(m)} dS_{(m)} - \sum_i \{U\}^T_i \{F\}^i \qquad (3.21)$$

Considering force terms and relating to shape functions (equation 3.15) we can similarly see that:

$$\sum_m \int_{V_m} \{u\}^T_{(m)} \{f\}^B_{(m)} dV_{(m)} = \{U\}^T \sum_m \int_{V_m} [S]^T_{(m)} \{f\}^B_{(m)} dV_{(m)} \qquad (3.22)$$

$$\sum_m \int_{S_m} \{u\}^T_{(m)} \{f\}^S_{(m)} dS_{(m)} = \{U\}^T \sum_m \int_{S_m} [S^S]^T_{(m)} \{f\}^S_{(m)} dS_{(m)} \qquad (3.23)$$

Where $[S^S]_{(m)}$ is the surface displacement interpolation matrix (i.e. surface shape function) for element m, and is obtained from the volume displacement interpolation matrix (shape function), $[S]_{(m)}$, in equation 3.15 by substituting the element surface coordinates. Putting equation 3.22 and 3.23 into equation 3.21 gives:

$$\Pi = \{U\}^T \left[\sum_m \frac{1}{2} \int_{V_m} [B]^T_{(m)} [D]_{(m)} [B]_{(m)} \, dV_{(m)} \right] \{U\} - \{U\}^T \sum_m \int_{V_m} [S]^T_{(m)} \{f\}^B_{(m)} dV_{(m)} - \{U\}^T \sum_m \int_{S_m} [S^S]^T_{(m)} \{f\}^S_{(m)} dS_{(m)} - \sum_i \{U\}^T_i \{F\}^i \qquad (3.24)$$

Which can be simplified to:

$$\Pi = \{U\}^T \left[\sum_m \frac{1}{2} \int_{V_m} [B]^T_{(m)} [D]_{(m)} [B]_{(m)} \, dV_{(m)} \right] \{U\} - \{U\}^T \left[\sum_m \int_{V_m} [S]^T_{(m)} \{f\}^B_{(m)} dV_{(m)} - \sum_m \int_{S_m} [S^S]^T_{(m)} \{f\}^S_{(m)} dS_{(m)} - \sum_i \{F\}^i \right] \qquad (3.25)$$

Equation 3.25 represents an equation for the potential energy of a finite element model which is entirely described by the parameters: $[S]_{(m)}$, $[B]_{(m)}$, $[D]_{(m)}$, $\{U\}$, $\{f^B\}$, $\{f^S\}$ and $\{f\}$. When performing a finite element analysis we choose to use a certain element type, this element type will have associated shape functions describing it's behaviour ad thus will determine $[S]_{(m)}$. The matrices $[B]_{(m)}$ and $[S^S]_{(m)}$ are determined by mathematically manipulating $[S]_{(m)}$. By assigning a particular material model to the finite elements we will describe $[D]_{(m)}$, which is essentially, the material property matrix. $\{U\}$ is the vector of nodal displacements which we are seeking to find – if we know what makes up $\{U\}$ then we have the answer to the problem. The applied forces $\{f^B\}$, $\{f^S\}$ and $\{f\}$ will be known as part of the problem description. Thus, equation 3.25 describes a general form for the formulation of a finite element analysis using potential energy methods like the minimum potential energy method.

It is common practice to assemble all of the force terms for each element into one global force vector which describes the entire loading scenario in the problem. Each $[B]^T[D][B]$ term for each element is also assembled into a global stiffness matrix which describes the physical behaviour of each element and subsequently the entire problem domain. Thus equation 3.25 can be rewritten as:

$$\Pi = \frac{1}{2}\{U\}^T [K]\{U\} - \{U\}^T \{F\} \tag{3.26}$$

where K is the global stiffness matrix, F is the global load vector and U is the global displacement vector.

If we now minimise equation 3.26, i.e. let: $\dfrac{\partial \Pi}{\partial U_i} = 0$ for i = 1,2,3,……..,n, we obtain:

$$[K]\{U\} = \{F\} \tag{3.27}$$

Which is the governing equation for all structural static finite element analysis, as discussed in chapter 1 (see equation 1.16).

The global stiffness matrix, K, is given by:

$$K = \sum_m \int_{V_m} [B]^T_{(m)} [D]_{(m)} [B]_{(m)} dV_{(m)} \tag{3.28}$$

The global force vector, F, is given by:

$$\{F\} = \{F\}^B + \{F\}^S + \{F\}^C \tag{3.29}$$

Where $\{F\}^B$ is the effect of element body forces and is given by:

$$\{F\}^B = \sum_m \int_{V_m} [S]^T_{(m)} \{f\}^B_{(m)} dV_{(m)} \tag{3.30}$$

$\{F\}^S$ is the effect of element surface forces:

$$\{F\}^S = \sum_m \int_{S_m} [S^S]^T_{(m)} \{f\}^S_{(m)} dS_{(m)} \tag{3.31}$$

and, $\{F\}^C$ are the concentrated loads or nodal point loads:

$$\{F\}^C = \sum_i \{F\}^i \tag{3.32}$$

Summary of the Above

The theory outlined above demonstrates how the potential energy method can be used to formulate the finite element method. The potential energy method is just one method of many that can be used to formulate a FEA. Other methods include: weighted residual methods like the least squares method, variational techniques such as the Rayleigh-Ritz method, the principal of virtual displacements etc. As this text is concerned with practical FEA we will confine our study of theory to the minimum potential energy method and use this to illustrate examples in subsequent chapters. There are many excellent textbooks on FEA theory using other methods available should the reader require this information. The important thing to understand about all the methods is that each will result in equations 3.27, 3.28 and 3.29 regardless of the formulation method used.

You may ask yourself the question "why do I need this complex theory when I can formulate a FEA based on the direct method outlined in example 1.1?" Unfortunately the direct method only works for simple problems with simple elements. When we use a more complex element (e.g. 2-d, 3-d, quadratic, cubic) assembling the global matrices becomes highly complex. The method above works for any level of complexity as it is based on using element shape functions. When a change in complexity occurs (e.g. going from 2D to 3D) then the element shape functions will change which will automatically update the entire problem formulation. Thus the above method is very general and is not tied to specific element types or problem types.

It is important to understand the above as we will use this methodology to describe how specific items (i.e. elements, material models, loads and boundary conditions) work in later chapters. Please take the time to read over this section a number of times and work through it on a sheet of paper. If you take the time to work through it you will find that it is not that complicated.

3.4.3. Dynamic Analysis and Non-linear Analysis

Equation 3.27 is a statement of the static equilibrium of the finite element mesh. In equilibrium considerations, applied forces may vary with time, in which case the displacements may also vary with time. In such a case equation 3.27 is a statement of equilibrium for a specific point in time. If, in reality, the loads are applied rapidly, inertia forces must be considered and a dynamic problem must be solved. Using d'Alembert's principle, the element inertia forces may be included as part of the body forces. In such a case equation 3.30 becomes:

$$\{F\}^B = \sum_m \int_{V(m)} [S]_{(m)}^T \left[\{f\}_{(m)}^B - \rho_{(m)} [S]_{(m)} \{\ddot{U}\} \right] dV_{(m)} \tag{3.33}$$

where $\{f\}_{(m)}^B$ no longer includes inertial forces, $\{\ddot{U}\}$ gives nodal point accelerations and $\rho_{(m)}$ is the mass density of element m. In this case the equilibrium equations are:

$$[M]\{\ddot{U}\} + [K]\{U\} = \{F\} \tag{3.34}$$

where [K] is the global stiffness matrix, [M] is the global mass matrix and {F} and {U} are time dependant. The global mass matrix is given by:

$$[M] = \sum_m \int_{V(m)} \rho_{(m)} [S]_{(m)}^T [S]_{(m)} dV_{(m)} \tag{3.35}$$

However, in a dynamic analysis some energy is dissipated during vibration, which in vibration analysis is usually taken account of by introducing velocity dependant damping forces. Introducing the damping forces as additional contributions to the body forces changes equation 3.33 as follows:

$$\{F\}^B = \sum_m \int_{V(m)} \left[\{S\}^T_{(m)} \left[\{f\}^B_{(m)} - \rho_{(m)} [S]_{(m)} \{\ddot{U}\} - c_{(m)} [S]_{(m)} \{\dot{U}\}\right] dV_{(m)}\right. \qquad (3.35)$$

where $\{\dot{U}\}$ is a vector of the nodal point velocities and $c_{(m)}$ is the damping property parameter of element m. In this case the equilibrium equations become:

$$[M]\{\ddot{U}\} + [C]\{\dot{U}\} + [K]\{U\} = \{F\} \qquad (3.36)$$

where [C] is the global damping matrix and can be written as:

$$[C] = \sum_m \int_{V(m)} c_{(m)} [S]^T_{(m)} dV_{(m)} \qquad (3.37)$$

Nonlinearities

In the above formulation it was assumed that the displacements of the finite element assembly are small, that the material is linearly elastic and that the boundary conditions remain unchanged during the application of loads. These assumptions have entered the equilibrium equation in the following manner:

(a)   The fact that all integrations have been performed over the original volume of the finite elements implies that displacements must be small. This affects the stiffness matrix, K, and the load vector, F.

(b)   The strain-displacement matrix, B, of each element was assumed to be constant and independent of element displacements.

(c)   The assumption of a linear elastic material is implied in the use of a constant stress-strain matrix, D.

(d)   Unchanged boundary conditions are implied by keeping constant constraint relations for the complete response.

These observations point to the different types of nonlinearity that may arise in a finite element analysis:

(1)   Non linearity due to large displacements, large rotations, but small strains.

(2)   Non linearity due to large displacements, large rotations and large strains.

(3)   Material non linearity.

(4)   Non linearity due to contact.

3.5. Summary of Chapter 3

After completing this chapter you should understand the various steps involved in setting up and solving a practical finite element analysis. You should also understand how the finite element method works and is formulated using the minimum potential energy method. In particular you should be able to:
  • Describe the important features of a particular problem and identify modelling strategies that can be used to make a mathematical model of the problem.

- Identify a suitable mathematical model by considering a hierarchy of available models for geometry, materials, loading and boundary conditions with reference to table 3.1.
- Use the hierarchy presented in table 3.1 to turn any stress analysis problem into an appropriate mathematical model for use in a finite element analysis.
- Describe how the minimum potential energy method can be used to formulate a finite element analysis, and understand how user selected features like element shape functions, material models and loads are implemented and how they will effect the overall formulation of a particular problem

3.6   Problems

P3.1. Use table 3.01 to determine a suitable mathematical model for the determination of maximum stress in a steel plate with a hole subjected to uniform in-plane tension as shown in the figure below. You may assume that the load is applied very slowly and does not cause permanent deformation of the plate.

Answer: Geometry level 5, plane stress. Material level 1, linear elastic isotropic. Loading level 2, static distributed force. Boundary conditions level 2, regions fixed in specific DOF

P3.2. Use table 3.01 to determine a suitable mathematical model for the analysis of stress distribution in a thick cylinder with open ends subjected to a uniform internal pressure as shown in the figure below. You may assume that the load in this case is large enough to cause yielding of the inner surface of the pressure vessel but it is applied very slowly.

Answer: Geometry level 6, axisymmetric. Material level 6, Elasto-plastic rate independent. Loading level 3, static pressure load. Boundary conditions level 2, regions fixed in specific DOF

P3.3. Use table 3.01 to determine a suitable mathematical model for the analysis of deformation of a rubber balloon which is inflated via a uniform internal pressure as shown in the figure below.

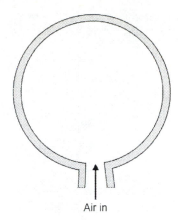

Air in

Answer: Geometry level 6, axisymmetric. Material level 2, non-linear elastic. Loading level 7, time-varying pressure load. Boundary conditions level 2, regions fixed in specific DOF

P3.4. Use table 3.01 to determine a suitable mathematical model for the analysis of deformation of a circular blank which is deformed into a cup using a shouldered cylindrical punch in a deep drawing process. As part of the analysis, the forming process will be carried out at various speeds to investigate the effect of deformation speed on the properties of the drawn cup.

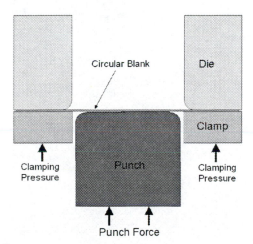

Answer: Geometry level 6, axisymmetric. Material level 7, elasto-plastic rate dependant. Loading levels 6 & 7, time-varying distributed force on the punch and time-varying pressure load on the blank holder. Boundary conditions levels 2 & 4, regions fixed in specific DOF and 2D contact surfaces.

P3.5. Use table 3.01 to determine a suitable mathematical model to determine the stress distribution in a turbine blade which is mounted on a disk rotating at 500 rpm as shown in the figure below. The turbine blade is made from a metal-matrix-composite material and has a highly complex geometry.

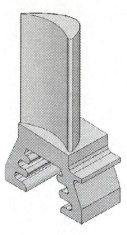

Answer: Geometry level 10, 3D solid. Material level 3, anisotropic linear elastic. Loading level 8, specified angular velocity. Boundary condition levels 2, base of the blade is fixed in specific DOF.

4

Elements

4.1.   Introduction

As discussed in chapter 3, the most fundamental decision to make before starting a finite element analysis is to decide what type of element should be used to model the problem under investigation. The geometry of the mathematical model, the material model to be used and type of loading to be applied, will all be dependent to some extent on the chosen element type and its behaviour. This chapter will begin with an overview of the general types of elements available for structural analysis and will then deal with each element type specifically. In each case we will describe how the element works, when and where it should be used and detail examples of appropriate use of the particular element in FEA.

4.2.   Basic Element Shapes and Behaviour

The types of elements available for structural analysis in most commercial finite element software packages can be summarised as shown in table 4.1.

Basic Shape	Subtypes	Representative Geometry
Point	Mass Element	
Line	Spar/Truss Element Spring Element Beam Element Pipe Elements 2D Contact/Gap Element 2D Surface Effect Element	
Area	2D Solid Element - Plane Stress - Plane Strain - Axisymmetric Plate Element	
Curved Area	Shell Element 3D Contact Element 3D Surface Effect Element	
Volume	3D Solid Element	

Table 4.1: Overview of Structural Finite Elements

We will now briefly describe each of the various basic element shapes from table 4.1 before going on to discuss the various subtypes in detail in the remainder of this chapter.

4.2.1. Point Elements

The only type of point element relevant to structural analyses are point mass elements. As you would expect, these are used to define a concentrated point mass at a particular point in a finite element model.

These elements are generally used in combination with spring elements to generate mass-spring models of engineering systems. Another typical use is to model pendulums in combination with spar or beam elements.

4.2.2. Line Elements

Line elements come in many different forms. Although each of the following subtypes appears as a line from one point in space to another, the formulation of the element linking those points in space is very different.

(i)   Spar/Truss Elements
These elements, which may be known as a spar, link or truss element, can extend or compress axially, but cannot bend or change their cross section. These elements may be thought of as a line which can change its length but not its direction.

(ii)   Spring Elements
Spring elements are essentially similar to spar/truss elements in that they can be compressed or extended but cannot be bent. These elements are typically used with point masses to build mass-spring models of engineering problems.

(iii)   Beam Elements
Beam elements can be used to model beam problems where tension, compression and bending of the element are all allowed. These elements allow the definition of section properties to model the beam cross sectional shape and hence accurately predict bending behaviour. In some cases the definition of tapered cross sections is available to model tapered beams. 3D beam elements may also allow torsional effects to be modelled.

(iv)   Pipe Elements
These elements are essentially a development of circular cross section beam elements and are used to model pipe networks, useful in building services engineering etc. Pipe elements will generally have tension, compression and bending effects available. 3D elements may allow for torsion and advanced pipe elements allow for hydrodynamic and buoyant effects in order to model submerged pipes etc.

(v)   2D Contact/Gap Elements
These elements are used to define contact between two structures (e.g. a circle impacting a rectangle) or a gap between two parts of a structure. Two types of line contact elements are available:

a) A 2D/3D point-to-point contact element is a line joining two points (i.e. nodes) on two separate structures and defines contact between those two points.

During loading, these points may break or maintain this contact and/or may slide relative to one another.

b) A 2D surface-to-surface contact element is used to model contact and sliding between two 2D structures.

(vi) 2D Surface Effect Elements
These surface effect elements are overlaid onto a face of a 2D solid element and used to apply various loads and surface effects.

4.2.3. Area Elements

Area elements come in two fundamental shapes: triangles and rectangles. Each of the various subtypes are available in both shapes. Sometimes, depending on the finite element software used, each of the subtypes are available as options within a particular element formulation.

(i) 2D Solid Element - Plane Stress
These elements are rectangles or triangles that are used to model thin sheet structures according to the plane stress assumption discussed in chapter 2 and 3. The elements will behave according to the governing equations for plane stress analysis and can deform in the x and y directions.

(ii) 2D Solid Element - Plane Strain
These elements are rectangles or triangles that are used to model thick structures where the cross-sectional properties don't vary significantly along the thickness according to the plane strain assumption discussed in chapter 2 and 3. The elements will behave according to the governing equations for plane strain analysis and can deform in the x and y directions.

(iii) 2D Solid Element - Axisymmetric
Axisymmetric elements are rectangles or triangles that are used to model axisymmetric problems as defined in chapter 2. These elements can deform in the radial and axial direction. In some cases advanced axisymmetric elements are available which can also model torsional effects.

(iv) Plate Element
Plate elements are basically plane stress elements which allow for the specification of plate thickness and are generally only used for thin plates where bending of the plate is not considered.

4.2.4 Curved Area Elements

These elements are essentially 3D elements that have no thickness.

(i) Shell Elements
Shell elements are curved plate elements that include both bending and membrane or stretching effects and are suitable for modelling 3D thin shell type structures. Typical examples of application include modelling thin walled pressure vessels and analysis of sheet metal forming processes. These elements can deform in the x, y and z-direction and element thickness is generally input during the element definition.

(ii) 3D Contact Elements
These elements are used to define contact between two structures (e.g. a sphere impacting a plate) or a gap between two parts of a structure. 3D surface-to-

surface contact elements are used to model contact and sliding between two 3D structures, e.g. shell to shell, solid to shell, shell to solid etc.

(iii)   3D Surface Effect Elements
These surface effect elements are overlaid onto a face of a 3D solid or shell element and used to apply various loads and surface effects.

4.2.5. Volume Elements – 3D Solid Elements

3D Solid elements come in two fundamental shapes bricks (hexahedrons) and pyramids (tetrahedrons). These elements are used for 3D modelling of solid structures. Many advanced types of 3D solids are available for modelling layered composite materials, reinforced concrete materials, anisotropic materials etc.

4.3.   Overview of Element Behaviour

In this section we will describe and define important element characteristics and concepts that we will need to understand when we discuss specific element types and their application later in the chapter. We will illustrate the concept of shape functions, demonstrate the difference between linear and quadratic elements and finally introduce the concept of natural element coordinate systems.

4.3.1. Shape Functions

In order to illustrate what shape functions are and how they work we will study a simple one dimensional stress analysis problem. Consider the steel bar shown in figure 4.01. The bar is rigidly fixed at its left hand edge and a load is applied to its right hand edge. The resulting displacement distribution in the bar is shown in figure 4.02.

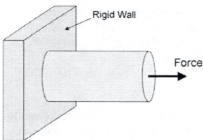

Figure 4.01: A One-Dimensional Structural Problem

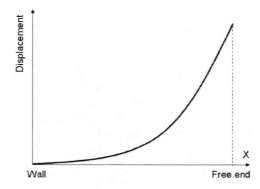

Figure 4.02: Distribution of Displacement along the Length of the Bar due to Applied Load

We will model the bar using 1D linear spar elements. As a first approximation we divide the bar into three elements and four nodes. The actual displacement distribution may be approximated by a combination of linear functions as shown in the figure 4.03. To better approximate the actual displacement gradient near the right hand edge of the bar in our FE model we have placed the nodes closer to each other in this region. Obviously we could further improve the accuracy of our approximation by also increasing the number of elements but for illustrative purposes we will proceed with our model as is.

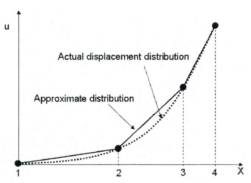

Figure 4.03: Modelling Displacement Distribution Using Linear Elements

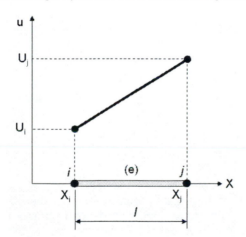

Figure 4.04: A Typical 1D Linear Spar Element

Let's consider a typical element from the model as shown in the figure 4.04. The displacement distribution along the element may be interpolated using a linear function as shown and the linear distribution may be expressed as:

$$u^{(e)} = c_1 + c_2 X \qquad (4.1)$$

Where X is the global x-coordinate and c_1 and c_2 are unknown constants. The element's end conditions will be given by the displacements of node i and node j according to:

$$U_i = c_1 + c_2 X_i$$
$$U_j = c_1 + c_2 X_j \qquad (4.2)$$

Where U_i is the displacement of node i and U_j is the displacement of node j.

Solving for the two unknowns c_1 and c_2 we obtain:

$$c_1 = \frac{U_i X_j - U_j X_i}{X_j - X_i} \quad c_2 = \frac{U_j - U_i}{X_j - X_i} \tag{4.3}$$

So, the distribution of displacement within the element in terms of its nodal values is:

$$u^{(e)} = \frac{U_i X_j - U_j X_i}{X_j - X_i} - \left(\frac{U_j - U_i}{X_j - X_i} \right) X \tag{4.4}$$

Grouping the like terms together, we obtain:

$$u^{(e)} = \left(\frac{X_j - X}{X_j - X_i} \right) U_i - \left(\frac{X - X_i}{X_j - X_i} \right) U_j \tag{4.5}$$

We can now define 1D linear shape functions, S_i and S_j according to the equations:

$$S_i = \frac{X_j - X}{X_j - X_i} = \frac{X_j - X}{l}$$
$$S_j = \frac{X - X_i}{X_j - X_i} = \frac{X - X_i}{l} \tag{4.6}$$

Where, l is the length of the element. So, we can now write the distribution of displacement within an element in terms of its shape functions as:

$$u^{(e)} = S_i U_i + S_j U_j \quad \text{or} \quad u^{(e)} = \begin{bmatrix} S_i & S_j \end{bmatrix} \begin{Bmatrix} U_i \\ U_j \end{Bmatrix} \tag{4.7}$$

So the shape functions relate the displacement of the element as a whole to the specific displacement of the nodes that make up the element, via the nodal coordinates X_i and X_j. The shape functions describe how an element behaves the only thing that makes one element type different from another is if it has different shape functions.

These 1D linear shape functions have unique properties that simplify the determination of certain integrals when we are implementing the potential energy method described in chapter 3. One of these properties is that a shape function will have a value of one at its corresponding node and a value of zero at the other adjacent node.

$$S_i \big|_{X=X_i} = \frac{X_j - X}{l} \bigg|_{X=X_i} = \frac{X_j - X_i}{l} = 1 \text{ and } S_i \big|_{X=Xj} = \frac{X_j - X}{l} \bigg|_{X=Xj} = \frac{X_j - X_j}{l} = 0$$

Or

$$S_j \big|_{X=X_i} = \frac{X - X_i}{l} \bigg|_{X=X_i} = \frac{X_i - X_i}{l} = 0 \text{ and } S_j \big|_{X=X_j} = \frac{X - X_i}{l} \bigg|_{X=X_j} = \frac{X_j - X_i}{l} = 1$$

$$\tag{4.8}$$

This property is illustrated in figure 4.05:

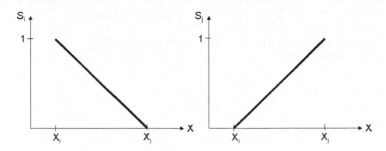

Figure 4.05: Properties of 1D Linear Shape Functions

Another important property of these 1D linear shape functions is that the shape functions add up to a value of unity:

$$S_i + S_j = \frac{X_j - X}{X_j - X_i} + \frac{X - X_i}{X_j - X_i} = 1 \qquad (4.9)$$

Another important property of shape functions is that the sum of derivatives of the functions is zero:

$$\frac{d}{dx}\left(\frac{X_j - X}{X_j - X_i}\right) + \frac{d}{dx}\left(\frac{X - X_i}{X_j - X_i}\right) = \frac{-1}{X_j - X_i} + \frac{1}{X_j - X_i} = 0 \qquad (4.10)$$

4.3.2. Linear Vs Quadratic Element Formulations

An obvious way to increase the accuracy of our finite element results is to use a displacement interpolation function that fits the actual displacement of the bar shown in figure 4.03. For example we can use a quadratic function instead of a linear function. Using a quadratic function requires the use of extra nodes – in this case three nodes instead of two nodes as was used in the linear 1D element above. We need three nodes to define an element because in order to fit a quadratic function we need three points.

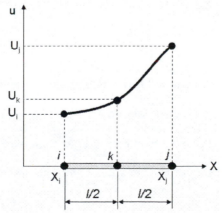

Figure 4.06: A Typical 1D Quadratic Spar Element

The third point can be created by placing a node in the middle of an element (designated as node "k") as shown in the figure 4.06 which illustrates a typical 1D quadratic element.

Using a quadratic approximation, the displacement distribution for a typical element can be represented by:

$$u^{(e)} = c_1 + c_2 X + c_3 X^2 \tag{4.11}$$

and the nodal values are:

$u = U_i$ at $X = X_i$
$u = U_k$ at $X = X_k$
$u = U_j$ at $X = X_j$

Substitution of these known nodal values into equation 4.11 results in three equations and three unknowns:

$$U_i = c_1 + c_2 X_i + c_3 X_i^2$$
$$U_k = c_1 + c_2 X_k + c_3 X_k^2 \tag{4.12}$$
$$U_j = c_1 + c_2 X_j + c_3 X_j^2$$

Solving for the unknown coefficients and rearranging gives us the elements displacement distribution in terms of the nodal values and the shape functions:

$$u^{(e)} = S_i U_i + S_j U_j + S_k U_k \quad \text{or} \quad u^{(e)} = \begin{bmatrix} S_i & S_j & S_k \end{bmatrix} \begin{Bmatrix} U_i \\ U_j \\ U_k \end{Bmatrix} \tag{4.13}$$

where the shape functions are given by:

$$S_i = \frac{2}{l^2}(X - X_j)(X - X_k)$$

$$S_j = \frac{2}{l^2}(X - X_i)(X - X_k) \tag{4.14}$$

$$S_k = \frac{-4}{l^2}(X - X_i)(X - X_j)$$

These quadratic shape functions will have properties similar to those of linear shape functions, namely:
   (1) a shape function has a value of one at its corresponding node and a value of zero at the other adjacent node
   (2) if we sum up the shape functions we will get a value of one.

However in this case the sum of the derivates of the quadratic shape functions is not zero.

Obviously, one could repeat the above process and use a cubic or even a fourth order displacement interpolation function in an attempt to increase accuracy. Most commercial FE packages however offer linear and quadratic formulations through their standard interfaces. Higher order formulations are available using a solution method known as the "P-method" which we will discuss later. For now it is sufficient to understand the main differences between linear and quadratic elements.

4.3.3. Global, Local and Natural coordinates

We need a global coordinate system to represent the location of each node, orientation of each element and to apply boundary conditions and loads to the entire problem. Also, the solution, giving nodal displacements, is generally represented (and more useful) in terms of the global coordinate system.

On the other hand we need to employ local and natural coordinates because they can offer certain advantages when we build the model or compute integrals during the solution process. The advantage becomes readily apparent when the integrals contain products of shape functions. For a one-dimensional element, the relationship between a global coordinate X and a local coordinate x is given by: $X = X_i + x$, as shown in figure 4.07.

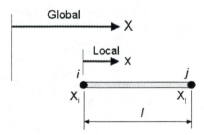

Figure 4.07: Relationship between Global and Local Nodal Coordinates for a Linear 1D Spar Element

Substituting for the global coordinate X in terms of local coordinate x into our equations for linear 1D shape functions developed earlier gives us:

$$S_i = \frac{X_j - X}{l} = \frac{X_j - (X_i + x)}{l} = 1 - \frac{x}{l}$$

$$S_j = \frac{X - X_i}{l} = \frac{(X_i + x) - X_i}{l} = \frac{x}{l}$$

(4.14)

where the local coordinate x varies from 0 to l.

Natural coordinates are basically local coordinates in a dimensionless form. We often need to use numerical methods to evaluate integrals for the purpose of calculating element stiffness matrices during the implementation of the minimum potential energy method and natural coordinates have the convenience of having −1 and 1 as the limits of integration. For example, if we establish a natural coordinate system with its origin at the midpoint of the element and let:

$$\xi = \frac{2x}{\ell} - 1$$

(4.15)

where x is the local coordinate, then we can specify the coordinates of node i as −1 and the coordinates of node j as 1. This relationship is shown in the figure 4.08.

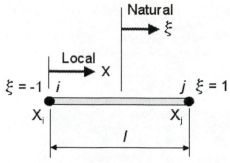

Figure 4.08: Relationship between Local and Natural Coordinates for a 1D Linear Spar Element

We can easily obtain the natural linear shape functions by substituting for x in terms of ξ into equation 4.14. This gives us:

$$S_i = \frac{1}{2}(1 - \xi)$$
$$S_j = \frac{1}{2}(1 + \xi)$$

(4.16)

Natural linear shape functions posses the same properties of linear shape functions, as we discussed earlier. As an example the displacement distribution over an element of our bar in figure 4.02 may be expressed by:

$$u^{(e)} = S_i U_i + S_j U_j = \frac{1}{2}(1 - \xi)U_i + \frac{1}{2}(1 + \xi)U_j$$

(4.17)

It is clear that at $\xi = -1$, $u = U_i$ and at $\xi = 1$, $u = U_j$.

It is important to note that the transformation from the global coordinate X or the local coordinate x to ξ can be made using the same shape functions S_i and S_j:

$$X = S_i X_i + S_j X_j = \frac{1}{2}(1 - \xi)X_i + \frac{1}{2}(1 + \xi)X_j$$
$$x = S_i x_i + S_j x_j = \frac{1}{2}(1 - \xi)x_i + \frac{1}{2}(1 + \xi)x_j$$

(4.18)

Comparing equations 4.18 and 4.17 shows that we have used a single set of parameters (S_i and S_j) to define the unknown variable u and we used the same parameters (S_i and S_j) to express the geometry. Finite element formulations that make use of this arrangement are called *isoparametric* formulations and an element expressed in such a manner is called an isoparametric element.

4.4. Mass Elements

4.4.1 Applications

Mass elements are used to apply concentrated point masses to a finite element model. Typical applications include modelling of mass-spring systems which are used to represent vibrating mechanical structures, as shown in figure 4.09.

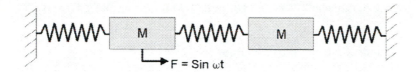

Figure 4.09: A Vibration Problem Modelled Using a Spring-Mass System

Mass elements are also very useful where we want to model the presence of an adjoining structure or its effects on the structure under consideration but we don't need to model the geometry of this adjoining structure, for example consider the problem of a mass held in a box subjected to a drop test as shown in figure 4.10.

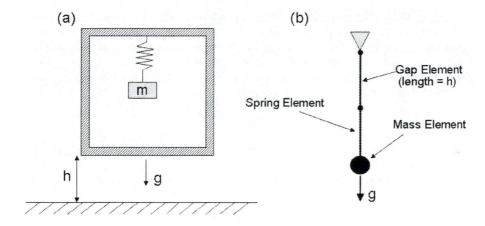

Figure 4.10: (a) Drop Test of a Box Containing a Heavy Object (b) FE Model Using a Mass Element

Other valid uses of mass elements include modelling of pendulum type structures, modelling of flywheels, or any type of structure that has a concentrated mass at a particular point within the structure.

Mass elements can also be useful during model updating. For example, suppose you had performed an experimental modal analysis to determine the natural frequencies and mode shapes of a particular structure and had subsequently built a representative finite element model. After post-processing the results from the model it may be apparent from the results that the model is not correctly representing the distribution of mass in the physical structure and, in such a case, it may be appropriate to add a number of point masses to the structure to improve the results.

4.4.2. Description

A mass element consists of a single node to which various mass and inertia terms can be applied. Mass elements usually allow the specification of different mass properties in each different coordinate direction. Mass elements can generally be applied to both 2D and 3D models.

4.4.3. Shape Functions

This element doesn't have any shape functions. The element formulation simply consists of an element mass matrix with terms down the diagonal. These terms are set during the element deformation and are related to mass in the x direction, mass in the y-direction, mass in the z-direction and the inertia terms I_{xx}, I_{yy} and I_{zz}.

4.5. 1D Quadratic Spar/Truss Elements

4.5.1. Applications

The practical applications of a 1D truss element are quite limited. This element can extend or compress axially and thus can only be subjected to tension or compression loads. This element is limited to 1D problems such as analysis of 1D spring networks that consist of springs in series or parallel. In most cases it is probably easier to carry out a paper based analysis of such systems than to build a FE model.

We will discuss 1D truss elements here as we will use them to introduce a number of key concepts which we will develop later when we look at more complex elements. We have already studied the application of 1D linear truss elements in example 1.1, we have discussed their shape functions in section 4.3.1 and we have shown how natural coordinates may be applied in section 4.3.3. Thus, we will confine this section to discussing the 1D quadratic truss element.

4.5.2. Description

An overview of the 1-D quadratic truss element in terms of natural coordinates is shown in figure 4.11. The element consists of 3 nodes, node i on the left hand edge, node j on the right hand edge and node k in the centre of the element.

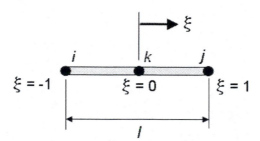

Figure 4.11: 1D Quadratic Spar Element in ξ Coordinates

The X-coordinate is mapped onto the ξ-coordinate system by the transformation:

$$\xi = \frac{2(X - X_k)}{X_j - X_i}$$

(4.19)

4.5.3. Shape Functions

In ξ-coordinates the 1D spar element quadratic shape functions are given by:

$$S_1 = -\frac{1}{2}\xi(1-\xi)$$

$$S_2 = \frac{1}{2}\xi(1-\xi)$$

$$S_3 = (1+\xi)(1-\xi)$$

(4.20)

These shape functions are known as *Lagrange Shape Functions* and are shown graphically in figure 4.12.

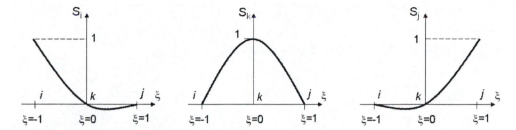

Figure 4.12: Lagrange Shape Functions for a 1D Quadratic Spar Element

The displacement field within the element is thus given by:

$$u^{(e)} = S_i U_i + S_j U_j + S_k U_k \tag{4.21}$$

as already explained in relation to equation 4.13 and now shown graphically in figure 4.13.

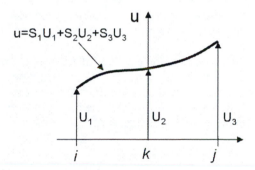

Figure 4.13: Interpolation of Element Displacement Using Quadratic Shape Functions

4.5.4. Calculation of Element Properties

The strain in the element is given by:

$$\varepsilon = \frac{du}{dx} = \frac{du}{d\xi}\frac{d\xi}{dx} = \frac{2}{X_j - X_i}\frac{du}{d\xi} \tag{4.22}$$

Equation 4.22 comes from the strain-displacement relation of equation 2.23, applying the chain rule and then substituting from equation 4.19 for dξ/dx.

Using equation 4.21 to substitute for du/dξ in equation 4.22 gives:

$$\varepsilon = \frac{2}{X_j - X_i}\left[\frac{dS_1}{d\xi}, \frac{dS_2}{d\xi}, \frac{dS_3}{d\xi}\right].U \tag{4.23}$$

Or: $$\varepsilon = \frac{2}{X_j - X_i}\left[-\frac{1-2\xi}{2}, \frac{1+2\xi}{2}, -2\xi\right].U \tag{4.24}$$

Equation 4.24 is of the form described of equation 3.17, which relates stress to displacement via a strain-displacement matrix [B], i.e.:

$$\{\varepsilon\}^{(e)} = [B]^{(e)}\{U\}$$

Where, in this case, the [B] matrix is given by:

$$[B] = \frac{2}{X_j - X_i}\left[-\frac{1-2\xi}{2}, \frac{1+2\xi}{2}, -2\xi\right] \qquad (4.25)$$

As previously discussed in chapter 3, using Hooke's law we can write element stress as:

$$\{\sigma\}^{(e)} = E[B]^{(e)}\{U\} \qquad (4.26)$$

Where E is the Young's modulus for the element.

At this point it should be noted that since the shape functions S_i, S_j and S_k are quadratic and the [B] matrix essentially contains derivatives of the shape functions, [B] will be linear in ξ. This means that displacement can vary quadratically in the element but stress and strain can vary linearly within the element.

At this stage we have expressions for $\{U\}$, $\{\varepsilon\}, \{\sigma\}$ for the element from the above equations. We can also determine $dx=(l/2)d\xi$ from equation 4.19. Since we are dealing with a 1D element we can safely assume that the cross sectional area of the element A_e, the element body load f^B and the element surface load f^S are constant within the element. Putting all this into the potential energy equation discussed in chapter 3 gives us:

$$\Pi = \sum_e \frac{1}{2}\int_e \{\sigma\}_{(e)}^T \{\varepsilon\}_{(e)}\,dx - \sum_e \int_e \{u\}_{(e)}^T f^B A\,dx - \sum_e \int_e \{u\}_{(e)}^T f^S dx - \sum_i U_i^T F^i \quad (4.27)$$

This equation is much simpler than those discussed in chapter 3 as in this case we are only dealing with one dimension and all the volume terms are reduced to Adx or dx terms.

Rewriting equation 4.27 in terms of known values and shape functions for the quadratic 1D element in terms of natural coordinates gives:

$$\Pi = \sum_e \frac{1}{2}U^T\left(E_e A_e \frac{l_e}{2}\int_{-1}^1 [B^T B]d\xi\right)U - \sum_e U^T\left(A_e \frac{l_e}{2}f^B\int_{-1}^1 S^T d\xi\right) - \sum_e U^T\left(\frac{l_e}{2}f^S\int_{-1}^1 S^T d\xi\right) - \sum_i U_i^T F^i \quad (4.28)$$

Comparing this to the general form of the potential energy equation:

$$\Pi = \sum_e \frac{1}{2}\{U\}^T[K]_e\{U\} - \sum_e \{U\}^T\{f^B\}_e - \sum_e \{U\}^T\{f^S\}_e - \sum_i U_i^T F_i$$

Gives the element stiffness matrix as: $\qquad k^e = \frac{E_e A_e l_e}{2}\int_{-1}^1 [B^T B]d\xi \qquad (4.29)$

Which may be evaluated, after substituting for [B] from equation 4.25 as:

$$k^e = \frac{E_e A_e}{3l_e} \begin{matrix} & i & j & k \\ \begin{bmatrix} 7 & 1 & -8 \\ 1 & 7 & -8 \\ -8 & -8 & 16 \end{bmatrix} & \begin{matrix} i \\ j \\ k \end{matrix} \end{matrix} \qquad (4.30)$$

Similarly, the element body force vector is given by:

$$\{f^B\}_e = \frac{A_e l_e f^B}{2} \int_{-1}^{1} S^T \, d\xi \qquad (4.31)$$

Which gives:

$$\{f^B\}_e = A_e l_e f^B \begin{Bmatrix} 1/6 \\ 1/6 \\ 2/3 \end{Bmatrix} \begin{matrix} i \\ j \\ k \end{matrix} \qquad (4.32)$$

And finally, the element surface force vector is given by:

$$\{f^S\}_e = \frac{l_e f^S}{2} \int_{-1}^{1} S^T \, d\xi = l_e f^S \begin{Bmatrix} 1/6 \\ 1/6 \\ 2/3 \end{Bmatrix} \begin{matrix} i \\ j \\ k \end{matrix} \qquad (4.33)$$

Now that we have an expression for each term in equation 4.28 we can now form the element equations for the element from equations 3.27 and 3.29.

4.5.5   Worked Example

The following worked example shows how the above equations are applied in practice and how the 1D quadratic Truss element is implemented in FEA software.

Example 4.1. Application of 1-D Quadratic Truss Element

Consider the stepped solid shaft shown in figure 4.14.

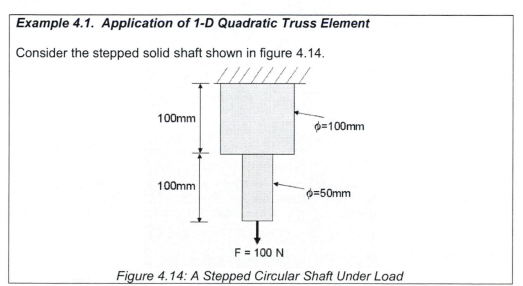

100mm

ϕ=100mm

100mm

ϕ=50mm

F = 100 N

Figure 4.14: A Stepped Circular Shaft Under Load

The shaft is of circular cross section and, in addition to its self weight, it supports a load of 100 N at its free end. It is held rigidly at the other end. Determine the stress distribution in the shaft using two 1D quadratic truss elements, assuming that the shaft is made from steel with E = 210 GPa and ρ = 7850 Kg/m^3.

Solution

A representative finite element model of the stepped shaft is shown in figure 4.15.

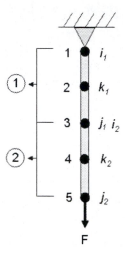

Figure 4.15: Finite Element Model of the Shaft Using 1D Quadratic Truss Elements

The model consists of two quadratic elements with three nodes in each element. The local node number scheme is shown on the right hand side of the model and the global node numbering scheme is shown on the left hand side. The element numbers are shown in circles.

We can form the element stiffness matrices from equation 4.30:

$$k^1 = \frac{210x10^9 \, \pi(0.05)^2}{3(0.1)} \begin{array}{c} \\ \end{array} \begin{array}{ccc} 1 & 3 & 2 \\ \left[\begin{array}{ccc} 7 & 1 & -8 \\ 1 & 7 & -8 \\ -8 & -8 & 16 \end{array}\right] & \begin{array}{c} 1 \\ 3 \\ 2 \end{array} \end{array}$$

and

$$k^2 = \frac{210x10^9 \, \pi(0.025)^2}{3(0.1)} \begin{array}{c} \\ \end{array} \begin{array}{ccc} 3 & 5 & 4 \\ \left[\begin{array}{ccc} 7 & 1 & -8 \\ 1 & 7 & -8 \\ -8 & -8 & 16 \end{array}\right] & \begin{array}{c} 3 \\ 5 \\ 4 \end{array} \end{array}$$

Where the numbers outside the matrix brackets indicate the global degrees of freedom that the matrix elements refer to. If we refer to figure 4.15 we will see that for element 1, the element is labelled i, j, k as usual, but this corresponds to 1, 3, 2 in the global numbering scheme.

Putting k^1 and k^2 together gives us the global stiffness matrix:

$$k^G = \frac{210x10^9 \, \pi(0.025)^2}{0.3} \begin{array}{c} \\ \\ \left[\begin{array}{ccccc} 28 & -32 & 4 & 0 & 0 \\ -32 & 64 & -32 & 0 & 0 \\ 4 & -32 & 35 & -8 & 1 \\ 0 & 0 & -8 & 16 & -8 \\ 0 & 0 & 1 & -8 & 7 \end{array}\right] \begin{array}{c} 1 \\ 2 \\ 3 \\ 4 \\ 5 \end{array} \end{array}$$

The force due to self weight for each element is given by:

$$f_w = mass \, x \, acceleartion = (\rho A L)g$$

Thus: $f_w = 7850\pi \, r^2 g$

And: $f_{w1} = 7850\pi (0.05)^2 g = 605 \, N$
$f_{w1} = 7850\pi (0.025)^2 g = 151N$

We can thus form the element body force vectors from equation 4.32:

$$\{f^B\}_1 = 0.007854 \, x \, 0.1 \, x \, 605 \begin{Bmatrix} 1/6 \\ 1/6 \\ 2/3 \end{Bmatrix} \begin{array}{c} i,1 \\ j,3 \\ k,2 \end{array}$$

$$\{f^B\}_1 = 0.001963 \, x \, 0.1 \, x \, 151 \begin{Bmatrix} 1/6 \\ 1/6 \\ 2/3 \end{Bmatrix} \begin{array}{c} i,1 \\ j,3 \\ k,2 \end{array}$$

Assembling $\{f^B\}_1$ and $\{f^B\}_2$ and the point load of 100,000 N into the global load vector, gives:

$$\{F\} = [0.07919 \quad 0.3167 \quad (0.07919 + 0.00494) \quad 0.01967 \quad (0.00494 + 100)]^T$$

We can now assemble the global problem equation: [K]{U}={F}

$$\frac{210x10^9 \, \pi(0.025)^2}{0.3} \begin{bmatrix} 28 & -32 & 4 & 0 & 0 \\ -32 & 64 & -32 & 0 & 0 \\ 4 & -32 & 35 & -8 & 1 \\ 0 & 0 & -8 & 16 & -8 \\ 0 & 0 & 1 & -8 & 7 \end{bmatrix} \begin{bmatrix} 0 \\ U_2 \\ U_3 \\ U_4 \\ U_5 \end{bmatrix} = \begin{bmatrix} 0.07919 \\ 0.31678 \\ 0.08413 \\ 0.01976 \\ 100.00494 \end{bmatrix}$$

Solving this equation gives: $\{U\} = 1x10^{-9}[0 \quad 3.04 \quad 6.08 \quad 18.2 \quad 30.3]^T$

The element strains can now be evaluated from the strain-displacement matrix as detailed in equation 4.25.

For element 1, $\{U\}_1 = [U_i \ \ U_j \ \ U_k] = [U_1 \ \ U_3 \ \ U_2] = 1 \times 10^{-9}[0 \ \ 6.08 \ \ 3.04]^T$

For element 2, $\{U\}_1 = [U_i \ \ U_j \ \ U_k] = [U_3 \ \ U_5 \ \ U_4] = 1 \times 10^{-9}[6.08 \ \ 30.3 \ \ 18.2]^T$

The strain-displacement equation for element 1 is given by:

$$\varepsilon_1 = \frac{2}{0.1}\left[-\frac{1-2\xi}{2} \quad \frac{1+2\xi}{2} \quad -2\xi\right]\begin{bmatrix} U_1 \\ U_3 \\ U_2 \end{bmatrix}$$

The strain at node 1 in element 1 is obtained by substituting $\xi=-1$ into the above equation. Similarly the strain at node 2 is obtained by using $\xi= 0$ and the strain at node 3 is obtained using $\xi= 1$.

Using this procedure for both element 1 and 2 we can obtain the strain distribution throughout the finite element model as:

$$\{\varepsilon\} = [6.08 \times 10^{-8} \ \ 6.08 \times 10^{-8} \ \ 3.03 \times 10^{-7} \ \ 2.42 \times 10^{-7} \ \ 2.42 \times 10^{-7}]^T$$

Now, knowing that $\{\sigma\} = E \{\varepsilon\}$ we can calculate the global stress vector, and hence obtain the stress distribution in the shaft as:

$$\{\sigma\} = [12733 \ \ 12733 \ \ 63655 \ \ 50932 \ \ 50932]^T \text{ Pa}$$

4.6. 2D Linear Truss Elements

4.6.1. Applications

2D Trusses are useful for analysing bridge structures, roof structures, small cranes, and pin jointed frames. Performing a 2D truss analysis is often a useful pre-analysis step to learn more about many beam problems before performing a more complex beam analysis or an analysis of a space frame structure. Examples of the type of structures to which analysis using 2D trusses is appropriate are shown in figure 4.16.

Figure 4.16: Examples of Structures which can be analysed Using 2D Truss Elements (roof structure, truss bridge and jib crane)

4.6.2   Description

2D Truss elements assume that only axial tension or compression loads are allowed and that all elements are connected by frictionless pin joints. If you are studying or have studied a course on engineering, then it is almost certain that you have already analysed truss structures in your study of basic statics and applied mechanics.

The major difference between 1D and 2D truss elements is that 2D truss elements can have different orientations. In order to take account of this element equations derived in the natural or local coordinate systems must be transformed to align with the global coordinate system before each individual element equation can be assembled into the global problem statement. A typical 2D truss element is shown in figure 4.17.

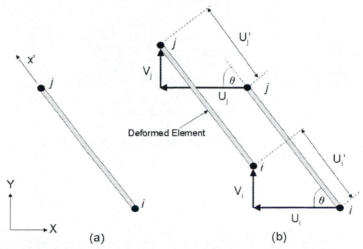

(a)   (b)

Figure 4.17: A 2D Truss Element in (a) its local coordinate system and (b) the global coordinate system.

In order to make clear the distinction between local and global coordinate systems, the local coordinate system is indicated by x' and the global coordinate system is indicated by X and Y. The local coordinate system x' runs along the length of the element from node i to node j. In the local coordinate system each node has one degree of freedom, i.e. it can move along the x' axis. In the global coordinate system each node has two degrees of freedom as it can move in both the X and Y directions.

If U_i' and U_j' are the displacements of node i and node j in the local coordinate system, then the nodal displacement vector in the local coordinate system (2 DOF), with reference to figure 4.17, is given by:

$$\{U'\} = \begin{bmatrix} U_i' \\ U_j' \end{bmatrix}$$

(4.34)

and, the nodal displacement vector in terms of global coordinates (4 DOF) is:

$$\{U\} = \begin{bmatrix} U_i & V_i & U_j & V_j \end{bmatrix}^T$$

(4.35)

Using elementary trigonometry with figure 4.17 we can show that:

$$U_i' = U_i \cos\theta + V_i \sin\theta \qquad (4.36)$$

And, $\quad U_j' = U_j \cos\theta + V_j \sin\theta$

If we now define the direction cosines l and m such that $l = \cos\theta$ and $m = \cos(90-\theta)$ then these direction cosines are the angles that the local x' axis makes with the respective global X and Y axes. Noting that $m = \cos(90-\theta) = \sin\theta$, we can now write equations 4.36 in matrix form as:

$$\{U'\} = [L]\{U\} \qquad (4.37)$$

Where the transformation matrix [L] is given by:

$$[L] = \begin{bmatrix} l & m & 0 & 0 \\ 0 & 0 & l & m \end{bmatrix} \qquad (4.38)$$

4.6.3. Calculation of Element Properties

It should be obvious from the above that the 2D truss element is a 1D element when viewed in its local coordinate system, thus we can use our previously derived element formulations for a 1D linear truss again for this element.

From example 1.1 we know that the element stiffness matrix for a truss element in the local coordinate system is:

$$[k'] = \frac{A_e E_e}{l_e} \begin{bmatrix} 1 & -1 \\ -1 & 1 \end{bmatrix} \qquad (4.39)$$

we now need to manipulate this equation to obtain an expression for the element stiffness matrix in terms of global coordinates. In order to do this, let's consider the strain energy in the element. The strain energy in local coordinates is given by:

$$\Lambda_e = \tfrac{1}{2}\{U'\}^T [k']\{U'\} \qquad (4.40)$$

Substituting equation 4.37 into equation 4.40, gives:

$$\Lambda_e = \tfrac{1}{2}\{U\}^T [L^T k' L]\{U\} \qquad (4.41)$$

Now, we know that the strain energy in global coordinates is given by:

$$\Lambda_e = \tfrac{1}{2}\{U\}^T [k]\{U\} \qquad (4.42)$$

By comparing equations 4.41 and 4.42 we find that the element stiffness matrix in global coordinates is given by:

$$[k] = [L^T k' L] = \frac{A_e E_e}{l_e} \begin{bmatrix} l^2 & lm & -l^2 & -lm \\ lm & m^2 & -lm & -m^2 \\ -l^2 & -lm & l^2 & lm \\ -lm & -m^2 & lm & m^2 \end{bmatrix} \qquad (4.43)$$

The element stiffness matrices are assembled in the usual manner to obtain the global stiffness matrix and hence solve the problem.

After solution, the stress in the truss element can be calculated from:

$$\sigma = E_e \varepsilon = E_e \frac{U_j^{'} - U_i^{'}}{l_e} = \frac{E_e}{l_e}[-1 \quad 1]\begin{Bmatrix} U_i^{'} \\ U_j^{'} \end{Bmatrix} = \frac{E_e}{l_e}[-1 \quad 1][L]\{U\} \quad (4.44)$$

Substituting for [L] from equation 4.38 gives:

$$\sigma = \frac{E_e}{l_e}[-l \quad -m \quad l \quad m]\{U\} \qquad\qquad (4.45)$$

Equation 4.45 allows the calculation of element stresses once the element global displacements {U} are known.

4.6.4. Worked Example

The following worked example shows how the above equations are applied in practice and how the 2D linear truss element is implemented in FEA software.

Example 4.2: Application of a 2D Linear Truss Element

Consider the truss framework shown in figure 4.18. The individual members are capable of expansion and contraction but cannot bend. A load of 10,000 N is supported as shown and the pin joints on the left hand side are held rigidly to prevent movement in any direction. Determine the stress distribution in each individual member and reaction forces at the supports, assuming that the members are made from steel with E = 210 GPa and each has a cross sectional area of 0.05 m^2.

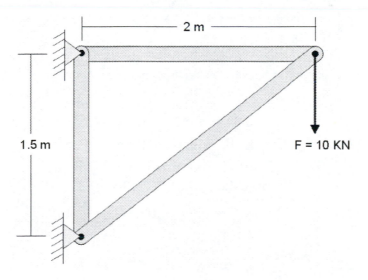

Figure 4.18: A 2D Truss Framework

A representative finite element model of the truss framework is shown in figure 4.19.

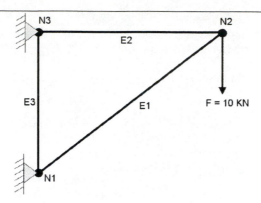

Figure 4.19: Finite Element Model of the Framework Using 2D Linear Truss Elements

In the finite element model, the nodes and elements are labelled in a clockwise direction. The tables below show the nodal coordinates and the relevant element data.

Node	X	Y
1	0	0
2	2	1.5
3	0	1.5

Element	Node i	Node j	l_e	l	m
1	0	0	2.5	0.8	0.6
2	2	1.5	2	1	0
3	0	1.5	1.5	0	-1

We can calculate the element length, l_e, and the direction cosines l and m, from the element coordinates via:

$$l_e = \sqrt{(X_2 - X_1)^2 + (Y_2 - Y_1)^2} \qquad l = \frac{X_2 - X_1}{l_e} \qquad m = \frac{Y_2 - Y_1}{l_e}$$

We can form the element stiffness matrices from equation 4.43:

$$k^1 = \frac{210x10^9(0.05)}{2.5} \begin{matrix} & 1 & 2 & 3 & 4 \\ \begin{bmatrix} 0.64 & 0.48 & -0.64 & -0.48 \\ 0.48 & 0.36 & -0.48 & -0.36 \\ -0.64 & -0.48 & 0.64 & 0.48 \\ -0.48 & -0.36 & 0.48 & 0.36 \end{bmatrix} & \begin{matrix} 1 \\ 2 \\ 3 \\ 4 \end{matrix} \end{matrix}$$

$$k^2 = \frac{210x10^9(0.05)}{2.5} \begin{matrix} & 3 & 4 & 5 & 6 \\ \begin{bmatrix} 1 & 0 & -1 & 0 \\ 0 & 0 & 0 & 0 \\ -1 & 0 & 1 & 0 \\ 0 & 0 & 0 & 0 \end{bmatrix} & \begin{matrix} 3 \\ 4 \\ 5 \\ 6 \end{matrix} \end{matrix}$$

$$k^3 = \frac{210x10^9(0.05)}{2.5} \begin{matrix} 5 & 6 & 1 & 2 \\ \begin{bmatrix} 0 & 0 & 0 & 0 \\ 0 & 1 & 0 & -1 \\ 0 & 0 & 0 & 0 \\ 0 & -1 & 0 & 1 \end{bmatrix} \begin{matrix} 5 \\ 6 \\ 1 \\ 2 \end{matrix} \end{matrix}$$

Putting k^1, k^2 and k^3 together gives us the global stiffness matrix:

$$K = 210x10^9(0.05) \begin{matrix} 1 & 2 & 3 & 4 & 5 & 6 \\ \begin{bmatrix} 0.4 & 0.858 & 0.1 & -0.192 & -0.5 & 0 \\ 0.192 & 0.194 & -0.192 & -0.144 & 0 & -0.666 \\ -0.4 & -0.192 & 0.4 & 0.192 & 0 & 0 \\ -0.192 & -0.144 & 0.192 & 0.144 & 0 & 0 \\ 0 & 0 & -0.5 & 0 & 0.5 & 0 \\ 0 & -0.666 & 0 & 0 & 0 & 0.666 \end{bmatrix} \begin{matrix} 1 \\ 2 \\ 3 \\ 4 \\ 5 \\ 6 \end{matrix} \end{matrix}$$

We can now assemble the global problem equation: [K]{U}={F}

$$210x10^9(0.05) \begin{bmatrix} 0.4 & 0.858 & 0.1 & -0.192 & -0.5 & 0 \\ 0.192 & 0.194 & -0.192 & -0.144 & 0 & -0.666 \\ -0.4 & -0.192 & 0.4 & 0.192 & 0 & 0 \\ -0.192 & -0.144 & 0.192 & 0.144 & 0 & 0 \\ 0 & 0 & -0.5 & 0 & 0.5 & 0 \\ 0 & -0.666 & 0 & 0 & 0 & 0.666 \end{bmatrix} \begin{Bmatrix} U_1 \\ V_1 \\ U_2 \\ V_2 \\ U_3 \\ V_3 \end{Bmatrix} = \begin{Bmatrix} 0 \\ 0 \\ 0 \\ -10,00 \\ 0 \\ 0 \end{Bmatrix}$$

Eliminating the trivial rows and columns from the global problem equation gives:

$$210x10^9(0.05) \begin{bmatrix} 0.4 & 0.192 & 0 & 0 \\ 0.192 & 0.144 & 0 & 0 \end{bmatrix} \begin{Bmatrix} 0 \\ -10,000 \end{Bmatrix} = \begin{Bmatrix} U_2 \\ V_2 \end{Bmatrix}$$

Solving this equation gives: $\{U\} = \begin{bmatrix} 0 & 0 & 8.818x10^{-6} & 1.837x10^{-5} & 0 & 0 \end{bmatrix}^T m$

The stress in each element can now be evaluated from equation 4.45 as follows:

$$\sigma_e = \frac{E_e}{l_e}[-l \quad -m \quad l \quad m]\{U\}$$

Giving:

$$\sigma_1 = \frac{210x10^9}{2.5}[-0.8 \quad -0.6 \quad 0.8 \quad 0.6] \begin{Bmatrix} 0 \\ 0 \\ 8.818x10^{-6} \\ 1.837x10^{-5} \end{Bmatrix} = 1,518,417 Pa$$

$$\sigma_2 = \frac{210x10^9}{2} \begin{bmatrix} -1 & 0 & 1 & 0 \end{bmatrix} \begin{Bmatrix} 8.818x10^{-6} \\ 1.837x10^{-5} \\ 0 \\ 0 \end{Bmatrix} = -370,356 \, Pa$$

$$\sigma_3 = \frac{210x10^9}{1.5} \begin{bmatrix} 0 & 1 & 0 & -1 \end{bmatrix} \begin{Bmatrix} 0 \\ 0 \\ 0 \\ 0 \end{Bmatrix} = 0 \, Pa$$

The reaction forces at each support can be obtained by considering the global reaction equation:

$$[R] = [K]\{U\} - \{F\}$$

$$\begin{Bmatrix} R_{1X} \\ R_{1Y} \\ R_{2X} \\ R_{2Y} \\ R_{3X} \\ R_{3Y} \end{Bmatrix} = 1.05x10^{10} \begin{bmatrix} 0.4 & 0.858 & 0.1 & -0.192 & -0.5 & 0 \\ 0.192 & 0.194 & -0.192 & -0.144 & 0 & -0.666 \\ -0.4 & -0.192 & 0.4 & 0.192 & 0 & 0 \\ -0.192 & -0.144 & 0.192 & 0.144 & 0 & 0 \\ 0 & 0 & -0.5 & 0 & 0.5 & 0 \\ 0 & -0.666 & 0 & 0 & 0 & 0.666 \end{bmatrix} \begin{Bmatrix} 0 \\ 0 \\ 8.818x10^{-6} \\ 1.837x10^{-5} \\ 0 \\ 0 \end{Bmatrix} - \begin{Bmatrix} 0 \\ 0 \\ 0 \\ -10,0 \\ 0 \\ 0 \end{Bmatrix}$$

This equation only needs to be evaluated at fixed degrees of freedom (i.e. where U=0 and F=0), so lines 3 and 4 of the above equation do not need to be considered.

Solving this equation gives: $\{R\} = \begin{bmatrix} -27,775 & -45,553 & 0 & 0 & -46,295 & 0 \end{bmatrix}^T N$

4.6.5. Case Study

You should now study case study A in chapter 10 which details a practical example of the use of 2-D truss elements in a FEA using a commercial FEA software package.

4.7. 3D Linear Truss Elements

4.7.1. Applications

Three dimensional truss elements are used to model truss type structures that cannot be satisfactorily represented using 2D elements. Typical examples include tower structures, space frames, bicycle frames and cranes. Examples of such problems are shown in figure 4.20.

4.7.2. Description

The 3D truss element is basically a simple development of the 2D truss element described in the previous section. Only axial tension or compression loads are allowed and all elements are connected via frictionless pin joints. The 3D truss element consists of two nodes which can occupy any points in the 3D coordinate system. A typical 3D truss element is shown in figure 4.21.

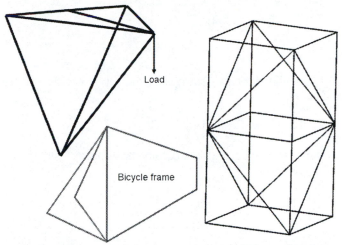

Figure 4.20: Examples of Structures Which May be Analysed Using 3D Truss Elements

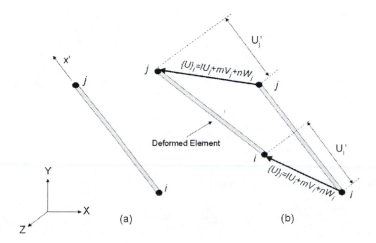

Figure 4.21: A 3D Truss Element in (a) its local coordinate system and (b) the global co-ordinate system.

The nodal displacement vector in local coordinates is given by:

$$\{U'\} = \begin{bmatrix} U'_i \\ U'_j \end{bmatrix}$$

and, the nodal displacement vector in terms of global coordinates (6 DOF) is:

$$\{U\} = \begin{bmatrix} U_i & V_i & W_i & U_j & V_j & W_j \end{bmatrix}^T \tag{4.46}$$

As before, the transformation between local and global coordinates is given by:

$$\{U'\} = [L]\{U\} \tag{4.47}$$

Where the transformation matrix [L] in this case is:

$$[L] = \begin{bmatrix} l & m & n & 0 & 0 & 0 \\ 0 & 0 & 0 & l & m & n \end{bmatrix} \tag{4.48}$$

where l, m and n are the direction cosines of the local x' axis with respect to the global X, Y and Z axes.

4.7.3. Calculation of Element Properties

From equations 4.41 and 4.42 we can obtain the element stiffness matrix in the global coordinate system as:

$$[k] = \frac{A_e E_e}{l_e} \begin{bmatrix} l^2 & lm & ln & -l^2 & -lm & -ln \\ lm & m^2 & mn & -lm & -m^2 & -mn \\ ln & mn & n^2 & -ln & -mn & -n^2 \\ -l^2 & -lm & -ln & l^2 & lm & ln \\ -lm & -m^2 & -mn & lm & m^2 & mn \\ -ln & -mn & -n^2 & ln & mn & n^2 \end{bmatrix} \tag{4.49}$$

The element stiffness matrices are assembled in the usual manner to obtain the global stiffness matrix and hence solve the problem. In this case the direction cosines can be calculated from:

$$l = \frac{X_i - X_j}{l_e} \qquad m = \frac{Y_i - Y_j}{l_e} \qquad n = \frac{Z_i - Z_j}{l_e} \tag{4.50}$$

Where the length of the element is given by:

$$l_e = \sqrt{(X_j - X_i)^2 + (Y_j - Y_i)^2 + (Z_j - Z_i)^2} \tag{4.51}$$

After solution, the stress in the 3D truss element can be calculated from:

$$\sigma = E_e \varepsilon = E_e \frac{U'_j - U'_i}{l_e} = \frac{E_e}{l_e} \begin{bmatrix} -1 & 1 \end{bmatrix} \begin{Bmatrix} U'_i \\ U'_j \end{Bmatrix} = \frac{E_e}{l_e} \begin{bmatrix} -1 & 1 \end{bmatrix} [L]\{U\} \tag{4.52}$$

Substituting for [L] from equation 4.48 gives:

$$\sigma = \frac{E_e}{l_e} \begin{bmatrix} -l & -m & -n & l & m & n \end{bmatrix} \{U\} \tag{4.53}$$

Equation 4.53 allows the calculation of element stresses once the element global displacements {U} are known.

4.7.3. Case Study

If you have completed case study A which deals with 2D trusses, you should now be able to perform a similar analysis with 3D truss elements. Give it a try!

4.8. Beam Elements

4.8.1. Applications

Beams are long and slender structural elements that primarily carry loads by bending. Beams are primarily used for carrying vertical loads but can also carry horizontal loads in certain cases. Typical examples of beams are the long horizontal members used in building construction and bridges, shafts supported by bearings, stiffening members in aircraft structures etc.

Beam elements are often used to analyse frame structures. Frame structures are similar in appearance to truss structures except in this case the members are rigidly connected and each member can be subjected to bending, tension or compression. Typical examples of frame structures are automobile chassis, aircraft structures, bicycle frames, etc.

The major difference between beam elements and truss elements is that bending is considered in beams and it is not in trusses. Thus beam elements can be used to give a more "realistic" answer to many truss problems.

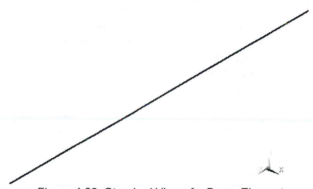

Figure 4.22: Standard View of a Beam Element

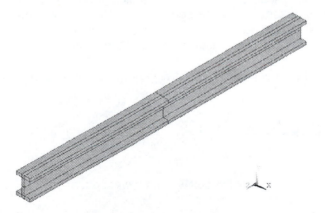

Figure 4.23: View of the Same Element with Display of Element Cross Sectional Shape Enabled

Beams are characterised by their cross section and applying section properties to beam elements is very important. Figure 4.22 shows a standard view of a beam element in a finite element software package. It can be seen from the figure that beam elements are essentially just lines. Figure 4.23 shows the same beam element as shown in figure 4.23 but this time the element shape has been turned on for display purposes. Figure 4.23 shows that the beam element has I beam section properties assigned to it and will thus behave in this manner. It is possible to apply a myriad of section properties or cross sectional shapes to beam elements some of the most common cross sections are shown in figure 4.24.

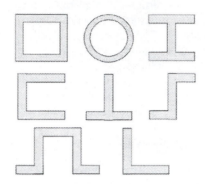

Figure 4.24: Common Cross Sections Used for Beam Elements

4.8.2. Description

A beam element generally consists of two nodes which link two points in space with a linear relationship describing the element behaviour. In many FE software packages both 2D and 3D beam elements are available. In some cases a third node is used as an "orientation node" this node is used to ensure that the beam cross section is orientated in the correct direction. Quadratic beam elements are also available

In order to describe beam elements we can make use of elementary beam theory:

$$\sigma = -\frac{My}{I} \qquad \varepsilon = \frac{\sigma}{E} \qquad \frac{d^2\upsilon}{dx^2} = \frac{M}{EI} \qquad (4.54)$$

where σ is the normal stress, M is the bending moment at the section under consideration, y is distance from the neutral axis, I is the moment of inertia of the section about the neutral axis, ε is the normal strain, υ is the deflection of the centroidal axis at x and E is the elastic modulus of the beam material.

Using the potential energy method: the strain energy $d\Lambda$ of a beam element of length dx is given by:

$$d\Lambda = \frac{1}{2}\int_A \sigma\varepsilon \, dA \, dx = \frac{1}{2}\left(\frac{M^2}{EI^2}\int_A y^2 dA\right) dx \qquad (4.55)$$

Since $\int_A y^2 dA$ is the moment of inertia I we can rewrite equation 4.55 as:

$$d\Lambda = \frac{1}{2}\frac{M^2}{EI} dx \qquad (4.56)$$

Substituting for M in terms of υ from equation 4.54 gives the total strain energy in the beam as:

$$\Lambda = \frac{1}{2} \int_0^L EI \left(\frac{d^2\upsilon}{dx^2} \right)^2 dx \qquad (4.57)$$

Thus, the potential energy of the beam is given by:

$$\Pi = \frac{1}{2} \int_0^L EI \left(\frac{d^2\upsilon}{dx^2} \right)^2 dx - \int_0^L f^d \upsilon \, dx - \sum_m F_i \upsilon_i - \sum_n M_n \upsilon'_k \qquad (4.58)$$

Where f^d is the distributed load per unit length, F_i is a point load at point i, M_n is the moment of the couple applied at point n, υ_i is the deflection at point i and υ'_n is the slope of the beam at point n.

4.8.3. Determination of Element Properties

4.8.3.1.   1D Beams

In order to illustrate how beam elements are formulated we will first consider the relatively simple case of a 1D linear beam element and we will subsequently develop the equations for the 1D case to more practical 2D problems. Thus each 1D beam element will have two nodes with two degrees of freedom at each node, vertical deflection and rotation in the x-y plane, as shown in figure 4.25.

Figure 4.25: A 1-D Linear Beam Element

So the displacement field in the element will be a function of the nodal vertical displacement and nodal rotation at each node, i.e.

$$U_e = f(v_i \, \theta_i \, v_j \, \theta_j) \qquad (4.59)$$

Shape functions for beam elements are different to those used for truss elements in the previous sections. For beam elements we need to describe nodal slopes as well as translations so we end up with four shape functions for a two node element. Hermite shape functions are used to satisfy nodal value and slope requirements, where, in natural coordinates:

$$S_i = \frac{1}{4}(1-\xi)^2(2+\xi)$$

$$S_{i\theta} = \frac{1}{4}(1-\xi)^2(\xi+1)$$

$$S_j = \frac{1}{4}(1+\xi)^2(2-\xi) \qquad (4.60)$$

$$S_{j\theta} = \frac{1}{4}(1+\xi)^2(\xi-1)$$

Practical Stress Analysis with Finite Elements

The Hermite shape functions can be used to write an expression for beam element deflection, υ as:

$$\upsilon(\xi) = S_i \upsilon_i + S_{i\theta} \left(\frac{d\upsilon}{d\xi} \right)_i + S_j \upsilon_j + S_{j\theta} \left(\frac{d\upsilon}{d\xi} \right)_j \qquad (4.61)$$

The coordinate transformation from X to ξ is given by:

$$X = \frac{1-\xi}{2} X_i + \frac{1+\xi}{2} X_j = \frac{X_i + X_j}{2} \xi + \frac{X_j - X_i}{2} \xi \qquad (4.62)$$

Since $X_j - X_i = l_e$ we can rewrite equation 4.62 as:

$$dx = \frac{l_e}{2} d\xi \qquad (4.63)$$

Using the chain rule and equation 4.63 we can write:

$$\frac{d\upsilon}{d\xi} = \frac{d\upsilon}{dx} \frac{dx}{d\xi} = \frac{l_e}{2} \frac{d\upsilon}{dx} \qquad (4.64)$$

Now, putting equation 4.64 into equation 4.61 for $d\upsilon/d\xi$ we have:

$$\upsilon(\xi) = S_i \upsilon_i + \frac{l_e}{2} S_{i\theta} \upsilon_{i\theta} + S_j \upsilon_j + \frac{l_e}{2} S_{j\theta} \upsilon_{j\theta} \qquad (4.65)$$

Using the total potential energy method, the element strain energy is given by:

$$\Lambda = \frac{1}{2} EI \int_e \left(\frac{d^2 \upsilon}{dx^2} \right)^2 dx \qquad (4.66)$$

From equation 4.64 we have:

$$\frac{d\upsilon}{dx} = \frac{2}{l_e} \frac{d\upsilon}{d\xi} \quad \text{thus:} \quad \frac{d^2 \upsilon}{dx^2} = \frac{4}{l_e^2} \frac{d^2 \upsilon}{d\xi^2} \qquad (4.67)$$

Substituting equation 4.67, 4.64 and 4.63 into equation 4.66 gives an equation of the form:

$$\Lambda = \frac{1}{2} \{U\}^T [k^e] \{U\} \qquad (4.68)$$

Where the element stiffness matrix is given by:

$$[k^e] = \frac{EI}{l_e^3} \begin{bmatrix} 12 & 6l_e & -12 & 6l_e \\ 6l_e & 4l_e^2 & -6l_e & 2l_e^2 \\ -12 & -6l_e & 12 & -6l_e \\ 6l_e & 2l_e^2 & -6l_e & 4l_e^2 \end{bmatrix} \qquad (4.69)$$

The load vector can be formed by considering the distributed load on the element. If we assume that the load is distributed uniformly over the element then:

$$\int_{l_e} f^d \upsilon \, dx = \left(\frac{f^d l_e}{2} \int_{-1}^{1} S \, d\xi \right) \{U\} = \{f\}^{e^T} \{U\} \tag{4.70}$$

where:

$$\{f\}^e = \left[\frac{f^d l_e}{2} \quad \frac{f^d l_e^2}{12} \quad \frac{f^d l_e}{2} \quad \frac{-f^d l_e^2}{12} \right]^T \tag{4.71}$$

Using the bending moment and shear force equations we can calculate the element bending moment and shear force equations:

$$M = EI \frac{d^2 \upsilon}{dx^2} = \frac{EI}{l_e^3} \left(6\xi \upsilon_i + (3\xi - 1)l_e \theta_i - 6\xi \upsilon_j + (3\xi + 1)l_e \theta_j \right) \tag{4.72}$$

$$V = \frac{6EI}{l_e^3} \left(2\upsilon_i + l_e \theta_i + 2\upsilon_j + l_e \theta_j \right) \tag{4.73}$$

4.8.3.2. 2D Beams

2D beams are a development of the 1D beams described above where in addition to bending and deflection deformations, in this case, axial deformations are also allowed. Figure 4.26 shows an overview of a 2D beam element.

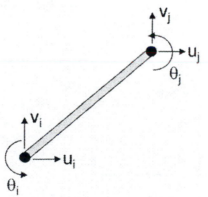

Figure 4.26: A 2D Beam Element

In this case we have two displacements and a rotational deformation for each node, thus the nodal displacement vector is given by:

$$\{U\} = \left[U_i \quad V_i \quad \theta_i \quad U_j \quad V_j \quad \theta_j \right]^T \tag{4.74}$$

If we define the local coordinate system x' and y' with direction cosies l and m, so that x' is orientated along the length of the element, then the nodal displacement vector in local coordinates is given by:

$$\{U'\} = \left[U'_i \quad V'_i \quad \theta'_i \quad U'_j \quad V'_j \quad \theta'_j \right]^T \tag{4.75}$$

Since θ is rotation with respect to the element it should be clear that $\theta_i = \theta'_i$ and $\theta_j = \theta'_j$ and the local to global transformation matrix is given by:

$$[L] = \begin{bmatrix} l & m & 0 & 0 & 0 & 0 \\ -m & l & 0 & 0 & 0 & 0 \\ 0 & 0 & 1 & 0 & 0 & 0 \\ 0 & 0 & 0 & l & m & 0 \\ 0 & 0 & 0 & -m & l & 0 \\ 0 & 0 & 0 & 0 & 0 & 1 \end{bmatrix}$$

(4.76)

If we now consider displacements in the local coordinate system we can notice that V'_i, θ'_i, V'_j and θ'_j are basically 1D beam element degrees of freedom while U'_i and U'_j are essentially 1D truss element displacements. Combining the relevant stiffness terms and placing them in the correct location in the element stiffness matrix gives the element stiffness matrix in local coordinates as:

$$[k']^e = \begin{bmatrix} \dfrac{EA}{l_e} & 0 & 0 & \dfrac{-EA}{l_e} & 0 & 0 \\ 0 & \dfrac{12EI}{l_e^3} & \dfrac{6EI}{l_e^2} & 0 & \dfrac{12EI}{l_e^3} & \dfrac{6EI}{l_e^2} \\ 0 & \dfrac{6EI}{l_e^2} & \dfrac{4EI}{l_e} & 0 & \dfrac{-6EI}{l_e^2} & \dfrac{2EI}{l_e} \\ \dfrac{-EA}{l_e} & 0 & 0 & \dfrac{EA}{l_e} & 0 & 0 \\ 0 & \dfrac{-12EI}{l_e^3} & \dfrac{-6EI}{l_e^2} & 0 & \dfrac{-12EI}{l_e^3} & \dfrac{-6EI}{l_e^2} \\ 0 & \dfrac{6EI}{l_e^2} & \dfrac{2EI}{l_e} & 0 & \dfrac{-6EI}{l_e^2} & \dfrac{4EI}{l_e} \end{bmatrix}$$

(4.77)

From equation 4.41 we know that:

$$\Lambda_e = \tfrac{1}{2}\{U\}^T \left[L^T k' L \right] \{U\}$$

And thus, $[k]^e = [L^T k' L]$. This allows us to calculate the element stiffness matrix in global coordinates from equations 4.76 and 4.77.

Similarly, the element distributed load vector in local coordinates is given by:

$$\{f'\}^e = \left[0 \quad \dfrac{f^d l_e^2}{12} \quad \dfrac{f^d l_e}{2} \quad 0 \quad \dfrac{f^d l_e}{2} \quad \dfrac{-f^d l_e^2}{12} \right]^T$$

(4.78)

And thus, the element distributed load vector in global coordinates is given by:

$$\{f\}^e = [L]^2 \{f'\}^e$$

(4.79)

4.8.4. Worked Example

The following example shows how the above equations are applied in practice and how the 2D beam element is implemented in FEA software.

Example 4.3: Application of a 2D Beam Element

A steel beam and two steel columns are used to support an opening in a small building as shown in figure 4.27(a). A representative finite element model of the beams is shown in figure 4.27(b). We are required to determine the displacements and rotations at the joints.

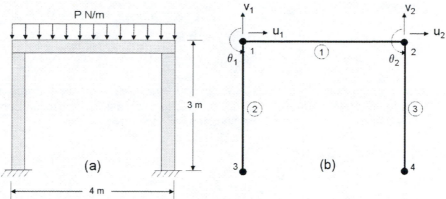

Figure 4.27: A 2-D Beam Problem

Given that E = 210 GPa, I = 0.05 m^4 and A = 0.005m^4, for element 1 we can determine that:

$$\frac{EA}{l} = \frac{(210x10^9)(0.005)}{4} = 2.625x10^8 \qquad \frac{EI}{l} = 2.625x10^9$$

$$\frac{EI}{l^2} = 6.562x10^8 \qquad \frac{EI}{l^3} = 1.64x10^8$$

Using equation 4.77 we can form the element stiffness matrix for each element. Element 1 is conveniently orientated so that it is aligned with the global coordinate system so for element 1:

$$
k'^1 = k^1 = 1x10^8
\begin{array}{cccccc}
U1 & V1 & \theta1 & U2 & V2 & \theta2 \\
\end{array}
\begin{bmatrix}
2.625 & 0 & 0 & -2.625 & 0 & 0 \\
0 & 19.68 & 39.37 & 0 & 19.68 & 39.37 \\
0 & 39.37 & 105 & 0 & -39.37 & 52.5 \\
-2.625 & 0 & 0 & 2.625 & 0 & 0 \\
0 & -19.68 & -39.37 & 0 & 19.68 & -39.37 \\
0 & 39.37 & 52.5 & 0 & -39.37 & 105
\end{bmatrix}
$$

For elements 2 and 3 we can determine that:

$$\frac{EA}{l} = \frac{(210x10^9)(0.005)}{3} = 3.5x10^8 \qquad \frac{EI}{l} = 3.5x10^9$$

$$\frac{EI}{l^2} = 3.166x10^9 \qquad \frac{EI}{l^3} = 3.88x10^8$$

Thus the local element stiffness matrix for elements 2 and 3 is given by:

$$
k'^2 = k'^3 = 1 \times 10^8
\begin{bmatrix}
3.5 & 0 & 0 & -3.5 & 0 & 0 \\
0 & 46.65 & 18.9 & 0 & 46.65 & 18.9 \\
0 & 18.9 & 140 & 0 & -18.9 & 70 \\
-3.5 & 0 & 0 & 3.5 & 0 & 0 \\
0 & -46.65 & -18.9 & 0 & 46.65 & -18.9 \\
0 & 18.9 & 70 & 0 & -18.9 & 140
\end{bmatrix}
$$

The transformation matrix [L] is used to transform the local element stiffness matrices into the global element stiffness matrices, where, in this case the direction cosines are given by l = 0 and m = 1 and thus from equation 4.76 we have:

$$
[L] =
\begin{bmatrix}
0 & 1 & 0 & 0 & 0 & 0 \\
-1 & 0 & 0 & 0 & 0 & 0 \\
0 & 0 & 1 & 0 & 0 & 0 \\
0 & 0 & 0 & 0 & 1 & 0 \\
0 & 0 & 0 & -1 & 0 & 0 \\
0 & 0 & 0 & 0 & 0 & 1
\end{bmatrix}
$$

The stiffness matrix in terms of global coordinates for elements 2 and 3 is obtained from $[k]^e = [L^T k' L]$ which gives:

Element 2:　U3　V3　θ3　U1　V1　θ1
Element 3:　U4　V4　θ4　U2　V2　θ2

$$
k^2 = k^3 = 1 \times 10^8
\begin{bmatrix}
46.65 & 0 & -18.9 & 46.65 & 0 & 18.9 \\
0 & 3.5 & 0 & 0 & 3.5 & 0 \\
-18.9 & 0 & 140 & 18.9 & 0 & 70 \\
46.65 & 0 & 18.9 & 46.65 & 0 & 18.9 \\
0 & -3.5 & 0 & 0 & 3.5 & 0 \\
-18.9 & 0 & 70 & 18.9 & 0 & 140
\end{bmatrix}
$$

Thus the global stiffness matrix can be formed:

$$
[K] = 1 \times 10^8
\begin{bmatrix}
49.275 & 0 & 18.9 & -2.625 & 0 & 0 \\
0 & 23.18 & 39.37 & 0 & 19.68 & 39.37 \\
18.9 & 39.37 & 245 & 0 & -39.37 & 52.5 \\
-2.625 & 0 & 0 & 49.275 & 0 & 18.9 \\
0 & -19.68 & -39.37 & 0 & 23.18 & -39.37 \\
0 & 39.37 & 52.5 & 18.9 & -39.37 & 245
\end{bmatrix}
$$

The applied distributed load can be resolved into vertical point loads in nodes 1 and 2 and bending moments about these nodes. In this case let's assume that the loads can be represented as shown in figure 4.28.

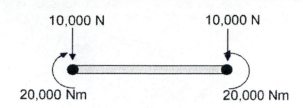

Figure 4.28: Loads on Element 1

From figure 4.28 we can easily form the global load vector:

$$\{F\} = \begin{Bmatrix} 0 \\ -10,000 \\ -20,000 \\ 0 \\ -10,000 \\ +20,000 \end{Bmatrix}$$

As always, the global problem equation is given by [K]{U} = {F}. Solving for {U} gives:

$$\{U\}^T = 1\text{x}10^{-7}\{8.98 \quad 42.1 \quad -23.3 \quad 1.13 \quad -49.8 \quad -1.71\}$$

4.8.5. Case Study

You should now examine case study B in chapter 10 which details a practical example of the use of 3D beam elements in a finite element analysis of a bicycle frame subjected to operational loads.

4.9. Pipe Elements

4.9.1 Applications

Pipe elements are generally used to analyse pipe networks in the civil engineering and building services engineering industries. The most basic pipe elements are essentially beam elements with a cylindrical cross section; however more advanced versions are curved to model pipe elbows and can include hydrostatic and hydrodynamic effects. The ability to model hydrostatic effects is important for pipes that may be submerged or partially submerged, e.g. undersea oil and gas pipelines. Both internal pressure, due to the fluid contained in the pipe, and external pressure, due to the external fluid, can be applied. Buoyant effects can also be included. Hydrodynamic effects are used to model the effect of ocean or river currents and wave motion on submerged or partially submerged pipes. This type of analysis is particularly important for floating pipes and pipes that bring fluid from under the seabed to the surface.

4.9.2. Description

Pipe elements are essentially a development of 3D beam elements with an annular cross section. A straight pipe element is identical to a 3D linear beam element with

Practical Stress Analysis with Finite Elements

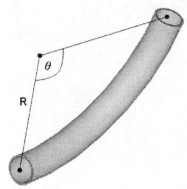

Figure 4.30: A Curved Pipe Element Used to Model Pipe Elbows

In some FE software a curved pipe element is available for modelling pipe elbows, as shown in figure 4.30. In this case the stiffness matrix in local coordinates is given by:

$$[k'] = [F']^{-1}$$

(4.80)

where $[F]^{-1}$ is the inverse of the pipe element flexibility matrix:

$$[F'] = \begin{bmatrix} F_{11} & 0 & F_{13} & 0 & F_{15} & 0 \\ 0 & F_{22} & 0 & F_{24} & 0 & F_{26} \\ -F_{13} & 0 & F_{33} & 0 & F_{35} & 0 \\ 0 & -F_{24} & 0 & F_{44} & 0 & F_{46} \\ -F_{15} & 0 & -F_{35} & 0 & F_{55} & 0 \\ 0 & -F_{26} & 0 & -F_{46} & 0 & F_{66} \end{bmatrix}$$

(4.81)

Where:

$$F_{11} = \frac{R^3 C_{fi}}{EI}\left(\frac{\theta I}{2} - \frac{3m\theta}{2} + \theta\right) + \frac{R}{2EA^W}(\theta I + m) + \frac{2R(1+v)}{EA^W}(\theta I - m)$$

$$F_{13} = \frac{R^3 C_{fi}}{EI}\left(I - 1 + \frac{\theta m}{2}\right) + \frac{R\theta I}{EA^W}\left(\frac{5}{2} + 2v\right)$$

$$F_{15} = \frac{R^2 C_{fi}}{EI}(I - \theta)$$

$$F_{22} = \frac{R^3(1+v)}{EI}(\theta - I) + \frac{R^3}{2EI}(1 + v + C_{f0})(\theta I - m) + \frac{R\theta(4(1+v))}{EA^W}$$

$$F_{24} = \frac{R^2}{2EI}(1 + v + C_{f0})(\theta I - m)$$

$$F_{26} = \frac{R^2}{EI}\left[(1 + v)(\cos(\theta - 1)) + \frac{\theta}{2}\sin\theta(1 + v + C_{f0})\right]$$

$$F_{33} = \left(\frac{\theta I}{2} - \frac{m}{2}\right)\left(\frac{R^3 C_{fi}}{EI} + \frac{R}{EA^W}\right) + \left(\frac{\theta I}{2} - \frac{m}{2}\right)\left(\frac{4R(1+v)}{EA^W}\right)$$

$$F_{44} = \frac{R}{2EI}(1+v+C_{f0})\theta l + \frac{R}{2EI}(1+v-C_{f0})m$$

$$F_{46} = \frac{R}{2EI}(1+v+C_{f0})\theta m$$

$$F_{55} = \frac{RC_{fi}\theta}{EI}$$

$$F_{66} = \frac{R}{2EI}(1+v+C_{f0})\theta l - \frac{R}{2EI}(1+v-C_{f0})m$$

In the above equations; R is the radius of curvature of the element, θ is the included angle of the element, E is the Young's modulus of the element material, v is the material Poisson ratio, I is the moment of inertia of the annular cross section, A^w is the cross sectional area of the pipe, C_{fi} is the in-plane flexibility of the pipe, C_{f0} is the out-of-plane flexibility of the pipe, $l = \cos\theta$ and $m = \sin\theta$.

4.9.3. Determination of Element Properties

Some pipe elements can model hydrostatic and hydrodynamic effects. Hydrostatic effects occur on the outside and inside of the element. Pressure on the outside of the element, due to immersion in a liquid, will tend to crush the pipe and buoyancy forces will tend to raise the pipe to the surface. Pressure on the inside of the pipe will tend to stabilise the cross section and resist crushing effects. Hydrodynamic effects occur because either a) the structure moves in a motionless fluid, b) the structure is fixed but there is fluid motion, or c) both the structure and fluid are moving. The fluid motion can consist of two components: current and wave motions. Including hydrodynamic effects in an analysis is obviously important for undersea pipes etc.

We will now consider hydrostatic effects and how these are incorporated in the element equations. The implementation of hydrodynamic effects is complex and beyond the scope of this text.

The buoyant force acting on a totally submerged element is given by:

$$\{f^b\} = C_b \rho_{fluid} \frac{\pi}{4} D_e^2 \{g\} \tag{4.82}$$

Where $\{f^b\}$ is the vector of body loads per unit length due to buoyancy, C_b is a buoyancy coefficient, ρ_{fluid} is the density of the fluid, D_e is the external diameter of the pipe and $\{g\}$ is the nodal acceleration vector.

The external, crushing, pressure at a node is given by:

$$P_o = -\rho_{fluid} gz + f^s \tag{4.83}$$

Where P_o is the crushing pressure at the node, g is acceleration due to gravity, z is the vertical coordinate of the node and f^s is the effect of any additional surface loads applied to the element(s) of which the node is part.

The internal, bursting, pressure at a node is given by:

$$P_i = -\rho_{int} g(z - z_{fs}) + f^s_{int} \tag{4.84}$$

Where P_i is the internal pressure at the node, ρ_{int} is the density of the fluid inside the pipe, g is acceleration due to gravity, z is the vertical coordinate of the node, z_{fs} is the vertical coordinate of the fluid free surface and f^s_{int} is the effect of any additional internal surface loads applied to the element(s) of which the node is part.

The stresses in the element can thus be calculated from equations 4.83 and 4.84 as follows:

$$\sigma_{axial} = \frac{P_i D_i^2 - P_o D_e^2}{D_e^2 - D_i^2} \tag{4.85}$$

$$\sigma_{hoop} = \frac{P_i D_i^2 - P_o D_e^2 + \left(\dfrac{D_i^2 D_e^2}{D^2}\right)(P_i - P_o)}{D_e^2 - D_i^2} \tag{4.86}$$

$$\sigma_{radial} = \frac{P_i D_i^2 - P_o D_e^2 - \left(\dfrac{D_i^2 D_e^2}{D^2}\right)(P_i - P_o)}{D_e^2 - D_i^2} \tag{4.87}$$

Where D_i is the internal pipe diameter and D is the diameter at which the stress state is required (i.e. the diameter of interest for results).

4.10. 2D Plane Stress and Plane Strain Solids

The formulation of elements for plane stress and plane strain is essentially the same in terms of how the element properties such as shape functions etc. are determined an implemented in FE software. The major difference occurs when the material property matrix [D] is introduced into the calculation of the stiffness matrices. Thus, in this section we will discuss plane stress and plane strain elements separately when describing applications and material property implementation, but when discussing element formulation a common discussion is required.

4.10.1. Applications of Plane Stress Elements

As discussed in chapter 3, a plane stress analysis is particularly suited to the analysis of thin, sheet like, structures where the through thickness dimension is much less than the other dimensions of the structure under consideration. Obvious applications include flat thin plates under tension. Some less obvious applications include the analysis of a wrench turning a nut and analysis of thin cantilevered beams. Many curved thin sheets can generally be assumed to be flat (i.e. in one plane) for the purposes of analysis and in particular to get a "first guess" or ballpark answer to a problem. Examples of such structures include pressure vessels of large radius and aircraft fuselage skin.

Plane stress elements are sometimes also available in "plane stress with thickness" or "plate" formulations. These are essentially 21/2D elements, as opposed to 2D elements, as they allow for specification of element thickness, thus allowing for the inclusion of different areas of thickness within the model.

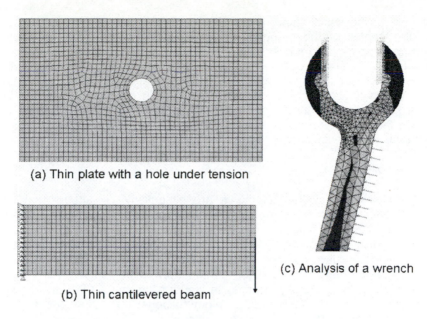

(a) Thin plate with a hole under tension

(b) Thin cantilevered beam

(c) Analysis of a wrench

Figure 4.31: Examples of Plane Stress Finite Element Models

4.10.2. Applications of Plane Strain Elements

Plain strain elements are particularly suited to the analysis of very long structures that have continuous unchanging loads along their length. Typical examples include splined shafts, long beams with constant loading, dams, pressure vessels, etc. Plain strain analysis is also very useful for certain metal forming processes such as rolling, drawing and forging. Figure 4.32 shows two examples of the use of plane strain elements to model typical long structural problems.

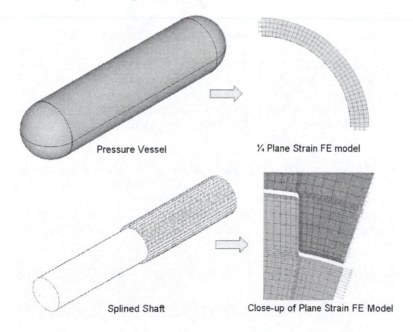

Pressure Vessel

¼ Plane Strain FE model

Splined Shaft

Close-up of Plane Strain FE Model

Figure 4.32: Examples of Plane Strain Finite Element Models

The top part of the figure shows a cylindrical pressure vessel modelled as a slice through the centre of the cylinder. Using symmetry and appropriate boundary conditions it is possible to model only one quarter of the slice, thus producing the plane strain representation of the problem shown in the top right hand side of figure 4.32. The bottom of the figure shows a splined shaft which is used to transfer rotational forces to another structure when inserted in a splined hole. The bottom right hand side of the figure shows a representative plane strain model of this problem. In this case cyclic symmetry has been used so that only one tooth of both the shaft and hole are considered.

4.10.3. Description

Plane stress and plane strain elements are essentially 2D sheets that can be triangular or rectangular in shape. Each shape is generally available in linear and quadratic formulations. Figure 4.33 summarises the types of plane elements generally found in most commercial FE software packages. In some cases these elements will be described as plane stress elements or plain strain elements, while in others they may be described as 2D solids or simply as "rectangles" and "triangles". In the latter case it is generally required to first specify that a 2D solid is required, then pick it's shape and finally pick an element behaviour option, usually from a choice of plane stress, plane strain or axisymmetric.

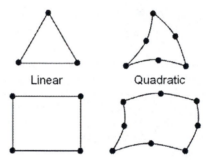

Figure 4.33: Overview of Plane Stress and Plain Strain Elements

Linear plane triangles consist of three nodes and have straight edges between these nodes. These elements are often known as "constant strain triangles" (CST) as their formulation results in each element having a constant strain after solution (i.e. it will be the same at each node). The ideal shape for these elements in a finite element mesh is as an equilateral triangle. The further the element is from this shape in a mesh, the less accurate it becomes.

Quadratic plane triangles have six nodes, one at each corner and one at the midpoint of each side of the triangle. Due to the quadratic formulation used, these elements can have curved edges and strain will not be constant within the element after solution (i.e. it can be different at each node) and, in fact, strain will vary linearly within the element and these elements are sometimes known as "linear triangles" (LT).

Linear plane rectangles have four nodes and straight edges between these nodes. The formulation of these elements results in strain varying in a bilinear fashion throughout the element, thus these elements are sometimes known as "bilinear quadrilaterals" (BQ). The ideal shape for these elements in a finite element mesh is as a square.

Quadratic plane rectangles have eight nodes, one at each corner and one at the midpoint of each side of the rectangle. Due to the quadratic formulation used, these elements can have curved edges and strain will vary in a quadratic manner within the element. We will designate these elements Q8 for discussions in later chapters.

4.10.4. Determination of Element Properties

4.10.4.1. 3 Node Plane Triangles (constant strain triangles)

As usual, we wish to represent the displacement of points inside the element in terms of the nodal point displacements using shape functions. For the three node triangle, with nodes i, j and k, three shape functions, S_i, S_j and S_k are required. S_i must have a value of 1 at node i and a value of 0 at the other nodes, and similarly for S_j and S_k. This results in three shape functions of the form shown in figure 4.34.

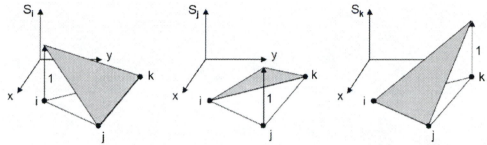

Figure 4.34: Area Shape Functions for a 3 Node Triangle

It is clear from figure 4.34 that these shape functions describe a plane surface that has a value of 1 at the dependant node and drops to zero at the other two nodes. Thus, $S_i+S_j+S_k$ will represent a plane at a height of 1 at each node which is parallel to the triangle. This means that:

$$S_i + S_j + S_k = 1 \tag{4.88}$$

These shape functions in terms of the natural coordinates ξ and η are given by:

$$S_i = \xi \qquad S_j = \eta \qquad S_k = 1 - \xi - \eta \tag{4.89}$$

The displacements within the element in terms of the shape functions of equation 4.89 and the unknown nodal displacements are:

$$u = S_i U_i + S_j U_j + S_k U_k$$
$$v = S_i V_i + S_j V_j + S_k V_k \tag{4.90}$$

Equation 4.90 can be rewritten in terms of natural coordinates using equation 4.89 to give:

$$u = (U_i - U_k)\xi + (U_j - U_k)\eta + U_k$$
$$v = (V_i - V_k)\xi + (V_j - V_k)\eta + V_k \tag{4.91}$$

Equation 4.90 can be expressed in matrix form: {u} = [S]{U}, where:

$$[S] = \begin{bmatrix} S_i & 0 & S_j & 0 & S_k & 0 \\ 0 & S_i & 0 & S_j & 0 & S_k \end{bmatrix} \tag{4.92}$$

Practical Stress Analysis with Finite Elements

The coordinate transformation from local to natural coordinates can be done using the same shape functions as described in equation 4.89. Thus we are again using an isoparametric formulation as previously described in section 4.3.3. Using equation 4.91 we have:

$$x = (x_i - x_k)\xi + (x_j - x_k)\eta + x_k$$
$$y = (y_i - y_k)\xi + (y_j - y_k)\eta + y_k$$

(4.93)

When determining element strains partial derivatives of u and v need to be evaluated with respect to x and y. From equations 4.90 and 4.93 we know that u, v, x and y are all functions of ξ and η. If we examine u and use the chain rule to evaluate partial derivatives we can write:

$$\frac{du}{d\xi} = \frac{du}{dx}\frac{dx}{d\xi} + \frac{du}{dy}\frac{dy}{d\xi}$$
$$\frac{du}{d\eta} = \frac{du}{dx}\frac{dx}{d\eta} + \frac{du}{dy}\frac{dy}{d\eta}$$

(4.94)

This can be written in matrix form as:

$$\begin{Bmatrix} \dfrac{du}{d\xi} \\ \dfrac{du}{d\eta} \end{Bmatrix} = \underbrace{\begin{bmatrix} \dfrac{dx}{d\xi} & \dfrac{dy}{d\xi} \\ \dfrac{dx}{d\eta} & \dfrac{dy}{d\eta} \end{bmatrix}}_{[J]} \begin{Bmatrix} \dfrac{du}{dx} \\ \dfrac{du}{dy} \end{Bmatrix}$$

(4.95)

Where, the 2x2 square matrix denoted as J in equation 4.95 is known as the *Jacobian* of the coordinate transformation. If we take the derivative of x and y in the Jacobian we obtain:

$$[J] = \begin{bmatrix} x_i - x_k & y_i - y_k \\ x_j - x_k & y_j - y_k \end{bmatrix}$$

(4.96)

If we invert equation 4.95 we obtain:

$$\begin{Bmatrix} \dfrac{du}{dx} \\ \dfrac{du}{dy} \end{Bmatrix} = [J]^{-1} \begin{Bmatrix} \dfrac{du}{d\xi} \\ \dfrac{du}{d\eta} \end{Bmatrix}$$

(4.97)

Where, $[J]^{-1}$ is the inverse of the Jacobian, [J] and is given by:

$$[J]^{-1} = \frac{1}{\det[J]} \begin{bmatrix} y_j - y_k & -y_i + y_k \\ -x_j + x_k & x_i - x_k \end{bmatrix}$$

(4.98)

Where: $\det[J] = (x_k - x_i)(y_j - y_k) - (x_j - x_k)(y_i - y_k)$ (4.99)

Comparing equation 4.99 with the equation for the area of a triangle it should be clear that det [J] is twice the area of the triangle. If the points i, j and k are ordered in an anticlockwise manner then det [J] will be positive in sign. Most FE software will use an anticlockwise ordering and will use det [J] to calculate the area of the element.

$$Area = \frac{1}{2}\left| \det[J] \right|$$

(4.100)

4.10.4.2.   4 Node Plane Quadrilaterals (Bilinear Quadrilaterals)

If we now consider the four node quadrilateral element shown in figure 4.35. The nodes are numbered i, j, k, and l in the local coordinate system as shown. Each node can be described in terms of its local coordinate x,y and each node has two degrees of freedom, i.e. displacement in the x and y directions.

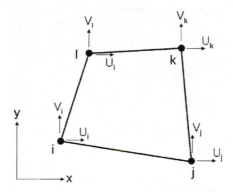

Figure 4.35: Quadrilateral Four-node Plane Element

As usual we must develop shape functions on an element which is described in terms of its natural coordinate system as shown in figure 4.36. For the four node quadrilateral, with nodes i, j, k and l, four shape functions, S_i, S_j, S_k and S_l are required. S_i must have a value of 1 at node i and a value of 0 at the other nodes, and similarly for S_j, S_k and S_l.

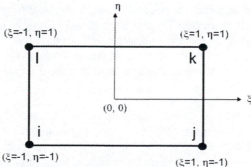

Figure 4.36: Quadrilateral Four-node Element in its Natural Coordinate System

The requirement that $S_i = 0$ at the three nodes i, j and k, effectively means that S_i must be zero along the edges $\xi = +1$ and $\eta = +1$. This means that S_i must be of the form:

$$S_i = c(1-\xi)(1-\eta)$$

(4.101)

Where c is a constant, which may be evaluated by considering that $S_i = 1$ at node i, which has coordinates $\xi = -1$ and $\eta = -1$. Thus c can be evaluated to be ¼. All four shape functions in terms of natural coordinates are:

$$S_i = \frac{1}{4}(1-\xi)(1-\eta)$$

$$S_j = \frac{1}{4}(1+\xi)(1-\eta)$$

(4.102)

$$S_k = \frac{1}{4}(1+\xi)(1+\eta)$$

$$S_l = \frac{1}{4}(1-\xi)(1+\eta)$$

The displacements within the element in terms of the shape functions of equation 4.102 and the unknown nodal displacements are:

$$u = S_i U_i + S_j U_j + S_k U_k + S_l U_l$$
$$v = S_i V_i + S_j V_{jy} + S_k V_k + S_l V_l$$

(4.103)

Equation 4.103 can be expressed in matrix form: $\{u\} = [S]\{U\}$, where:

$$[S] = \begin{bmatrix} S_i & 0 & S_j & 0 & S_k & 0 & S_l & 0 \\ 0 & S_i & 0 & S_j & 0 & S_k & 0 & S_l \end{bmatrix}$$

(4.104)

As we are using an isoparametric formulation we can use the same shape functions in equation 4.102 to describe the position of any point in the element in terms of nodal coordinates using:

$$x = S_i x_i + S_j x_j + S_k x_k + S_l x_l$$
$$y = S_i y_i + S_j y_j + S_k y_k + S_l y_l$$

(4.105)

As usual, the displacement field is related to the components of strains (ε_{xx}, ε_{yy} and γ_{xy}) and subsequently to the nodal displacements using shape functions. In deriving the element stiffness matrix from strain energy, we need to take the derivatives of the components of the displacement field with respect to the x and y coordinates which in turn means taking the derivatives of the appropriate shape functions with respect to x and y.

Currently, however, we have the shape functions expressed in terms of natural coordinates (ξ and η). So in general it is required to establish relationships that allow the derivatives of a function f=f(x,y) to be taken with respect to x and y and to express them in terms of derivatives of the function f=f(x,y) with respect to ξ and η. Using the chain rule we can write:

$$\frac{\partial f}{\partial \xi} = \frac{\partial f}{\partial x}\frac{\partial x}{\partial \xi} + \frac{\partial f}{\partial y}\frac{\partial y}{\partial \xi}$$

$$\frac{\partial f}{\partial \eta} = \frac{\partial f}{\partial x}\frac{\partial x}{\partial \eta} + \frac{\partial f}{\partial y}\frac{\partial y}{\partial \eta}$$

(4.106)

Expressing this equation in matrix form gives:

$$\begin{Bmatrix} \dfrac{\partial f}{\partial \xi} \\[2mm] \dfrac{\partial f}{\partial \eta} \end{Bmatrix} = \overbrace{\begin{bmatrix} \dfrac{\partial x}{\partial \xi} & \dfrac{\partial y}{\partial \xi} \\[2mm] \dfrac{\partial x}{\partial \eta} & \dfrac{\partial y}{\partial \eta} \end{bmatrix}}^{[J]} \begin{Bmatrix} \dfrac{\partial f}{\partial x} \\[2mm] \dfrac{\partial f}{\partial y} \end{Bmatrix} \qquad (4.107)$$

Where [J] is the Jacobian matrix. Equation 4.107 can be inverted to give:

$$\begin{Bmatrix} \dfrac{\partial f}{\partial x} \\[2mm] \dfrac{\partial f}{\partial y} \end{Bmatrix} = \dfrac{1}{\det[J]} \begin{bmatrix} J_{22} & -J_{12} \\[2mm] -J_{21} & J_{11} \end{bmatrix} \begin{Bmatrix} \dfrac{\partial f}{\partial \xi} \\[2mm] \dfrac{\partial f}{\partial \eta} \end{Bmatrix} \qquad (4.108)$$

The stiffness matrix for the four node quadrilateral element can be derived from the strain energy in the body:

$$\Lambda = \int_V \frac{1}{2}\{\sigma\}^T\{\varepsilon\}dV = \sum_e t_e \int_e \frac{1}{2}\{\sigma\}^T\{\varepsilon\}dA \qquad (4.109)$$

Where, t_e is the thickness of element e.

The strain displacement relations for the element are:

$$\{\varepsilon\} = \begin{Bmatrix} \varepsilon_x \\ \varepsilon_y \\ \gamma_{xy} \end{Bmatrix} = \left\{ \dfrac{\partial u}{\partial x} \quad \dfrac{\partial v}{\partial y} \quad \left(\dfrac{\partial u}{\partial y} + \dfrac{\partial v}{\partial x} \right) \right\}^T \qquad (4.110)$$

By letting f = u in equation 4.108 we get:

$$\begin{Bmatrix} \dfrac{\partial u}{\partial x} \\[2mm] \dfrac{\partial u}{\partial y} \end{Bmatrix} = \dfrac{1}{\det[J]} \begin{bmatrix} J_{22} & -J_{12} \\[2mm] -J_{21} & J_{11} \end{bmatrix} \begin{Bmatrix} \dfrac{\partial u}{\partial \xi} \\[2mm] \dfrac{\partial u}{\partial \eta} \end{Bmatrix} \qquad (4.111)$$

Similarly, by letting F = v, we get:

$$\begin{Bmatrix} \dfrac{\partial v}{\partial x} \\[2mm] \dfrac{\partial v}{\partial y} \end{Bmatrix} = \dfrac{1}{\det[J]} \begin{bmatrix} J_{22} & -J_{12} \\[2mm] -J_{21} & J_{11} \end{bmatrix} \begin{Bmatrix} \dfrac{\partial v}{\partial \xi} \\[2mm] \dfrac{\partial v}{\partial \eta} \end{Bmatrix} \qquad (4.112)$$

Brining together equations 4.110, 4.111 and 4.112 gives:

$$\{\varepsilon\} = [A] \left\{ \dfrac{\partial u}{\partial \xi} \quad \dfrac{\partial u}{\partial \eta} \quad \dfrac{\partial v}{\partial \xi} \quad \dfrac{\partial v}{\partial \eta} \right\}^T \qquad (4.113)$$

Where the [A] matrix is given by:

$$[A] = \frac{1}{\det[J]} \begin{bmatrix} J_{22} & -J_{12} & 0 & 0 \\ 0 & 0 & -J_{21} & J_{11} \\ -J_{21} & J_{11} & J_{22} & -J_{12} \end{bmatrix} \tag{4.114}$$

From the interpolation equations (equation 4.103) we know that:

$$\left\{ \frac{\partial u}{\partial \xi} \quad \frac{\partial u}{\partial \eta} \quad \frac{\partial v}{\partial \xi} \quad \frac{\partial v}{\partial \eta} \right\}^T = [G]\{U\} \tag{4.115}$$

Where, the [G] matrix is given by:

$$[G] = \frac{1}{4} \begin{bmatrix} -(1-\eta) & 0 & (1-\eta) & 0 & (1+\eta) & 0 & -(1+\eta) & 0 \\ -(1-\xi) & 0 & -(1+\xi) & 0 & (1+\xi) & 0 & (1-\xi) & 0 \\ 0 & -(1-\eta) & 0 & (1-\eta) & 0 & (1+\eta) & 0 & -(1+\eta) \\ 0 & -(1-\xi) & 0 & -(1+\xi) & 0 & (1+\xi) & 0 & (1-\xi) \end{bmatrix} \tag{4.116}$$

Putting together equations 4.113 and 4.115 gives us our standard equation:

$$\{\varepsilon\} = [B]\{U\} \tag{4.117}$$

Where: $\qquad [B] = [A][G] \tag{4.118}$

Thus, the strain in the quadrilateral element can be expressed in terms of its nodal displacement, via the [A] and [G] matrices, which are functions of the shape functions.

The stress in the element is now given by:

$$\{\sigma\} = [D][B]\{U\} = [D][A][G]\{U\} \tag{4.119}$$

Where [D] is the (3x3) material property matrix which will be given by equation 2.34 in the case of plane stress and by equation 2.35 in the case of plane strain. The strain energy of the element, from equation 4.109, can now be rewritten as:

$$\Lambda = \sum_e \frac{1}{2}\{U\}^T \left[t_e \int_{-1}^{1}\int_{-1}^{1} [B]^T [D][B]\det[J]\,d\xi\,d\eta \right]\{U\} = \sum_e \frac{1}{2}\{U\}^T [k^e]\{U\} \tag{4.120}$$

Where the (8x8) element stiffness matrix for the four node quadrilateral is given by:

$$[k^e] = t_e \int_{-1}^{1}\int_{-1}^{1} [B]^T [D][B]\det[J]\,d\xi\,d\eta \tag{4.121}$$

Similarly, it can be shown that the (8x1) element body force vector is given by:

$$\{f_e^B\} = t_e \left[\int_{-1}^{1}\int_{-1}^{1} [S]^T \det[J]\,d\xi\,d\eta \right] \begin{Bmatrix} f_x^B \\ f_y^B \end{Bmatrix} \tag{4.122}$$

Also, the element surface load vector can be determined, for example for a pressure load applied on edge j-k of the quadrilateral element, we have:

$$\{f_e^s\} = \frac{t_e \ell_{j-k}}{2} \begin{bmatrix} 0 & 0 & f_x^s & f_y^s & f_x^s & f_y^s & 0 & 0 \end{bmatrix}^T \tag{4.123}$$

Where, ℓ_{j-k} is the length of the edge j-k.

4.10.4.3 Six Node Plane Triangles (Linear Strain Triangles)

The six node triangle, shown in figure 4.37, is an improvement on the constant strain triangle that is able to model more complex shapes and with greater accuracy.

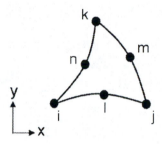

Figure 4.37: Six Node Plane Triangle Element

The shape functions for this element are given by:

$$\begin{array}{lll} S_i = \xi(2\xi - 1) & S_j = \eta(2\eta - 1) & S_k = \lambda(2\lambda - 1) \\ S_l = 4\xi\eta & S_m = 4\lambda\eta & S_i = 4\xi\lambda \end{array} \tag{4.124}$$

Where $\lambda = 1 - \xi - \eta$.

The determination of element properties follows the same procedure as used above for other 2D elements:
1. We express the global element displacement in terms of nodal displacements using the shape functions, $\{u\} = [S]\{U\}$
2. We express strains in terms of nodal displacements using a strain-displacement matrix which is derived from the shape functions, $\{\varepsilon\} = [B]\{U\}$
3. We evaluate the element stiffness matrix from 4.121.

An important consideration regarding the placement of the mid-side node in this element and all other isoparametric elements that contain mid-side nodes is that the node should not be placed between ¼ and ¾ of the length of the element edge. Failing to do so may result in det [J] having a value of zero and thus rendering all calculations of element properties impossible.

4.10.4.4. Eight Node Plane Quadrilaterals

The eight node plane quadrilateral is a higher order version of the four node quadrilateral discussed in section 4.10.4.2. In this case the element has curved boundaries and a mid-side node placed on each edge of the element. This element is more suited to modelling complex shapes with greater accuracy.

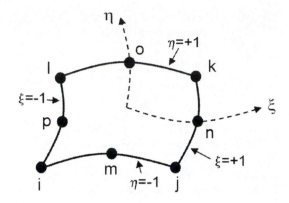

Figure 4.38: Eight Node Plane Quadrilateral Element

The shape functions for this element are given by:

$$S_i = -\frac{1}{4}(1-\xi)(1-\eta)(1+\xi+\eta) \qquad S_m = \frac{1}{2}(1-\xi^2)(1-\eta) \qquad (4.125)$$

$$S_j = -\frac{1}{4}(1+\xi)(1-\eta)(1-\xi+\eta) \qquad S_n = \frac{1}{2}(1+\xi)(1-\eta^2)$$

$$S_k = -\frac{1}{4}(1+\xi)(1+\eta)(1-\xi-\eta) \qquad S_o = \frac{1}{2}(1-\xi^2)(1+\eta)$$

$$S_l = -\frac{1}{4}(1-\xi)(1+\eta)(1+\xi-\eta) \qquad S_p = \frac{1}{2}(1-\xi)(1-\eta^2)$$

Using the steps outlined in the previous section we can calculate element properties based on these shape functions and hence calculate the element stiffness matrix using equation 4.121.

4.10.5 Case Studies

Case study C in chapter 10 compares the use of three node triangles, four node quadrilaterals, six node triangles and eight node quadrilaterals to model a plane stress cantilevered beam problem and compares the accuracy of the results produced by each element type.

Case Study D in chapter 10 discusses the use of plane strain elements to perform an analysis of a long pipe subjected to compressive loads.

4.11. 2D Axisymmetric Solids

4.11.1. Applications

These elements are used to analyse three dimensional axisymmetric solids subjected to axisymmetric loading. Such problems can be reduced to a 2D representation as both geometry and the applied loading is unchanging in the circumferential direction. If we take, for example, the axisymmetric problem shown in figure 4.38(a), it is clear that because of symmetry about the z axis all deformations and stresses will be independent of the circumferential coordinate, θ. Thus the problem can be examined as a 2D problem in the rz plane, as shown in figure 4.38(b).

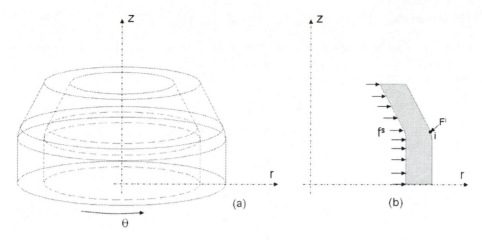

Figure 4.39: An Example Axisymmetric Problem

Common applications of axisymmetric elements include the analysis of pressure vessels subjected to internal and/or external pressure (e.g. gas cylinders, diver's air cylinders, submarines etc.), analysis of shafts and flywheels, analysis of Belleville springs (i.e. a conical disc spring), analysis of circular deep drawing metal forming processes and analysis of other axisymmetric metal forming processes such as sheet bulging and bore expansion. Examples of some of these applications are shown in figure 4.40.

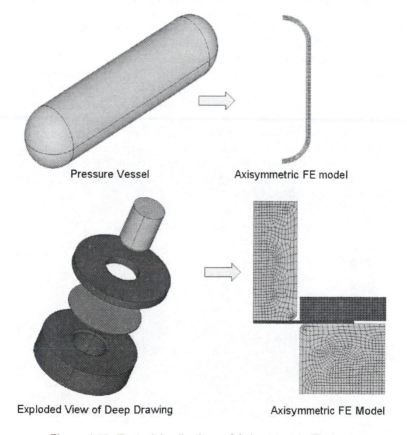

Pressure Vessel   Axisymmetric FE model

Exploded View of Deep Drawing   Axisymmetric FE Model

Figure 4.40: Typical Applications of Axisymmetric Elements

4.11.2. Description

Axisymmetric elements are a form of 2D element and are thus available in the same shapes as other 2D elements as shown again in figure 4.41. Both triangular and quadrilateral shapes are generally available in both linear and quadratic formulations.

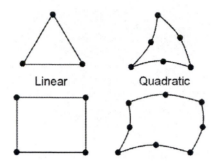

Linear   Quadratic

Figure 4.41: Types of Axisymmetric Elements

The major difference between axisymmetric and 2D plane elements is in how they are formulated, and hence how they are "seen" by the finite element solver. 2D elements are generally "seen" as being in two dimensions whereas axisymmetric elements are essentially "seen" as 3D elements by the FE solver.

Some FE software packages require axisymmetric models to be placed on the global ZX plane and others on the XY plane it is important to check in which plane your software requires your model to be placed. In either case the axis about which the area will be revolved will be the Z or Y plane.

A common mistake by novice FE users when applying axisymmetric elements is to not place the model in the correct location in the global coordinate system. For example, if one wishes to model a hollow cylinder the correct procedure is to place a rectangular model a specified positive distance away from the global Y axis. If one placed the rectangle on the global Y axis an effective model of a solid cylinder would result. This is summarised in figure 4.42, where figure 4.42(a) shows the correct procedure for modelling a hollow cylinder by placing the rectangle at a distance r_{in}, equivalent to the inner radius of the cylinder, away from the global Y axis. Figure 4.42(b) shows the consequence of forgetting to take account of the inner radius, where a solid cylinder is modelled instead.

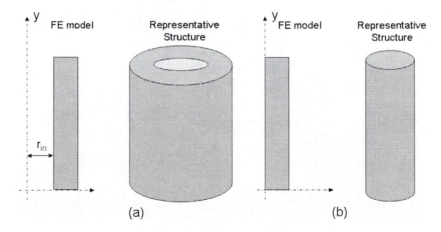

Figure 4.42: The Importance of Correct Model Position When Using Axisymmetric Elements

4.11.3. Determination of Element Properties

In order to illustrate how element properties are determined and how axisymmetric elements are formulated we will refer to the problem shown in figure 4.39 and assume the Z axis is the axis about which the area shall be revolved. If we consider an elemental volume of the problem to be modelled, as shown in figure 4.43, we can write the potential energy of the elemental volume, as:

$$\Pi = \frac{1}{2}\int_{0}^{2\pi}\int_{A}\{\sigma\}^{T}\{\varepsilon\}r\,dAd\theta - \int_{0}^{2\pi}\int_{A}\{u\}^{T}\{f^{B}\}r\,dAd\theta - \int_{0}^{2\pi}\int_{L}\{u\}^{T}\{f^{S}\}r\,dld\theta - \sum_{i}\{u^{i}\}^{T}\{F^{i}\} \quad (4.126)$$

Where, from figure 4.43, $dV = r.d\theta.dr.dz = r.d\theta.dA$ and $r.dl.d\theta$ is the surface area of the elemental volume over which the surface load $\{f^{S}\}$ acts, and the point load $\{F^{i}\}$ represents a line load distributed around a circle. As all variables in the integrals in equation 4.126 are independent of θ, we can rewrite the equation as:

$$\Pi = 2\pi\left(\frac{1}{2}\int_{A}\{\sigma\}^{T}\{\varepsilon\}r\,dA - \int_{A}\{u\}^{T}\{f^{B}\}r\,dA - \int_{L}\{u\}^{T}\{f^{S}\}r\,dl\right) - \sum_{i}\{u^{i}\}^{T}\{F^{i}\} \quad (4.127)$$

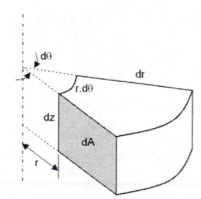

Figure 4.43: An Elemental Volume of the Problem Shown in Figure 4.38(a)

In this case, as we are working in the rz plane:

$$\{u\} = \{u \quad w\}^{T} \qquad \{f^{B}\} = \{f_{r}^{B} \quad f_{z}^{B}\}^{T} \qquad \{f^{S}\} = \{f_{r}^{S} \quad f_{z}^{S}\}^{T}$$

It can be shown that the strain displacement relationships are given by:

$$\{\varepsilon\} = \{\varepsilon_{r} \quad \varepsilon_{z} \quad \gamma_{rz} \quad \varepsilon_{\theta}\}^{T} = \left\{\frac{\partial u}{\partial r} \quad \frac{\partial w}{\partial z} \quad \left(\frac{\partial u}{\partial z} + \frac{\partial w}{\partial r}\right) \quad \frac{u}{r}\right\}^{T} \quad (4.128)$$

The stress vector is given by:

$$\{\sigma\} = \{\sigma_{r} \quad \sigma_{z} \quad \tau_{rz} \quad \sigma_{\theta}\}^{T} \quad (4.129)$$

Stress is related to strain in the usual manner, via the material property matrix [D]:

$$\{\sigma\} = [D]\{\varepsilon\} \quad (4.130)$$

Where the [D] matrix in this case is given by:

$$[D] = \frac{E(1-v)}{(1+v)(1-v)} \begin{bmatrix} 1 & \dfrac{v}{1-v} & 0 & \dfrac{v}{1-v} \\ \dfrac{v}{1-v} & 1 & \dfrac{v}{1-v} & 0 \\ 0 & 0 & \dfrac{1-2v}{2(1-v)} & 0 \\ \dfrac{v}{1-v} & \dfrac{v}{1-v} & 0 & 1 \end{bmatrix}$$ (4.131)

We will now detail the steps used to formulate a three node triangular axisymmetric finite element based on the details above, as shown in figure 4.44. The methodology used will follow the steps used to formulate the 2D Constant Strain Triangle detailed in section 4.10.4.1, however, in this case the r and z coordinates replace the x and y coordinates respectively.

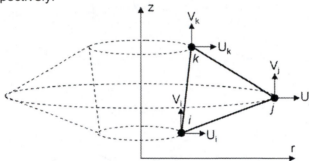

Figure 4.44: Three Node Axisymmetric Triangular Element

Using the three shape functions previously defined for the constant strain triangle, we can relate element displacement to nodal displacements via shape functions, i.e.

$$\{u\} = [S]\{U\}$$ (4.132)

Where:
$$[S] = \begin{bmatrix} S_i & 0 & S_j & 0 & S_k & 0 \\ 0 & S_i & 0 & S_j & 0 & S_k \end{bmatrix}$$ (4.133)

And
$$\{U\} = \{U_i \quad W_i \quad U_j \quad W_j \quad U_k \quad W_k\}^T$$ (4.134)

As before, we can now define the shape functions:
$$S_i = \xi \qquad S_j = \eta \qquad S_k = 1 - \xi - \eta$$ (4.135)

Combining equations 4.135 and 4.132 gives:

$$u = \xi U_i + \eta U_j + (1 - \xi - \eta)U_k$$
$$v = \xi W_i + \eta W_j + (1 - \xi - \eta)W_k$$ (4.136)

Using an isoparametric formulation gives:

$$r = \xi r_i + \eta r_j + (1 - \xi - \eta) r_k$$
$$z = \xi z_i + \eta z_j + (1 - \xi - \eta) z_k \qquad (4.137)$$

Using the chain rule we can determine:

$$\begin{Bmatrix} \dfrac{\partial u}{\partial \xi} \\ \dfrac{\partial u}{\partial \eta} \end{Bmatrix} = [J] \begin{Bmatrix} \dfrac{\partial u}{\partial r} \\ \dfrac{\partial u}{\partial z} \end{Bmatrix} \quad \text{and} \quad \begin{Bmatrix} \dfrac{\partial w}{\partial \xi} \\ \dfrac{\partial w}{\partial \eta} \end{Bmatrix} = [J] \begin{Bmatrix} \dfrac{\partial w}{\partial r} \\ \dfrac{\partial w}{\partial z} \end{Bmatrix} \qquad (4.138)$$

Where the Jacobian in this case is given by:

$$[J] = \begin{bmatrix} r_i - r_k & z_i - z_k \\ z_j - z_k & r_j - r_k \end{bmatrix} \qquad (4.139)$$

If we invert equation 4.138 we obtain:

$$\begin{Bmatrix} \dfrac{\partial u}{\partial r} \\ \dfrac{\partial u}{\partial z} \end{Bmatrix} = [J]^{-1} \begin{Bmatrix} \dfrac{\partial u}{\partial \xi} \\ \dfrac{\partial u}{\partial \eta} \end{Bmatrix} \quad \text{and} \quad \begin{Bmatrix} \dfrac{\partial w}{\partial r} \\ \dfrac{\partial w}{\partial z} \end{Bmatrix} = [J]^{-1} \begin{Bmatrix} \dfrac{\partial w}{\partial \xi} \\ \dfrac{\partial w}{\partial \eta} \end{Bmatrix} \qquad (4.140)$$

Where, $[J]^{-1}$ is the inverse of the Jacobian, $[J]$ and is given by:

$$[J]^{-1} = \frac{1}{\det[J]} \begin{bmatrix} z_j - z_k & -r_i + r_k \\ -r_j + r_k & z_i - z_k \end{bmatrix} \qquad (4.141)$$

Bringing these transformation relationships into the strain displacement equations in equation 4.128 and using equation 4.136 for u and w, gives:

$$\{\varepsilon\} = [B]\{U\} \qquad (4.142)$$

Where:

$$[B] = \frac{1}{\det[J]} \begin{bmatrix} z_j - z_k & 0 & z_k - z_i & 0 & z_i - z_j & 0 \\ 0 & r_k - r_j & 0 & r_i - r_k & 0 & r_j - r_i \\ r_k - r_j & z_j - z_k & r_i - r_k & z_k - z_i & r_j - r_i & z_i - z_j \\ \dfrac{S_i \det[J]}{r} & 0 & \dfrac{S_j \det[J]}{r} & 0 & \dfrac{S_k \det[J]}{r} & 0 \end{bmatrix} \quad (4.143)$$

We can now calculate element properties in the usual manner by substituting equations 4.143 and 4.132 into equation 4.127. Evaluating this gives the element stiffness matrix as:

$$[k^e] = 2\pi \int_e [B]^T [D][B] r \, dA \qquad (4.144)$$

Often, in order to simplify the evaluation of element properties when using axisymmetric elements [B] and r are evaluated at the centroid of the triangular element and used as a representative value for that particular element. At the centroid:

$$S_i = S_j = S_k = \frac{1}{3}$$

(4.145)

And $\quad \bar{r} = \dfrac{r_1 + r_2 + r_3}{3}$

(4.146)

Where, $\bar{r}$ is the radius of the centroid. If we now denote $\left[\bar{B}\right]$ as the element strain-displacement matrix [B] evaluated at the centroid of the element, then equation 4.144 becomes:

$$\left[k^e\right] = 2\pi \bar{r} \left[\bar{B}\right]^T \left[D\right]\left[\bar{B}\right] \int_e dA = 2\pi \bar{r} A_e \left[\bar{B}\right]^T \left[D\right]\left[\bar{B}\right]$$

(4.147)

It should be noted that $2\pi \bar{r} A_e$ is the volume of the ring-shaped element shown in figure 4.43. As usual, the area of the element A_e can be determined from half the determinant of the Jacobian, as shown in equation 4.100.

Similarly the body force term can be determined as:

$$\left\{f_e^B\right\} = \frac{2\pi \bar{r} A_e}{3} \left[\overline{f_r^B} \quad \overline{f_z^B} \quad \overline{f_r^B} \quad \overline{f_z^B} \quad \overline{f_r^B} \quad \overline{f_z^B}\right]^T$$

(4.148)

The surface load on, for example, the i-j edge of the element can be determined as:

$$\left\{f_e^S\right\}_{i-j} = 2\pi \ell_{i-j} \left\{U_i \quad W_i \quad U_j \quad W_j\right\} \begin{Bmatrix} af_r^S \\ af_z^S \\ bf_r^S \\ bf_z^S \end{Bmatrix}$$

(4.149)

Where: a = 1/6(2r$_i$+r$_j$) and   b = 1/6(r$_i$+2r$_j$)

4.11.4. Case Studies

Please now study case study E in chapter 10 which details an axisymmetric analysis of a pressure vessel using axisymmetric elements.

4.12.   2D Plate Elements

These elements are applicable to the analysis of membrane problems where bending of the plate is not considered. In addition, the geometry of the plate must not vary in the thickness direction. Plate elements are essentially plane stress elements in which the thickness of the plate is taken into account during the calculation of element properties. In some FE software packages these elements may be known as "plane stress with thickness" elements or "2D thin shell" elements. In many FEM texts and FE software plate elements are not considered separately from plane stress elements and are generally considered part of the plane stress family of elements.

4.13. 3D Shell Elements

4.13.1. Applications

Shell elements are used to analyse 3D structures where both bending and membrane effects are expected. Obvious examples include thin walled pressure vessels, structures made from sheet metal, aircraft structures and analysis of sheet metal forming processes. Some relevant examples are shown in figure 4.45, where figure 4.45(a) shows a shell element model of a stiffened panel from an aircraft wing and figure 4.45(b) shows a shell element model of a metal forming operation which creates a T-branch from a thin walled straight tube using internal pressure.

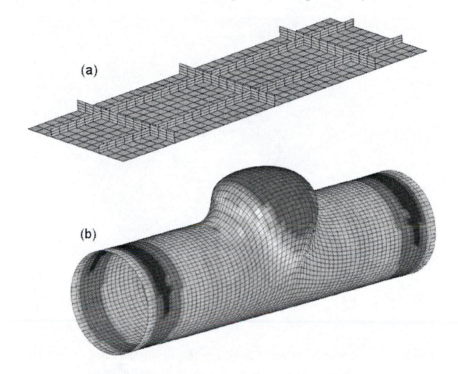

(a)

(b)

Figure 4.45: Typical Applications of 3D Shell Elements

Shell type structures are much more common than expected and occur in all branches of engineering. Reinforced concrete shell roofs are commonly used in new large buildings, particularly in landmark type buildings such as museums and airports. Modern automotive body structures are almost exclusively manufactured from thin sheet steel or aluminium which has been pressed into specific shapes. Shell elements are useful in the analysis of both the manufacture and the in service behaviour of these structures. Aircraft and spacecraft structures obviously lend themselves to analysis with shell elements as they essentially consist of a load bearing sheet metal skin which is supported by stiffeners made from bent sheet metal. In many cases, both ship and submarine structures also lend themselves to analysis using shell elements.

Shell elements are also available in a layered composite option which allows for the analysis of thin walled structures made from carbon fibre reinforced plastic (CFRP) or Fibreglass (GFRP). These elements can be useful for the analysis of aircraft structures, boat hulls, racing car structures etc. These elements are discussed in section 4.16.

When attempting to determine if a structure or problem is suitable for analysis with shell elements the main criteria to be satisfied is that the wall thickness of the structure must be significantly less than the other dimension of the problem. A structure may be generally described as a thin-walled structure when its surface area to wall thickness ratio is 10:1 or greater.

4.13.2. Description

Although shell elements have a thickness associated with them, they are generally represented as a curved area placed midway through the wall thickness of the structure, as shown in figure 4.46.

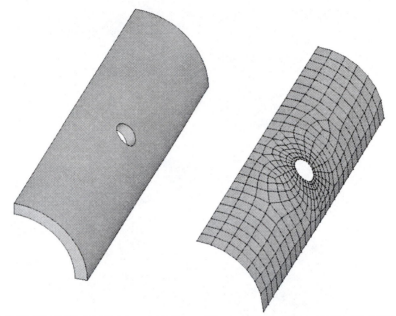

Figure 4.46: Shell Elements are created on a Plane Midway Through the Wall Thickness.

Shell elements may generally be considered as 3D curved versions of the 2D plane elements discussed above, and thus, shell elements are available in similar shapes and element orders, as shown in figure 4.47.

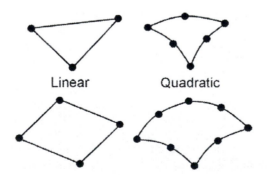

Figure 4.47: Types of Shell Elements

Since shell elements are generally used to model curved surfaces in many applications, the importance of using higher order elements (i.e. quadratic instead of linear) is illustrated in figure 4.48. It can be clearly see from figure 4.48(b) that the quadratic element follows curvature in both directions and thus provides for a more realistic and accurate representation of the problem.

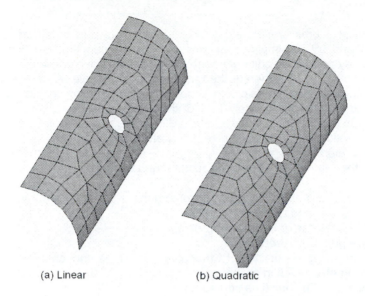

(a) Linear   (b) Quadratic

Figure 4.48: Quadratic Shell Elements Capture Curvature with Greater Accuracy

It is important to note that shell elements have many layers, usually a top, middle and bottom. This means that during post-processing it is important to select the layer of interest. For example, if an analysis of a pressure vessel was been carried out, the stress on the inside of the vessel may be of most interest, and thus the corresponding shell layer should be selected. This will be discussed further in chapter 9.

Recent advances in shell element formulations have seen the introduction of "solid-shell" elements. These elements are seen as being somewhere between shells and 3D solids as they can be used to model relatively thick shell type structures and, although they look like 3D solid elements (i.e. they have the same topology), they behave more like shell elements by modelling bending effects with greater accuracy than 3D solid elements.

4.13.3. Determination of Element Properties

The determination of element properties for shell elements is highly complex as the shape functions have to take account of both membrane and bending effects in a 3D stress regime. Many different methods of formulating shell elements are available, the most simple of which attempt to combine the membrane formulation used for 2D elements with plate bending theory. More complex formulations include the Belytschko-Tsay shell formulation and the Hughes-Liu shell formulation, which are used in more demanding situations such as non-linear analyses. Discussion of the formulation of these elements is beyond the scope of this text and the interested reader is referred to the bibliography section.

4.13.4. Case Study

Please now study case study F in chapter 10 which details an analysis of an aircraft fuselage panel using thin shell elements.

4.14. 3D Solid Elements

4.14.1 Applications

All real world problems are, in reality, three dimensional. In most cases we make assumptions and try to simplify problems so that we can that we can use truss, beam, 2D, axisymmetric or shell elements. Where this is not possible, and a full 3D stress regime must be considered, then three-dimensional elements are required. In most cases it will still be possible to simplify the problem by taking advantage of symmetry and thus modelling only part of the problem, with appropriate boundary conditions, but in any case a 3D model will still be required. In many cases an analysis with solid elements may be required after carrying out an analysis using a less complex element in order to gain more information about a problem, as discussed in the previous chapter in relation to model hierarchies.

3D solid elements are applicable to any problem that falls into the above category. In particular they are suited to problems that:
1. Are not truss or framework problems.
2. Are not beam problems.
3. Can not be satisfactorily modelled in 2D as the cross section of the geometry varies in all three dimensions.
4. Are not thin shell type structures
5. Do not exhibit axisymmetry.

There are many examples of such problems in stress analysis such as: analysis of turbine blades, analysis of complex metal forming processes, analysis of hip prosthesis behaviour, etc. Some examples of finite element models using 3D elements are shown in figure 4.49.

Figure 4.49: Examples of FE Models Using 3D Solid Elements: Femur and Turbine Blade

A clear advantage of solid elements is that the FE model will look like the actual problem and assumptions regarding loads will generally not be required. So we will create a 3D FE model of the problem (or part of the problem) to which we can directly apply pressures, forces or displacements (without any need to take account of plane stress or axisymmetry etc.). These advantages, however, are balanced by the disadvantages associated with solid elements such as: difficulty in producing solid models, difficulty in meshing in 3D, more computing effort required to solve the problem and more post-processing effort required to interrogate the results. A significant disadvantage of solid elements is the fact that as the mesh is refined the computational time required to obtain a solution increases drastically. In order to offset this, it may be necessary to use a coarser mesh than would be used with an equivalent 2D mode, thus causing accuracy problems.

It is generally recommended, because of the above problems, that 3D solid elements are not used during the design stage, but are saved for the verification and analysis stage of product development. A particular problem with novice users is where a problem is correctly identified as requiring a 3D analysis but where the use of shell elements is not considered and 3D solids are incorrectly used. A typical example of this is the analysis of thin walled structures, where solid elements are not appropriate as they are overly stiff in bending.

3D Solid elements are also available in a composite material option which can be used to model structures such as reinforced concrete or laminated composites. These elements are dealt with in section 4.16.

4.14.2 Description

Just as is the case with 2D elements, 3D elements come in two basic shapes with two different basic formulations of element behaviour for each shape, as shown in figure 4.50.

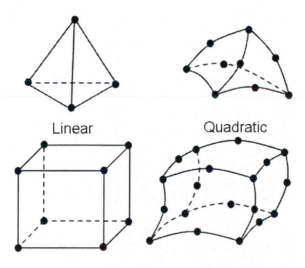

Figure 4.50: Types of 3D Solid Elements

Tetrahedral elements are the equivalent of 2D triangles and are basically shaped like a pyramid. Hexahedral elements are the equivalent of 2D quadrilateral elements and are brick shaped. Traditionally, hexahedral elements were preferred for use in finite element models as they were more robust that tetrahedrals and gave more accurate results for stress analysis. Tetrahedral elements, however, allow for easier meshing of complex curved structures. In the past this caused a lot of confusion as when presented with a complex shape for analysis the gut reaction was to use tetrahedrals, however if accuracy was a major issue then using tetrahedrals was frowned upon. In recent years much effort has been put into improving the formulation of tetrahedral

Practical Stress Analysis with Finite Elements

elements and much of the above is no longer a concern. Theoretically, tetrahedral elements are too stiff during bending and hexahedral elements are preferred. If this is an issue in your analysis, then it is important to check the help system of your FE package for more information.

4.14.3 Determination of Element Properties

4.14.3.1 Four Node Linear Tetrahedral Solid Element

This element consists of four nodes, numbered i, j, k and l. Lagrange type shape functions are used so that the shape function S_i has a value of 1 at node i and a value of zero at all other nodes, and similarly for the other three shape functions, as usual. We can define the shape functions in terms of the 3D natural coordinate system (ξ, η, ζ) as:

$$S_i = \xi \qquad S_j = \eta \qquad S_k = \zeta \qquad S_k = 1 - \xi - \eta - \zeta \qquad (4.150)$$

Following the usual procedure, we can express the element displacement in terms of the nodal point displacements via the shape functions, i.e.

$$\{u\} = [S]\{U\} \qquad (4.151)$$

Where:
$$[S] = \begin{bmatrix} S_i & 0 & 0 & S_j & 0 & 0 & S_k & 0 & 0 & S_l & 0 & 0 \\ 0 & S_i & 0 & 0 & S_j & 0 & 0 & S_k & 0 & 0 & S_l & 0 \\ 0 & 0 & S_i & 0 & 0 & S_j & 0 & 0 & S_k & 0 & 0 & S_l \end{bmatrix} \qquad (4.152)$$

And
$$\{U\} = \{U_i \ V_i \ W_i \ U_j \ V_j \ W_j \ U_k \ V_k \ W_k \ U_l \ V_l \ W_l\}^T \qquad (4.153)$$

Using an isoparametric formulation, we have the transformation:

$$\begin{aligned} x &= S_i x_i + S_j x_j + S_k x_k + S_l x_l \\ y &= S_i y_i + S_j y_j + S_k y_k + S_l y_l \\ z &= S_i z_i + S_j z_j + S_k z_k + S_l z_l \end{aligned} \qquad (4.154)$$

or

$$\begin{aligned} x &= x_l + \xi(x_i - x_l) + \eta(x_j - x_l) + \zeta(x_k - x_l) \\ y &= y_l + \xi(y_i - y_l) + \eta(y_j - y_l) + \zeta(y_k - y_l) \\ z &= z_l + \xi(z_i - z_l) + \eta(z_j - z_l) + \zeta(z_k - z_l) \end{aligned} \qquad (4.155)$$

$$[J] = \begin{bmatrix} \dfrac{\partial x}{\partial \xi} & \dfrac{\partial y}{\partial \xi} & \dfrac{\partial z}{\partial \xi} \\ \dfrac{\partial x}{\partial \eta} & \dfrac{\partial y}{\partial \eta} & \dfrac{\partial z}{\partial \eta} \\ \dfrac{\partial x}{\partial \zeta} & \dfrac{\partial y}{\partial \zeta} & \dfrac{\partial z}{\partial \zeta} \end{bmatrix} = \begin{bmatrix} (x_i - x_l) & (y_i - y_l) & (z_i - z_l) \\ (x_j - x_l) & (y_j - y_l) & (z_j - z_l) \\ (x_k - x_l) & (y_k - y_l) & (z_k - z_l) \end{bmatrix} \qquad (4.156)$$

Using the chain rule for evaluation partial derivatives, as per the usual procedure, we obtain three equations similar to equations 4.138. The (3x3) Jacobian 3D matrix is given by equation 4.156.

The volume of the element can be calculated from the Jacobian as follows:

$$V_e = \left| \int_0^1 \int_0^{1-\xi} \int_0^{1-\xi-\eta} \det[J] \, d\xi \, d\eta \, d\zeta \right| = \frac{1}{6} \left| \det[J] \right| \tag{4.157}$$

Inversing the relationship of the 3D version of equation 4.138, we obtain the 3D version of 4.140, where the (3x3) matrix [A], is the inverse of the 3D Jacobian matrix given in 4.156 above.

Using the 3D strain displacement relationships, the fact that the Jacobian relates derivatives of x,y,z to derivatives of ξ,η,ζ and equation 4.151 we obtain the element strain displacement relationships:

$$\{\varepsilon\} = [B]\{U\} \tag{4.158}$$

Where:

$$[B] = \begin{bmatrix} A_{11} & 0 & 0 & A_{12} & 0 & 0 & A_{13} & 0 & 0 & A_{C1} & 0 & 0 \\ 0 & A_{21} & 0 & 0 & A_{22} & 0 & 0 & A_{23} & 0 & 0 & A_{C2} & 0 \\ 0 & 0 & A_{31} & 0 & 0 & A_{32} & 0 & 0 & A_{33} & 0 & 0 & A_{C3} \\ 0 & A_{31} & A_{21} & 0 & A_{32} & A_{22} & 0 & A_{33} & A_{23} & 0 & A_{C3} & A_{C2} \\ A_{31} & 0 & A_{11} & A_{32} & 0 & A_{12} & A_{33} & 0 & A_{13} & A_{C3} & 0 & A_{C1} \\ A_{21} & A_{11} & 0 & A_{22} & A_{12} & 0 & A_{23} & A_{13} & 0 & A_{C2} & A_{C1} & 0 \end{bmatrix} \tag{4.159}$$

Where the calculated matrix elements in [B] are given by:
$$A_{C1} = -(A_{11} + A_{12} + A_{13}) \quad A_{C2} = -(A_{21} + A_{22} + A_{23}) \quad A_{C1} = -(A_{31} + A_{32} + A_{33})$$

We can now calculate element properties in the usual manner by substituting equations 4.158 and 4.159 into equation 3.25. Evaluating this gives the element stiffness matrix as:

$$[k^e] = V_e [B]^T [D][B] \tag{4.160}$$

Where V_e is the volume of the element, given by equation 4.157. Similarly the body force vector can be determined as:

$$\{f_e^B\} = \frac{V_e}{4} \{f_x^B \quad f_y^B \quad f_z^B \quad f_x^B \quad f_y^B \quad f_z^B \quad f_x^B \quad f_y^B \quad f_z^B \quad f_x^B \quad f_y^B \quad f_z^B\}^T \tag{4.161}$$

The element surface load vector on, for example, the surface given by nodes i, j and k is given by:

$$\{f_e^S\} = \frac{A_e}{4} \{f_x^B \quad f_y^B \quad f_z^B \quad f_x^B \quad f_y^B \quad f_z^B \quad f_x^B \quad f_y^B \quad f_z^B \quad 0 \quad 0 \quad 0\}^T \tag{4.162}$$

Where A_e is the area of the surface defined by nodes i, j and k.

Once equation 3.25 has been solved using equations 4.160, 4.161, and 4.162 inserting the appropriate boundary conditions and point loads, giving the element nodal point displacement vector {U}, element strains may be calculated from equation 4.158 and element stress subsequently calculated using equation 4.163.

$$\{\sigma\} = [D][B]\{U\} \tag{4.163}$$

As the element strains will be constant throughout the element, it follows that stress will also be constant within the element. The element stress vector will be given by:

$$\{\sigma_e\} = \{\sigma_x \quad \sigma_y \quad \sigma_z \quad \tau_{xy} \quad \tau_{yz} \quad \tau_{zx}\}^T \tag{4.164}$$

The applications of this element are quite limited due to the fact that stress and strain are constant in the element; however, it illustrates the basic procedure that is used to formulate all 3D solid finite elements.

4.14.3.2. Eight Node Linear Hexahedral Solid Element

This element consists of eight nodes, numbered i, j, k, l, m, n, o and p as shown in figure 4.51. Lagrange type shape functions are again used so that the shape function S_i has a value of 1 at node i and a value of zero at all other nodes, and similarly for the other three shape functions, as usual.

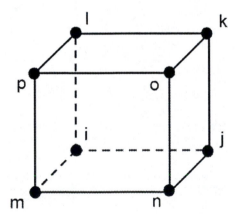

Figure 4.51: The 8-Node Hexahedral Element

We can define the shape functions in terms of the 3D natural coordinate system (ξ, η, ζ) as:

$$S_i = \frac{1}{8}(1-\xi)(1-\eta)(1-\zeta) \quad S_j = \frac{1}{8}(1-\xi)(1+\eta)(1-\zeta)$$

$$S_k = \frac{1}{8}(1-\xi)(1+\eta)(1+\zeta) \quad S_l = \frac{1}{8}(1-\xi)(1-\eta)(1+\zeta)$$

$$S_m = \frac{1}{8}(1+\xi)(1-\eta)(1-\zeta) \quad S_n = \frac{1}{8}(1+\xi)(1+\eta)(1-\zeta) \tag{4.165}$$

$$S_o = \frac{1}{8}(1+\xi)(1+\eta)(1+\zeta) \quad S_p = \frac{1}{8}(1+\xi)(1-\eta)(1+\zeta)$$

In this case, the element nodal displacement vector {U} will have 24 rows to accommodate the three degrees of freedom at each of the eight nodes. Following the same procedure as used in the previous section, for the 4-node tetrahedral element, we can develop an isoparametric formulation, using the shape functions to define displacement of any point in the element in terms of its nodal values and also to describe the location of any point within the element. This will yield similar equations to equations 4.154 and 4.155 but with 8 terms in each equation in this case.

By relating element strains to nodal displacements via: $\{\varepsilon\} = [B]\{U\}$

We can obtain the element stiffness matrix from:

$$[k^e] = \int_{-1}^{+1}\int_{-1}^{+1}\int_{-1}^{+1}[B]^T[D][B]\,|\det[J]|\,d\xi\,d\eta\,d\zeta \qquad (4.166)$$

Where, [J] is the (3x3) Jacobian matrix given in equation 4.156. Element body forces, surface loads and stress calculations are carried out in a manner similar to that explained in the previous section which deals with the 4-node tetrahedral.

4.14.3.3.   Higher Order 3D Solid Elements

The formulation of the higher order, quadratic tetrahedral and quadratic hexahedral elements is beyond the scope of this book. These elements are formulated using the same methods as discussed for higher order 2D elements earlier in this chapter. Hence, if you understand how these are formulated then you should easily be able to understand how higher order 3D elements are dealt with.

The formulation of these elements is covered in great detail in many of the references provided in the bibliography at the end of this book and the interested reader is referred there. If these elements are provided in your FE software then it is most likely that the formulation and theoretical background used for these elements are provided in the help system or documentation provided with your FE software.

4.14.4 Case Study

Please now study case study G in chapter 10 which details an example using 3D solid elements to model a human femur and an artificial hip replacement.

4.15.   Contact Elements

4.15.1 Applications

In many types of structural analysis contact can occur when the structure of interest either makes contact with itself (e.g. a cantilever beam is bent so much that it touches itself) or another structure. In certain complex problems such as analysis of metal forming or impact analysis, the presence of contact is fundamental to the problem. Contact problems in FEA can be generally classified into either a "rigid-to-flexible" contact or a "flexible-to-flexible" contact. Metal forming is a typical example of a "rigid-to-flexible" contact problem, where the forming tools are treated as rigid as they have a much higher stiffness relative to the deformable body they contact (i.e. the sheet being formed). In general, any time a soft material comes in contact with a hard material, the problem may be assumed to be rigid-to-flexible. Other examples of rigid-to-flexible contact include a soft bullet impacting a wall or a hard bullet hitting a soft target. Flexible-to-flexible contact where all bodies are deformable, due to their similar stiffnesses, is more common. Examples of flexible-to-flexible contact include bolted joints, a bird impacting an aircraft structure, interference fits, stent deployment, etc.

Figure 4.52 shows a typical example of the application of contact elements, where contact is defined between the punch and the blank and the dies and the blank in an axisymmetric finite element model of a deep drawing process. This axisymmetric model has already been explained in figure 4.40.

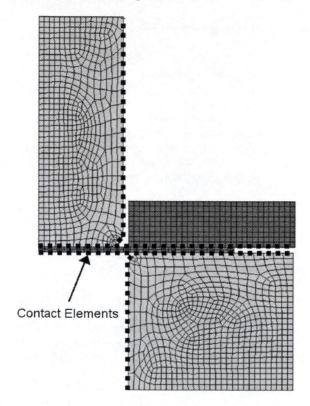

Contact Elements

Figure 4.52: Example of Contact Element Application

4.15.2 Description

In order to model contact in FEA the possible interactions of bodies must be analysed before the model is built. A significant problem is that, in general, the regions of contact will not be known until you've run the problem. Depending on the loads, material, boundary conditions, and other factors, surfaces can come into and go out of contact with each other in a largely unpredictable and abrupt manner. Thus, identifying all possible regions of interaction before running the model is important. If one of these interactions is at a point, the corresponding component of the FE model is a node, and contact can be created between this node and another entity. If one of the interactions is at a surface, the corresponding component of your model is an element (i.e. a beam, shell, or solid element). Contact elements are used to define contact between an entity (either a node or an element) on one body and an entity on the other. These contact elements are overlaid on the parts of the model that are being analyzed for interaction. In general, the following types of contact elements are available:

- Node to node contact elements

Node-to-node contact elements are typically used to model point-to-point contact applications. Obviously, this means that we will know the location of such precise contact in advance of running the model. This type of contact problem usually involves little relative sliding between contacting surfaces (in contrast to surface-to-surface problems where large sliding surfaces are present). Examples of node-to-node contact applications are modelling spot welds, analysis of pipe whip and any other application where contact between specific discrete points is required.

In some specific cases node-to-node contact elements can also be used to solve a surface-to-surface problem. This is only valid if the nodes of the two surfaces line up, the relative sliding deformation is negligible, and deflections (rotations) of the two surfaces remain small. In reality this will only occur with simple geometries. Interference fit problems are a typical example of such a case where what first appears as a surface-to-surface problem can be sufficiently modelled using node-to-node.

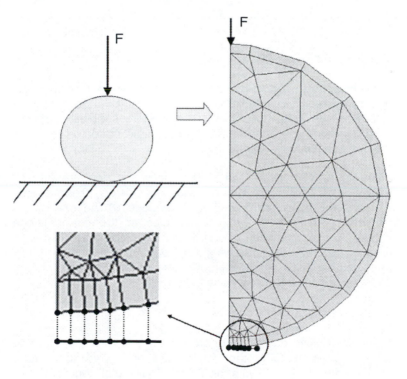

Figure 4.53: Example of Node-to-node Contact Used to Model Contact of an Axisymmetric Model of a Sphere Pressed against a Rigid Flat Plate

- Line to line contact elements

This type of contact is used to model contact between two beams or pipes modelled using beam or pipe elements. Obviously, if the beams come into contact then the contact surface will be defined via a line. Typical examples include crossing beams (e.g. woven fabric, tennis racquet strings) or a pipe sliding inside another pipe.

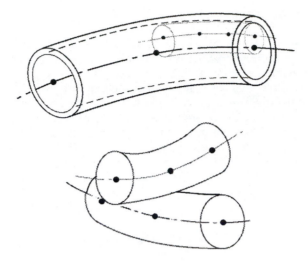

Figure 4.54: Examples of Line-to-Line Contact: Pipe sliding inside another pipe (top) and Woven Fabric Modelled Using Beams (bottom)

- Node to surface contact elements

These elements are used to model point to surface contact applications. These elements can be used to model point-to-point, point-to-surface or edge-to-surface type contact. This form of contact is much more simple to implement as the nodes on one body do not have to line up with the nodes on the other body (i.e. meshes do not have to be compatible).

The main difference between the node-to-node and node-to-surface techniques is that the node-to-surface method allows for large deformations and large relative part sliding, without requiring knowledge of the exact location of the contact areas beforehand. Typical examples of the use of node-to-surface contact are: contact between two cylinders, the corners of snap-fit parts and draw-beads used in deep drawing. In some cases node-to-surface contact elements can be used to model surface-to-surface contact, if the contacting surface is defined by a group of nodes.

- Surface to surface contact elements

These elements have no restrictions on the shape of the contact surfaces and represent the most useful method for the solution of real-world problems. Using this method both "rigid-to-flexible" or "flexible-to-flexible" contact can be modelled. These contact elements use a target surface and a contact surface to form a contact pair. These elements are well suited to problems where large areas of contact exist and there is large relative sliding between bodies. Typical applications include interference fit assembly contact, contact during deep-drawing, forging and other metal forming processes, contact between two disks etc.

In straightforward cases asymmetric contact, where all contact elements are on one surface and all target elements are on the other, is used. In some cases, however, this does not produce satisfactory results and it is required to use symmetric contact, where two contact pairs are generated with each surface acting as both a target and a contact surface. Symmetric contact is useful when it is not clear which surface is the target and which is the contact and is almost a prerequisite for some types of metal forming analysis.

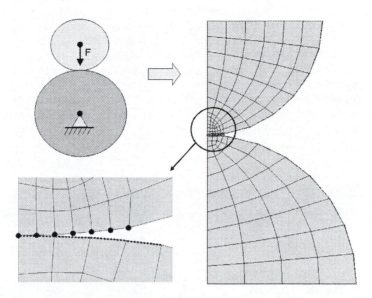

Figure 4.55: Example of Node-to-Surface Contact, Plane Stress Model of Two Cylinders in Contact

These elements provide more useful contact results such as normal pressure and friction stress contour plots which offer greater insight into the problem for engineers and designers. When using the surface-to-surface approach there are no restrictions on the shape of the contact surfaces and discontinuities are generally allowed.

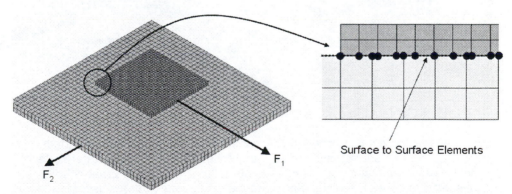

Surface to Surface Elements

Figure 4.56: Example of Surface-to-Surface Contact; relative sliding between two plates

Surface-to-surface contact elements, however, are not well-suited for point-to-point, point-to-surface, edge-to-surface, or 3-D line-to-line contact applications, such as those described in the sections above. In such cases node-to-node or node-to-surface elements should be used, where appropriate. For bodies with many corners or edges it is possible to use surface-to-surface contact elements for most of the contact regions and use a few node-to-surface contact elements near contact corners.

- Advanced and Specialised Contact Options

In some cases more specialised and advanced contact options are available in some commercial FE codes. Specific types of contact such as spot welds, draw-beads, rivets, etc. are modelled using specially developed contact elements which are generally named after the type of contact they model. In advanced FE software different types of contact may be available such as tied, tied with failure and eroding contact. Tied contact is used to permanently tie two bodies together at the contact location specified. Tied with failure ties two bodies together until a specified failure limit is reached and then the contact is released. Eroding contact is useful in impact analysis and ensures that when a projectile comes in contact with a target, elements in the target are eroded or removed from the model once a specified failure criterion has been reached.

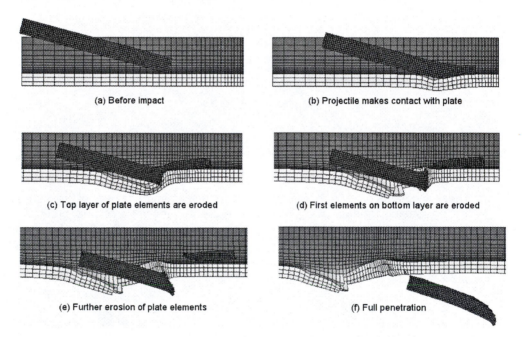

(a) Before impact   (b) Projectile makes contact with plate

(c) Top layer of plate elements are eroded   (d) First elements on bottom layer are eroded

(e) Further erosion of plate elements   (f) Full penetration

Figure 4.57: Example of Eroding Contact: Projectile Penetrates a Plate

4.15.3. How Contact Elements Work

When examining contact between two bodies, it is usual to take the surface of one body as a contact surface and the surface of the other body as a target surface. For a rigid-flexible contact problem (such as metal forming) the deformable body (blank) is assumed to be the contact surface and the rigid surface (die) is assumed to be the target surface. For a flexible-flexible contact problem it is usually arbitrary which body is selected as the target or contact. The contact and target surfaces for a specific region of contact are normally referred to as a "Contact Pair".

There are many different approaches available for formulating contact elements in FEA, however the most common methods are based on the penalty approach. In this method when one surface comes into contact with another a spring of stiffness K_N is placed between the two contacting entities (e.g. nodes). In general the K_N value is determined automatically from the stiffness of the contact and target surfaces or it can be user specified to allow for more control over contact behaviour. Since the spring will be very stiff it will impart a restoring force on the contact entity which has penetrated the target entity. The restoring force will cause the contact entity to move back outside the target. Figure 4.58 (a) illustrates a sphere in contact with a plate, figure 4.58 (b) shows the springs connecting the nodes of the sphere with those of the plate upper surface and figure 4.58 (c) shows the springs contracting due the restoring force and thus forcing the sphere nodes back to the surface of the plate. K_N can be varied and thus used to specify acceptable levels of penetration.

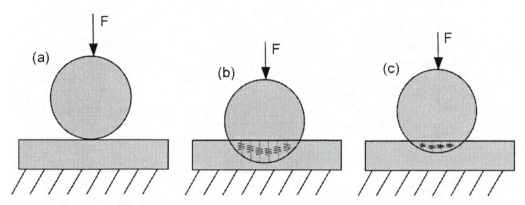

Figure 4.58: Overview of Penalty Approach for Contact Elements

When a penetrating node stays in contact with the target, it may either stick to the surface or slip along the surface. A number of methods are available for modeling friction in contact analyses, however the most commonly used methods are based on a Coloumb friction model. In this model the two contacting surfaces are permitted to carry shear stresses across their interface, up to a defined value, before they begin sliding relative to each other. The equivalent shear stress at which sliding begins is defined as:

$$\tau_{slide} = \mu P_{Con} + CSR \qquad (4.167)$$

Where, μ is the coefficient of friction, P_{Con} is the contact pressure and CSR is a specification of the cohesion sliding resistance. In some cases, the coefficient of friction depends on the relative velocity of the surfaces in contact. In such cases the coefficient of friction can be specified in terms of a static coefficient of friction and a dynamic coefficient of friction, where, for example:

$$\mu = \mu_d \left(1 + \frac{e^{-DCV_{REL}}}{(\mu_s / \mu_d)} \right) \qquad (4.168)$$

Where, μ_d is the dynamic friction coefficient, μ_s is the static friction coefficient, DC is a decay coefficient and V_{REL} is the relative velocity of the contact surfaces. Equation 4.168 essentially defines an exponential decay function whereby the coefficient of friction starts at μ_s, when both surfaces are at rest, and reduces exponentially to μ_d as relative motion continues.

It is important to note that contact problems involving friction result in an un-symmetric contact element stiffness matrix, in contrast to frictionless or bonded contact where the contact element stiffness matrix will be symmetrical. This means, that in general, contact problems involving friction will be more computationally expensive than those without friction.

4.15.4. Case Study

Please now study case study H in chapter 10 which details an analysis of a forging metal forming process using 2D plane strain elements and 2D surface to surface contact elements.

4.16. Composite Elements

4.16.1. Applications

Composite elements are obviously suited for the analysis of structures made from composite materials. These include: structures made from CFRP such as aircraft structures and consumer products, structures made from GFRP such as boat hulls and aircraft structures, structures made from Kevlar such as bullet proof vests and protective helmets, and structures made from reinforced concrete.

4.16.2. Description

Composite elements are generally available in two types: layered elements and reinforced concrete elements. Layered elements are available as shells, 3D solids and combination "solid-shell" type elements. Reinforced concrete elements are available as 3D solids.

4.16.2.1.   Layered Elements

Layered solids and shells are defined in a similar manner. The input parameters required include the number of layers in the element, the material properties associated with each layer and the orientation of each layer in relation to the element coordinate system. These elements are particularly suited to the analysis of laminated composite materials, where the reinforcing fibres may be orientated in different directions in each layer as illustrated in figure 4.59.

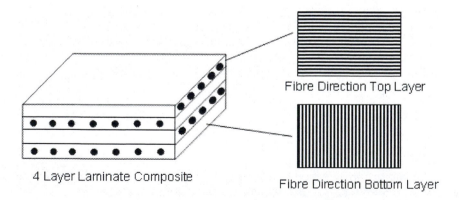

4 Layer Laminate Composite

Fibre Direction Top Layer

Fibre Direction Bottom Layer

Figure 4.59: Schematic of a Laminated Composite Material

In many cases, different materials are permitted for each layer of the element and a large number of layers may be specified. As each layer consists of fibre reinforced laminate, it is important that an orthotropic material model is used to take account of the laminates material behaviour.

Layered elements may also be used to model non-fibre reinforced laminates such as sandwich type structures. These structures are popular in aircraft manufacture and may consist, for example, of two thin layers of aluminium sheet with a soft material sandwiched in between. In such cases the orientation of each layer is not important as an isotropic material model can be used to model the behaviour of each material. This type of approach can also be used to model bimetallic strips etc.

These elements are also available as a "solid-shell" type element, which is essentially a shell element that can be used to model relatively thick shell type structures, in which a number of layers of element can be used through the thickness.

4.16.2.2 Reinforced Concrete Elements

These elements are available as 3D solids and are capable of modelling cracking of concrete (due to tensile loads) and crushing (due to compressive loads). Reinforcement of the concrete matrix via rebars is provided in a similar model to the layered element described above. In this case a limited number of rebar material types are allowed and the orientation of the rebars relative to the element coordinate system may be specified by the user. These elements are obviously also useful for modelling other reinforced composite materials such as fibreglass and can be used without the reinforcement option to model rock or un-reinforced concrete. These elements typically have many non-linear capabilities in order to accurately model the behaviour of concrete such as, cracking, crushing, creep and plastic deformation.

4.16.3. Determination of Element Properties

The properties for these elements are determined in a manner largely similar to laminate theory used for the analysis of laminated composite materials. This theory essentially develops a stiffness matrix for each lamina and subsequently develops an equation for the overall laminate behaviour based on the properties of each layer. Properties such as volume fraction, which relates the volume of the reinforcement material to the volume of the matrix material, may also be used in the determination of element properties. Discussion of these methods is beyond the scope of this book and it is quite likely that the reader who requires to use composite elements will already be familiar with this theory and will understand how element properties can be derived from it.

4.16.4. Case Study

You should now study case study I in chapter 10 which details an analysis of the loading on a reinforced concrete strip foundation from a small house which aims to determine how the weight of the house is transferred to the subsoil via the foundation.

4.17. Surface Effect Elements

4.17.1 Applications

Surface effect elements are essentially elements that are overlaid on the surface of other elements and hence used to apply loads to the selected surface of that element. 2D surface effect elements are curved or straight lines and 3D surface effect elements are curved or flat areas. Figure 4.60 shows typical applications of 2D and 3D surface effect elements where the surface element is used to apply a pressure load in a particular direction.

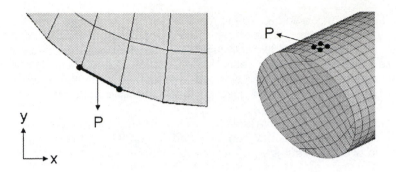

Figure 4.60: 2D and 3D Surface Effect Elements

4.18. Summary of Chapter 4

This chapter describes each of the element types typically available for structural analysis in a commercial finite element code. In each case the element type has been described, a number of typical applications have been highlighted and the theoretical basis of the element has been discussed, where appropriate. This chapter should be used together with section 3.3 when building a finite element model. By working through table 3.01 and referencing the appropriate material in this chapter the most appropriate element type for the analysis being undertaken should be clear.

4.19. Problems

P4.1. Use a one dimensional quadratic truss element to formulate a finite element analysis of the problem shown in the figure below. A stepped shaft is fixed at the step junction and supports a non-symmetrical compressive load. Assume that the shaft is made from Aluminium with E = 70 GPa and ρ = 2700 Kg/m³.

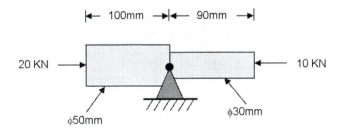

Use the methodology from example 4.1 to determine the stiffness matrix for the problem and hence determine the displacement of each of the free ends of the shaft and the stress distribution in the shaft. You may ignore the effects of self weight.

Answers:

$$k^G = 1 \times 10^8 \begin{bmatrix} 32.06 & -36.64 & 4.58 & 0 & 0 \\ -36.64 & 73.28 & -36.64 & 0 & 0 \\ 4 & -36.64 & 43.61 & -13.2 & 1.65 \\ 0 & 0 & -13.2 & 26.4 & -13.2 \\ 0 & 0 & 1.65 & -13.2 & 11.55 \end{bmatrix} \begin{matrix} 1 \\ 2 \\ 3 \\ 4 \\ 5 \end{matrix}$$

$$\{U\} = 1 \times 10^{-5} \begin{bmatrix} 1.46 & 0.728 & 0 & -1 & -2 \end{bmatrix}^T$$

Displacement of right hand end = -2 x 10⁻⁵ m
Displacement of left hand end = 1.46 x 10⁻⁵ m

P4.2. Use a one dimensional linear truss element to formulate a finite element analysis of the problem shown in the figure below. A tapered bar is fixed at one end and supports a tensile load at its free end. Assume that the shaft is made from Steel with E = 210 GPa and ρ = 7800 Kg/m³. You must use five equal length elements to model the bar and you must also consider the bar's self weight.

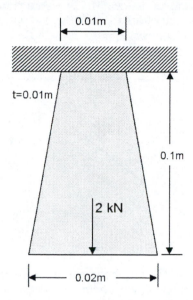

You are required to determine the displacement distribution and stress distribution in the bar.

Answers: $\{U\} = \{0 \quad 0.173 \quad 0.319 \quad 0.447 \quad 0.559 \quad 0.659\}^T \times 10^{-5}\, m$

$\{\sigma\} = \{18.1 \quad 15.3 \quad 13.3 \quad 11.7 \quad 10.5\}^T\, MPa$

P4.3. A small jib crane is used to lift a load of 10 kN as shown in the figure below. You are required to determine the displacement of point A, the stress in member B, and the reaction at the crane base using a finite element analysis of a 2D spar model of the crane.

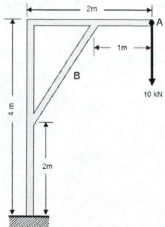

You may assume that the crane is made from steel with E = 210 GPa and that each of the crane structure members has a cross sectional area of 0.05 m^2.

Answers: Defection of point A =
          Stress in member B =
          Reaction at crane base =

P4.4. Two steel I-beams have been welded together and used to support a 10 kN load as shown in figure on the left below. The cross sectional area of the I-beams is 0.05m^2 and the second moment of area is 2.036 x 10^{-4} m^4. Member E2 has a length of 2m and member E1 has a length of 2.5 m.

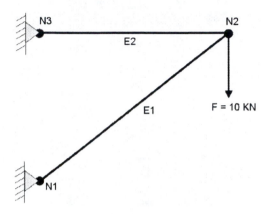

You are required to determine the displacement and rotation of the point N2 using a finite element model consisting of two beam elements. Note that because we are using beam elements points N1 and N3 are restrained from displacement but are allowed to rotate! Compare your results to example 4.2 which analyses a similar truss model.

Answers:

P4.5. A cylindrical bushing fits perfectly into a cylindrical hole as shown in the figure below. During service the bushing is subjected to an internal pressure of 2 x 10^6 Pa. You are required to determine the displacement of the internal wall of the bushing using two triangular axisymmetric elements. You may assume that the bushing is made of steel with E = 200 GPa and v = 0.3. Consider a finite length of the bushing in your analysis, e.g. 5mm.

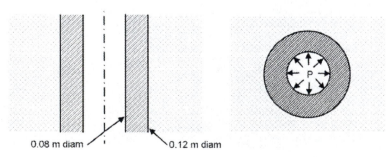

0.08 m diam   0.12 m diam

Answer: displacement of inner wall = 1.35 x 10^{-7} m

5

Material Models

5.1.   Introduction

In the previous chapter we discussed the selection of element type which will determine the shape function matrix [S] that will be used in the minimum potential energy equation 3.25. From equation 3.28, we can recall that the stiffness matrix of each finite element will be determined by a strain displacement matrix [B] and a material property matrix [D]. As we know, the strain-displacement matrix [B] is simply a derivative of the shape function matrix [S], thus once we know [S] we can determine [B]. Consequently, the next stage in our analysis is to establish the material property matrix [D]. This chapter is concerned with discussing the various material models available during finite element analysis of stress problems and describing how these models are implemented in FE software to eventually become the [D] matrix in equation 3.28. Each material model will be discussed in increasing order of complexity in reference to table 3.01 in chapter 3.

5.2.   Linear Elastic Models

5.2.1.   Linear Elastic Isotropic

An isotropic linear elastic material model is the most basic and fundamental material model available for stress analysis problems. As discussed in chapter 2, this model assumes that the material obeys Hooke's law with stress being linearly proportional to strain and the slope of the stress-strain line is given by the Young's modulus for the material. It also assumes that the material is isotropic and hence has the same properties in all directions.

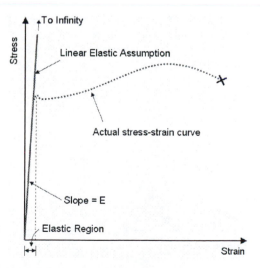

Figure 5.01: Overview of Linear Elastic Material Model

As we discussed in chapter 3, very few materials are linearly elastic in practice, however most metals behave in a near linear elastic manner before yielding. Thus, if we are confident that yielding will not occur then a linear elastic material model is a valid assumption for most metals. In many cases even if a linear elastic assumption is not valid it is usual to run a linear elastic model as a "first guess" in order to check how the FE model is responding to the applied loading. It is important to remember also, that the linear elastic portion of the stress-strain curve usually only accounts for a very small part of the strain history (up to approximately 0.2%). This means that very small displacements should be expected. If large displacements are experienced then either the linear elastic assumption is not valid or other inputs to the FE model are incorrect (normally the magnitude of the loads). Please see sections 2.3.3 and 2.5 for further discussion.

When defining a linear elastic material model the parameters needed are the Young's modulus, E, for the material and, usually, the Poisson's ratio and density of the material. In order to use this model with confidence, however, it is important to know what the yield stress for the material is.

When using this model results should be constantly checked to ensure that the yield stress has not been exceeded and that displacements are small. If this is not the case then the model is no longer valid and a more complex model must be used.

5.2.2. Linear Elastic Orthotropic

An orthotropic material is a material which has mechanical properties that are different in three mutually perpendicular directions at a point in the body, and that have three mutually perpendicular planes of material symmetry. To be classed as orthotropic the material must have at least two orthogonal places of symmetry, with material properties independent of direction within each plane. If we compare this to an isotropic material, where every plane is a plane of symmetry and an anisotropic material, where there are no planes of symmetry, we can see that an orthotropic material is somewhere in between isotropic and anisotropic. The mechanical properties of an orthotropic material will depend on the direction in which they are measured.

A typical example of an orthotropic material with two axes of symmetry is a lamina of CFRP. The strength of stiffness of the lamina will be much greater in the direction parallel to the fibres compared to the transverse direction. Such materials are often referred to as being "transversely isotropic".

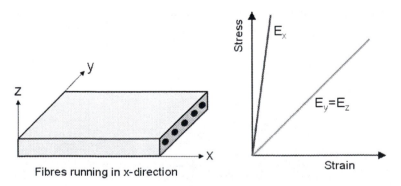

Fibres running in x-direction

Figure 5.02: Orthotropic CFRP Composite Material

Wood is an example of an orthotropic material with three mutually perpendicular axes. In this case material properties along the grain and in each of the two perpendicular directions are different. Another, less obvious, but very important example is sheet metal which has been rolled during its processing. The properties in the rolling direction and each of the two transverse directions will be different due to the anisotropic grain structure that develops during rolling.

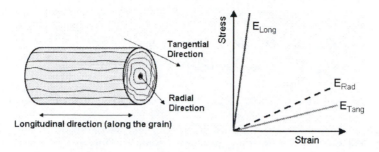

Figure 5.03: Orthotropic Material Model for Wood

An isotropic material requires only 2 elastic constants in its definition (i.e. Young's modulus and Poisson's ratio) whereas an orthotropic material requires 9 elastic constants in its definition. The nine constants are: The Young's modulus in each direction (E_x, E_y and E_z), the three Poisson ratio values (υ_{xy}, υ_{yz} and υ_{zx}) and the three shear modulus values (G_{xy}, G_{yz} and G_{zx}). The orthotropic form of Hooke's law is more easily expressed using the compliance matrix $[D]^{-1}$ (which is the inverse of the stiffness matrix and which relates strain to stress) as follows:

$$
\{\varepsilon\} = [D]^{-1}\{\sigma\} =
\begin{bmatrix}
\dfrac{1}{E_x} & \dfrac{-V_{xy}}{E_y} & \dfrac{-V_{zx}}{E_z} & 0 & 0 & 0 \\[2ex]
\dfrac{-V_{xy}}{E_y} & \dfrac{1}{E_y} & \dfrac{-V_{yz}}{E_y} & 0 & 0 & 0 \\[2ex]
\dfrac{-V_{zx}}{E_z} & \dfrac{-V_{yz}}{E_y} & \dfrac{1}{E_z} & 0 & 0 & 0 \\[2ex]
0 & 0 & 0 & \dfrac{1}{2G_{yz}} & 0 & 0 \\[2ex]
0 & 0 & 0 & 0 & \dfrac{1}{2G_{zx}} & 0 \\[2ex]
0 & 0 & 0 & 0 & 0 & \dfrac{1}{2G_{xy}}
\end{bmatrix}
\begin{Bmatrix}
\sigma_x \\ \sigma_y \\ \sigma_z \\ \tau_{xy} \\ \tau_{yz} \\ \tau_{zx}
\end{Bmatrix}
\qquad (5.01)
$$

When using this model it is important to be aware of the uniaxial yield or failure stress in each direction in order to determine when or if the linear elastic assumption breaks down.

5.2.3. Linear Elastic Anisotropic

In direct contrast to isotropic and orthotropic materials is a fully anisotropic material which does not have any planes of symmetry. Definition of a linear elastic anisotropic material model would appear to require the specification of all 36 members of the material stress-strain matrix [D]. It can be shown, however, that in most cases the matrix is symmetric about the diagonal and so only requires the specification of the 21 elastic constants shown in equation 5.02.

$$[D] = \begin{bmatrix} D_{11} \\ D_{21} & D_{22} \\ D_{31} & D_{32} & D_{33} \\ D_{41} & D_{42} & D_{34} & D_{44} \\ D_{51} & D_{52} & D_{35} & D_{45} & D_{55} \\ D_{61} & D_{62} & D_{36} & D_{46} & D_{56} & D_{66} \end{bmatrix} \qquad (5.02)$$

Determining each of the members D_{ij} of [D] requires a great deal of material testing.

Even though, in this case, the material behaviour will be different in every direction, the material will still behave in a linear elastic fashion. Determining when the linear elastic assumption breaks down is difficult as it requires knowledge of yielding/failure properties in each direction.

5.3. Non-Linear Elastic Models

Non-linear elastic material models are very useful for modelling elastomers (i.e. rubber like materials), foams, certain plastics and biomaterials. These materials are able to undergo large reversible elastic deformations and thus, unless damage has occurred, they will return to their original shape after load removal. If we take the example of rubber, it has unique properties that make it behave different to most other engineering materials:

- Large deformations, to approximately 500% strain are possible
- The stress-strain curve is highly non-linear
- It has unique damping properties
- Its behaviour is time dependant and temperature dependant
- It is almost incompressible (i.e. volume does not change significantly with increasing stress)

Characterisation (i.e. material testing) of rubber materials tends to base itself on "stretch ratio" rather than strain, where the stretch ratio, λ, is defined by:

$$\lambda = L/L_0 \qquad (5.03)$$

Where, L is the deformed length of the sample and L_0 is the original length.

If we consider a 3D block of rubber-like material; it is clear that if we apply a load in one (or more) directions we will also cause deformation (i.e. stretch) to occur in the other two directions. Thus we can determine three principal stretch ratios, λ_x λ_y and λ_z or λ_1 λ_2 and λ_3, in much the same manner as we determined principal strains in chapter 2.

All FE material models used for non-linear elastic materials are based on strain energy density functions, which are usually given the symbol, W. If we take the derivative of W with respect to strain, we obtain stress. Most of the models available are based on strain invariants which are functions of the three principal stretch ratios and are given by:

$$I_1 = \lambda_1^2 + \lambda_2^2 + \lambda_3^2$$
$$I_2 = \lambda_1^2\lambda_2^2 + \lambda_2^2\lambda_3^2 + \lambda_3^2\lambda_1^2 \qquad (5.04)$$
$$I_3 = \lambda_1^2\lambda_2^2\lambda_3^2$$

If the material is perfectly incompressible (remember that most rubbers are <u>almost</u> incompressible) then $I_3 = 1$.

5.3.1. Neo-Hookean Model

The simplest form of hyperelastic material model is the Neo-Hookean model. This model is based on a strain energy density function defined by:

$$W = C_{10}(I_1 - 3) + \frac{K}{2}(J - 1)^2 \qquad (5.05)$$

Where the constant C_{10} is defined as half the initial shear modulus of the material, K is the initial bulk modulus of the material and J the ratio of the deformed volume over the original volume of material. If the material is assumed to be perfectly incompressible then J = 1, and equation 5.05 reduces to:

$$W = C_{10}(I_1 - 3) \qquad (5.06)$$

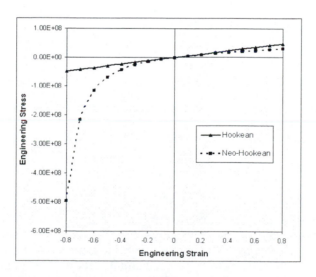

Figure 5.04: Comparison of Hookean and Neo-Hookean Material Models for Identical Shear Modulii and v = 0.5.

Figure 5.04 shows a comparison of a linear elastic model (Hookean) and a Neo-Hookean material model, where the same value of shear modulus, G, was used in each case. In the case of the linear elastic model a Poisson's ratio of 0.5 was used. The graphs are defined by equations 5.07 and 5.08.

Linear Elastic (Hookean) $\qquad \sigma = E\varepsilon = 2G(1+v)\varepsilon \qquad (5.07)$

Neo-Hookean $\qquad \sigma = G\left((1+\varepsilon) - (1+\varepsilon)^{-2}\right) \qquad (5.08)$

It is clear from figure 5.04 that the behaviour of the Neo-Hookean model under tension and compression is very different. In this case the linear elastic model provides a reasonable estimation of tensile behaviour but is highly inaccurate under compression.

The Neo-Hookean model has been shown to have good correlation with experimental results up to 40% strain in uniaxial tension and 90% strain in simple shear. If you expect deformations beyond these values then please choose a more suitable model from those listed below.

5.3.2. Mooney-Rivlin Model

The Mooney-Rivlin material model is a development of the Neo-Hookean model and comes in a number of forms. It is generally available in a two parameter, three parameter, five parameter or nine parameter form. It is good practice to begin an analysis by using the two parameter model and move up the more complex models in sequence. By examining the resulting stress-strain curves and comparing to the material test data the most suitable number of parameters can be chosen. As a general rule if the material stress-strain curve has single curvature (i.e. no inflection points) then a two parameter model is suitable, if one inflection point is present then a five parameter model is suitable and if two inflection points are present then the nine parameter model should be used. This is summarised in figure 5.05.

Figure 5.05: Typical Hyperelastic Material Stress Strain Curves

The simplest form is the two parameter Mooney-Rivlin model which is given by:

$$W = C_{10}(I_1 - 3) + C_{01}(I_2 - 3) + \frac{K}{2}(J - 1)^2 \tag{5.07}$$

Where C_{10} and C_{01} are constants which are obtained from curve fitting the experimentally measured stress-strain curve for the material. When defining the material model in FE software it is usual to specify the material density and Poisson's ratio as well as C_{10} and C_{01}. With hyperelastic materials it is desirable to specify a high value of Poisson's ratio, but remember not to exceed 0.499 due to reasons mentioned in section 2.4.2. This two parameter model has proven accurate up to 100% strain in tension and 30% strain in compression. Using a higher parameter model may provide better results at higher strains.

It should already be clear that hyperelastic material behaviour is much more complicated than the typical behaviour of metallic materials. Hyperelastic stress-strain relationships will normally be significantly different for tension, compression, and shear modes of deformation. Consequently when determining the constants C_{10} and C_{01} test data that encompasses all possible modes of deformation (tension, compression, and shear) should be available.

The three parameter Mooney-Rivlin model is given by:

$$W = C_{10}(I_1 - 3) + C_{01}(I_2 - 3) + C_{11}(I_1 - 3)(I_2 - 3) + \frac{K}{2}(J - 1)^2 \qquad (5.08)$$

Where, again, C_{10}, C_{01} and C_{11} are constants which are obtained from curve fitting the experimentally measured stress-strain curve for the material.

The five parameter Mooney-Rivlin model is given by:

$$W = C_{10}(I_1 - 3) + C_{01}(I_2 - 3) + C_{20}(I_1 - 3)^2 + C_{11}(I_1 - 3)(I_2 - 3) + C_{02}(I_1 - 3)^2 + \frac{K}{2}(J - 1)^2$$

$$(5.09)$$

Where, again, $C_{10}, C_{01}, C_{20}, C_{11}$ and C_{02} are constants which are obtained from curve fitting the experimentally measured stress-strain curve for the material.

Finally, the nine parameter Mooney-Rivlin model is given by:

$$W = C_{10}(I_1 - 3) + C_{01}(I_2 - 3) + C_{20}(I_1 - 3)^2 + C_{11}(I_1 - 3)(I_2 - 3) + C_{02}(I_1 - 3)^2$$

$$+ C_{30}(I_2 - 3)^3 + C_{21}(I_1 - 3)^2(I_2 - 3) + C_{12}(I_1 - 3)(I_2 - 3)^2 + C_{03}(I_2 - 3)^2 + \frac{K}{2}(J - 1)^2 \qquad (5.10)$$

Where, as before, $C_{10}, C_{01}, C_{20}, C_{11}, C_{02}, C_{30}, C_{21}, C_{12}$ and C_{03} are constants which are obtained from curve fitting the experimentally measured stress-strain curve for the material. Various forms of the Mooney-Rivlin model have been used to model many materials, including many biomaterials such as arterial tissue.

5.3.3. The Yeoh Model

The Yeoh method of modelling hyperelastic materials has proven popular because it depends on only the first strain invariant, I_1, and it has been shown to satisfactorily model various modes of deformation based only on data obtained from a uniaxial tensile test. This has the advantage of reduced requirements for materials testing. The Yeoh model has, however, been shown to be inaccurate at modelling deformations involving small strains.

Similarly to the Mooney-Rivlin model, the Yeoh model is available in a number of different forms. The form of the Yeoh strain energy density function is given by:

$$W = \sum_{i=1}^{N} C_{i0}(I_1 - 3)^i + \sum_{k=1}^{N} \frac{1}{d_k}(J - 1)^{2k} \qquad (5.11)$$

Where, C_{i0} and d_k are material constants. In particular, d_k is obtained from the initial bulk modulus where $K = 2/d_1$. It should be clear that setting $N=1$ in equation 5.11 is equivalent to the Neo-Hookean formulation.

As an example, the two parameter Yeoh model is given by:

$$W = C_{10}(I_1 - 3) + C_{20}(I_1 - 3)^2 + \frac{1}{d_2}(J - 1)^4 \qquad (5.12)$$

As before, if the material is considered to be perfectly incompressible then J = 1 and the final term in equation 5.12 is eliminated. The Yeoh model has proven popular for modelling the behaviour of natural rubber.

5.3.4. The Arruda-Boyce Model

This model is a development of the Yeoh model that attempts to overcome the inaccuracies of modelling deformation at small strains. The Arruda-Boyce model does this by considering the physics of the deformation of the underlying molecular structure of the hyperelastic material. The strain energy density function for the Arruda-Boyce model is:

$$W = G_i \left[\begin{array}{c} \dfrac{1}{2}(I_1 - 3) + \dfrac{1}{20\lambda_L}(I_1^2 - 9) + \dfrac{11}{1050\lambda_L^4}(I_1^3 - 27) \\ + \dfrac{19}{7000\lambda_L^6}(I_1^4 - 81) + \dfrac{519}{6733750\lambda_L^8}(I_1^5 - 243) \end{array} \right] + \dfrac{K_i}{2}\left(\dfrac{J^2 - 1}{2} - \ln J \right) \qquad (5.13)$$

Where, G_i is the initial shear modulus, λ_L is the limiting network stretch (where "network" refers to the underlying molecular network), K_i is the initial bulk modulus and J is the ratio of the deformed volume over the original volume of material. As always, if the material is assumed to be perfectly incompressible then J = 1. If λ_L approaches infinity then equation 5.13 will describe the Neo-Hookean model.

This model is valuable as a standard tensile test produces sufficient data to satisfactorily model all modes of deformation at all strain levels.

5.3.5. The Gent Model

The Gent model is method that attempts to accurately model behaviour at both low and high strains. In this case the strain energy density function is given by:

$$W = \dfrac{GJ_m}{2}\ln\left(1 - \dfrac{I_1 - 3}{J_m} \right)^{-1} + \dfrac{1}{d}\left(\dfrac{J^2 - 1}{2} - \ln J \right)^4 \qquad (5.14)$$

Where, G is the initial shear modulus, J_m is the limiting value of I_1-3, d is the initial bulk modulus divided by two and J is the determinant of the elastic deformation gradient. As J_m approaches infinity this model becomes equivalent to the Neo-Hookean model.

5.3.6. The Ogden Model

The Ogden model has been shown to provide good correlation with material test data at larger strain levels. The applicable strain level can be up to 700%. The strain energy density function for this model is given in equation 5.15.

$$W = \sum_{i=1}^{N} \dfrac{\mu_i}{\alpha_i}\left(\lambda_1^{\alpha_i} + \lambda_2^{\alpha_i} + \lambda_3^{\alpha_i} - 3 \right) + \sum_{k=1}^{N} \dfrac{1}{d_k}(J - 1)^{2k} \qquad (5.15)$$

If α is greater than 2 then the material stiffens with increasing strain. Similarly, if α is less than 2 the material will soften with increasing strain. The Ogden model has has been successfully used in the analysis of O-rings, seals, and many other industrial products.

5.3.7. The Ogden Compressible Foam Model

Some hyperelastic materials, particularly rubber foams, are highly compressible and significant volume changes can take place with relatively little stress. The Ogden compressible foam model was designed to model such materials and its strain energy density function is given by:

$$W = \sum_{i=1}^{N} \frac{\mu_i}{\alpha_i} \left(J^{\alpha_i/3} \left(\lambda_1^{\alpha_i} + \lambda_2^{\alpha_i} + \lambda_3^{\alpha_i} \right) - 3 \right) + \sum_{k=1}^{N} \frac{\mu_i}{\alpha_i \beta_i} \left(J^{-\alpha_i \beta_i} - 1 \right) \tag{5.16}$$

Where, α_i, β_i and μ_i are material constants, $\lambda^{\alpha i}$ are the deviatoric principal stretches, and the second term in 5.16 describes the volume change.

5.3.8. The Blatz-Ko Model

The Blatz-Ko hyperelastic model is a much more simple method of modelling highly compressible hyperelastic materials (like rubber foams). This model has a strain energy density function which is given by:

$$W = \frac{K}{2} \left(\frac{I_2}{I_3} + 2\sqrt{I_3} - 5 \right) \tag{5.17}$$

Definition of the initial shear modulus, K, is all that is required to implement this material model. The Blatz-Ko model is popular for modelling polyurethane foams.

5.3.9. Example of Hyperelastic Material Modelling

Figure 5.06 shows data obtained from an experimental test of a hyperelastic material which comprises natural rubber reinforced with 55pph carbon black. This experimental data was obtained from www.polymerfem.com. It can be seen from the figure that three types of test were carried out: a standard tensile test (uniaxial tension), a biaxial tension test and a planar tension test.

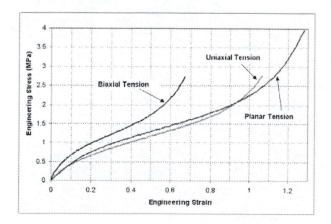

Figure 5.06: Experimental Testing of Natural Rubber filled with 55 pph Carbon Black (courtesy of www.polymerfem.com)

We will now compare how the different models described above can be used to model the experimental response of this material. For clarity we will only consider the uniaxial test data here and attempt to fit four of the above models to this data. In reality a model would be chosen based on its ability to fit to all of the available test data.

Most commercial FE software that offers the ability to model hyperelastic materials will also have a curve fitting function that allows you to enter experimental stress-strain data and will thus calculate the required material constants for the chosen hyperelastic material model. Figure 5.07 shows how the Neo-Hookean, 3rd order Mooney-Rivlin, 3rd Order Yeoh and 3rd order Ogden models compare in terms of their ability to model the uniaxial material test data from figure 5.06. It can be seen from figure 5.07, that in this case, the 3rd order Ogden model provides the best fit to the uniaxial experimental data. Further investigation, however, shows that the Ogden model is poor at representing the biaxial test data and, in fact, it is the Yeoh model that provides the best fit to all three experimental data sets. This illustrates the importance of having as much test data available as possible and of experimenting with each hyperelastic material model in order to ensure accurate modelling of material behaviour.

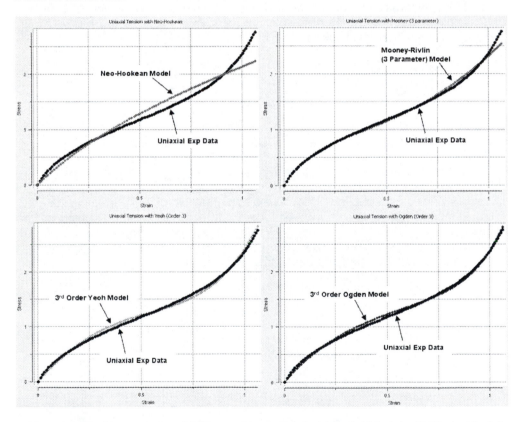

Figure 5.07: Comparison of Different Hyperelastic Material Models Ability to Model Uniaxial Tension of Natural Rubber with 50 pph CB.

5.4. Visco-Elastic Material Models

As discussed in chapter 3, a visco-elastic material is a material that exhibits both elastic (recoverable) and viscous (non-recoverable) deformation. When a load is applied to a viscous material the elastic deformation is instantaneous whereas the viscous deformation occurs over time (i.e. strain rate is dependant on time, which is known as creep).

Hysteresis is a common feature of visco-elastic materials subjected to cyclical loading. Unlike elastic materials, a visco-elastic material will lose some energy when a load is applied and then removed. Hysteresis is thus observed in the stress-strain curve, with the area of the loop being equal to the energy lost during the loading cycle. Another feature of visco-elastic materials is stress relaxation. If a visco-elastic material is given a prescribed strain, stress will decrease with time and this phenomenon is known as stress relaxation. In most visco-elastic materials the creep, stress relaxation and hysteresis properties are all dependant on temperature, making the modelling of such behaviour highly complex.

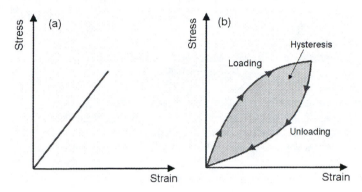

Figure 5.08: Stress-Strain Curves for a Linear Elastic Material (a) and a Visco-Elastic Material (b) Showing Hysteresis

Rubber can be modelled as a visco-elastic material as it exhibits each of the above properties. Many plastic and synthetic polymers are also visco-elastic. Human tissues such as blood vessels, cartilage, heart muscles, saliva, mucus etc. can be satisfactorily modelled using visco-elastic material models. Glass and glass like materials are commonly modelled using visco-elastic material models

There are a number of material models available for describing this type of behaviour, we will discuss the following: Maxwell, Kelvin-Voight, standard linear solid and generalised Maxwell.

5.4.1. Maxwell Visco-Elastic Model

The Maxwell model attempts to model visco-elasticity as been analogous to a spring and a dashpot in series, where the spring models the elastic portion of deformation and the dashpot accounts for the viscous portion of deformation, as shown in figure 5.09(a). Figure 5.09(b) shows that a sudden application of force produces an instantaneous deformation of the spring which is followed by creep of the dashpot (slow increase in deformation). Conversely, as shown in figure 5.09(c), a sudden application of deformation causes an immediate reaction in the spring which is followed by stress relaxation (as the dashpot expands) which predicts that stress decays exponentially with time. The Maxwell model can be represented by equation 5.18.

$$\frac{d\varepsilon_{total}}{dt} = \frac{d\varepsilon_{damp}}{dt} + \frac{d\varepsilon_{spring}}{dt} = \frac{\sigma}{\eta} + \frac{1}{E}\frac{d\sigma}{dt} \qquad (5.18)$$

Where, ε_{damp} is the stain in the dashpot, ε_{spring} is the strain in the spring, σ is the applied stress, E is the stiffness of the material and η is the viscosity of the material.

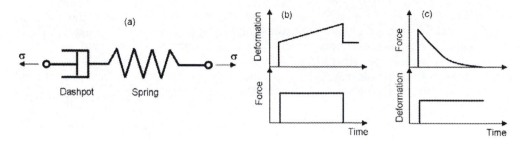

Figure 5.09: Overview of Maxwell Model of Visco-Elasticity (a) Creep Function (b) and Relaxation Function (c)

It is important to note the Maxwell model is limited as, although it describes stress relaxation behaviour quite accurately, it predicts that under constant stress that strain will increase linearly with time (see figure 5.09(b)). Clearly this is inaccurate, as most visco-elastic materials will creep with strain rate decreasing with time.

5.4.2. The Kelvin-Voight Visco-Elastic Model

This model again uses the assumption that the viscous behaviour of the material can be modelled using a dashpot and the elastic behaviour modelled using an elastic spring, however, in this case the spring and dashpot are assumed to be in series, as shown in figure 5.10(a).

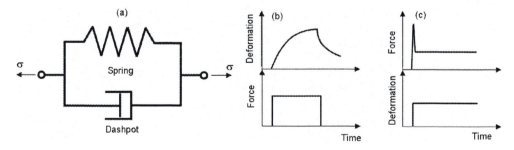

Figure 5.10: Overview of Kelvin-Voight Model of Visco-Elasticity (a) Creep Function (b) and Relaxation Function (c)

In this case the constitutive equation can be expressed as a linear first order differential equation, as shown in equation 5.19.

$$\sigma(t) = E\varepsilon(t) + \eta\frac{d\varepsilon(t)}{dt} \qquad (5.19)$$

Figure 5.10(b) shows that under application of a constant stress the material deforms at a decreasing strain rate (i.e. the slope is constantly reducing) until it reaches the steady-state strain. Deformation builds up gradually as the dashpot will not move instantaneously. After the stress is released the material gradually relaxes exponentially to the un-deformed state. In contrast to the Maxwell model, this model of creep is a quite realistic model of the actual behaviour of most visco-elastic materials. The Kelvin-Voight model is, however, inaccurate at modelling stress relaxation as shown in figure 5.10(c). This is due to the fact that there is instantaneous stress relaxation due to the dashpot being in parallel with the spring.

5.4.3. Standard Linear Solid Visco-Elastic Model

This model combines the Maxwell model with a parallel spring as shown in figure 5.11(a). The Standard Linear Solid visco-elastic material model can be represented by equation 5.20.

$$\frac{d\varepsilon}{dt} = \frac{1}{(E_1 + E_2)}\left[\frac{E_2}{\eta}\left(\frac{\eta}{E_2}\frac{d\sigma}{dt} + \sigma - E_1\varepsilon\right)\right] \tag{5.20}$$

This model is essentially an attempt to combine the Maxwell and Kelvin-Voight models in order to obtain the benefits of both. In this case, when subjected to a constant stress, as shown in figure 5.11(b) the material will immediately deform to a prescribed strain, equivalent to the elastic portion of strain, and then will continue to deform at a constantly reducing strain rate until the steady state strain rate is reached. Upon removal of stress, there is a partial instantaneous relaxation followed by a gradual exponential decay of strain rate. Figure 5.11(c) shows that stress relaxation effects are satisfactorily predicted by this model.

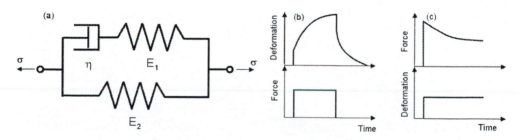

Figure 5.11: Overview of the Standard Linear Solid Model of Visco-Elasticity (a) Creep Function (b) and Relaxation Function (c)

5.4.4. Generalised Maxwell Model

The generalised Maxwell model attempts to take account of the fact that relaxation does not occur at a single instant in time but rather at several different times during the loading cycle. If we consider the microstructure of the material, it is clear that different molecular segments will have different lengths, thus shorter molecular segments will contribute less to relaxation then longer segments and hence their elaxation will vary with time. This model uses a number of spring-dashpot Maxwell models in parallel to represent the time dependency of relaxation, as shown in figure 5.12. The constitutive equation in this case is an exponential or Prony series representation of the stress relaxation function. By appropriately setting parameters in the generalised Maxwell model it can be made equivalent to the Maxwell, Kelvin-Voight or Standard Linear Solid models.

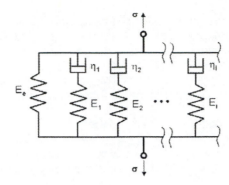

Figure 5.12: Overview of Generalised Maxwell Model of Visco-Elasticity

5.5. Strain-rate Independent Elasto-Plastic

5.5.1. Strain Rate Effects

Before we go into the detail of discussing Elasto-plastic models it is important to understand the difference between strain rate dependant and strain rate independent material models. A strain-rate-dependant material model is one that takes account of the rate of deformation of the material. Thus, when a strain-rate-dependant material is deformed at a quickly it will behave differently that if the deformation took place at a slow speed. A strain-rate-independent material will behave in the same way regardless of what speed the deformation takes place at. As an example, consider figure 5.13, which shows a ¼ symmetry finite element model of a cylinder impacting a rigid wall at high velocity.

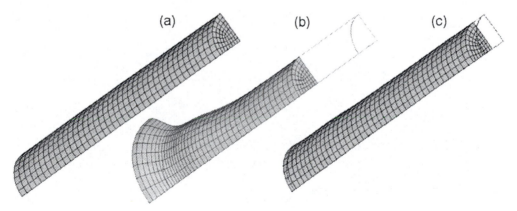

Figure 5.13: (a) FE model of cylinder impacting a rigid wall at high velocity (b) strain rate dependant (c) strain rate independent.

Figure 5.13(a) shows the original un-deformed model: the nodes on the bottom left hand edge have been constrained to simulate the rigid wall and a prescribed velocity is applied to remaining nodes. Figure 5.13(b) shows the resulting deformation of the cylinder when a strain-rate-sensitive material model was used and figure 5.13(c) shows the results obtained when a strain-rate-independent model was used. It should be noted that the same material parameters were used for both material models! Obviously when deformation takes place at a high velocity, a material model that includes strain rate effects is going to give a more accurate answer; in this case figure 5.13(b) provides the more accurate results.

Strain-rate-independent material models can however give excellent results for many problems that are not sensitive to strain rate effects. Before deciding which model to use it is important to think about the problem and determine if strain rate effects are likely to be important, if so then it is essential to use a strain-rate-dependant material model.

5.5.2. Bilinear Elasto-Plastic Material Model

The bilinear elasto-plastic material model is the most simple material model available for modelling plastic deformation. As the name suggests this model uses two lines to represent the stress strain curve: one line for the elastic portion and one line for the plastic portion. Figure 5.14 shows a typical stress-strain curve for a metallic material and illustrates how the tangent modulus (E_{Tan}) can be used to define the slope of the second line.

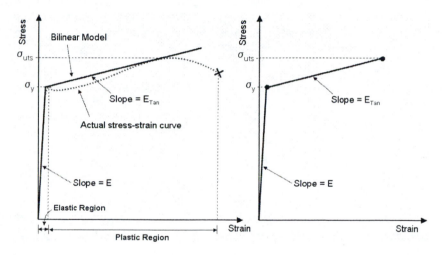

Figure 5.14: Bilinear Elasto-plastic Material Model

In this simple model a straight line is used to approximate the highly curved plastic portion of the stress strain curve. The slope of this straight line is known as the "tangent modulus". As its name suggests the tangent modulus may be obtained by drawing a tangent to the plastic portion of the stress-strain curve and measuring its slope. In practice, however, a bilinear model is often used when reliable experimental stress-strain data is not available so the tangent modulus must be obtained in another manner. In such cases the tangent modulus may be calculated using published material data and equation 5.21:

$$E_{Tan} = \frac{\sigma_{UTS} - \sigma_{yield}}{\varepsilon_{fail} - \varepsilon_{proof}}$$

(5.21)

Where, σ_{UTS} is the ultimate tensile stress for the material, σ_{yield} is the uniaxial yield stress for the material, ε_{fail} is the material failure strain (sometimes referred to as "elongation at break") and ε_{proof} is the proof strain (usually 0.002). At first glance this approximation may seem rather crude, as it is clear that the UTS and the failure strain do not occur at the same point. In practice, however, most metals fail very soon after the UTS is reached and, in the absence of any better data, this approximation usually gives reasonable results.

It should be obvious from the above, that the bilinear model is quite crude and should really only be used when sufficient material test data is not available to use a more complex model. Having said that, the bilinear model is often used as a "first guess" as it is so easy to implement and can give valuable information about the performance of a FE model.

Hardening Laws

The bilinear model is available in two forms which use different hardening rules: bilinear isotropic hardening and bilinear kinematic hardening. The hardening rule basically describes how the yield surface will change as yielding continues, so that the stress states for subsequent yielding can be established. Any stress state (i.e. combination of principal stresses) inside the yield surface is elastic and outside the yield surface is plastic. In isotropic hardening, the yield surface will remain centered about its initial centerline and as plastic strain develops the yield surface will expand in size about this centreline, as illustrated in figure 5.15(a). In contrast, kinematic hardening assumes that the yield surface will not change in size but will change location with progressive yielding, as shown in figure 5.15(b).

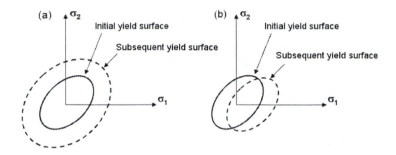

Figure 5.15: Isotropic (a) and Kinematic (b) Hardening Laws in 2D

The bilinear kinematic model works on the assumption that the total stress range (i.e. difference between the UTS and the maximum compressive stress) is equal to twice the yield stress. This is sometimes known as the "Bauschinger effect" and is illustrated in figure 5.16(a). The kinematic form is most suitable for materials that obey the von-Mises yield criterion applied to problems in which small strains are expected. For large strain applications the bilinear isotropic model is more suitable. This version again uses the von-Mises yield criterion coupled with an isotropic hardening law and assumes that the total stress range is equal to twice the UTS, as illustrated in figure 5.16(b).

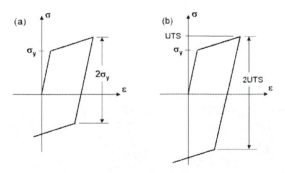

Figure 5.16: Stress-Strain Behaviour of (a) Bilinear Kinematic Model and (b) Bilinear Isotropic Model

5.5.3. Multi-linear Elasto-Plastic Material Model

An obvious extension of the bilinear material models is the multi-linear model. Clearly, if we use more than two lines we will get a closer approximation to the experimentally measured stress-strain curve for the material. The more lines we use, the closer we get to the highly curved stress-strain curve.

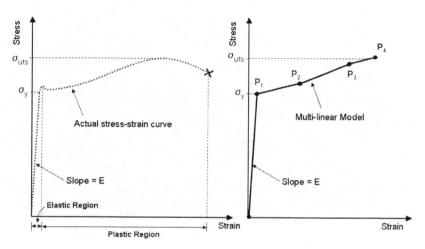

Figure 5.17: Multi-Linear Elasto-Plastic Material Model

As many lines as required can be used to ensure a good fit to the experimentally derived stress-strain curve, however, only lines of positive slope may be used. This essentially means that stress-strain behaviour beyond the UTS cannot be included as, in general, most stress-strain curves will become negative after the UTS as stress decreases with increasing strain in the lead up to failure. Figure 5.17 shows an example of a multi-linear model using four lines and four points on the stress-strain curve. Point P_1 is essentially the yield point and point P_4 is the point corresponding to the UTS. The other points are measured directly off the experimentally derived material stress-strain curve.

Similarly to the bilinear model, the multi-linear model is available with both isotropic and kinematic hardening laws. As before, the kinematic law is more suited to low strains and the isotropic law to large strains. In some cases advanced options are available that allow the combination of Isotropic and Kinematic hardening for simulation of cyclic loading or stress softening.

In general both the bilinear and multi-linear models by default use the von-Mises yield criterion to predict when yielding will occur. In some cases, the Hill yield criterion may be available as an alternative method of predicting yield, this method is described in section 5.5.5.

5.5.4. Power Law Elasto-Plastic Material Model

A power law elasto-plastic material model assumes that the stress-strain curve for the material can be modelled using an equation of the form:

$$\sigma = K\varepsilon^n \qquad\qquad (5.22)$$

Where, K is the material strength coefficient and n is the hardening parameter. This method of modelling stress-strain curves has already been described in section 2.4.3.

As with the bilinear and multi-linear models, the Power law model is available in both isotropic and kinematic forms. In the isotropic form a Voce hardening law is used, given by:

$$\sigma_{eqv} = \sigma_y + R_0 \bar{\varepsilon}^{pl} + R_\infty (1 - e^{b\bar{\varepsilon}^{pl}})$$ (5.23)

Where, σ_{eqv} is equivalent stress, σ_y is the initial yield stress, $\bar{\varepsilon}^{pl}$ is the equivalent plastic strain and R_0, R_∞ and b, are parameters used to change the shape of the plastic stress-strain curve. R_0 is essentially the tangent modulus of the stress-strain curve, $(R_\infty + \sigma_y)$ is the location on the stress (i.e. vertical) axis where the tangent crosses the stress axis and b is used to alter the shape of the curve to ensure a better fit.

In the kinematic form of the power law model models the Bauschinger effect by describing the translation of the yield surface in stress space through a "back stress" which results in strain in one direction causing a reduction in stress in the opposite direction. The back stress, α, is given by:

$$\alpha = \frac{C}{\gamma}(1 - e^{-\gamma\bar{\varepsilon}^p})$$ (5.24)

Where, C and γ are material parameters that are obtained from cyclic test data. C is effectively a tangent modulus and γ defines the rate at which the kinematic tangent modulus decreases as plastic deformation continues.

In certain cases an advanced hardening option, known as Chaboche hardening, may be available. This option essentially combines isotropic and kinematic hardening into one model, to describe a yield surface that both moves and changes size in stress space.

5.5.5. Anisotropic Hill Elasto-Plastic Material Model

Each of the elasto-plastic material models described above assumes that the material is isotropic. For many metals this will not be the case as it will have been processed using techniques such as rolling which will ensure that the material has different material properties in each direction. This is particularly true of sheet metals, where if one conducts a tensile test in three different directions (i.e. parallel to the rolling direction, perpendicular to the rolling direction and at 45° to the rolling direction) then results similar to those shown in figure 5.18 would be obtained.

The above models each use von-Mises stress to calculate yielding for isotropic materials. We may recall, from chapter 2, that the von-Mises yield criterion is given by:

$$\frac{1}{2}\sqrt{(\sigma_1 - \sigma_2)^2 + (\sigma_2 - \sigma_3)^2 + (\sigma_3 - \sigma_1)^2} = \sigma_y$$ (5.24)

Where, σ_1, σ_2 and σ_3 are the principal stresses and σ_y is the uniaxial yield stress.

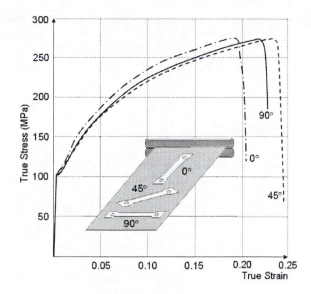

Figure 5.18: Effect of Anisotropy on Stress-Strain Results in Rolled Aluminium Sheet

In the case of anisotropic materials the Hill anisotropic yield criterion is used to calculate yielding and is given by:

$$F(\sigma_3 - \sigma_2)^2 + G(\sigma_3 - \sigma_1)^2 + H(\sigma_1 - \sigma_2)^2 + 2L\tau_{23}^2 + 2M\tau_{31}^2 + 2N\tau_{12}^2 = \frac{2}{3}(F + G + H)\sigma_y^2 \quad (5.25)$$

In the case of rolled sheet metal the directions 1,2 and 3 usually represent the rolling (sheet length), transverse (sheet width) and normal (sheet thickness) directions respectively. The constants F, G, H, L, M and N are constants which are obtained by conducting material tests in different directions on the sheet. The user is usually required to enter the yield stress and tangent modulus in each direction and this data is used by the software to calculate the constants.

For general planar anisotropy, such as that illustrated in figure 5.18, equation 5.25 reduces to:

$$\sigma_1{}^2 - \frac{2r_0}{1 + r_0}\sigma_1\sigma_2 + \frac{r_0(1 + r_{90})}{r_{90}(1 + r_0)}\sigma_2^2 = \sigma_y^2 \quad (5.26)$$

Where r_0 and r_{90} are the anisotropy parameters in the rolling and transverse directions respectively. The r factor is the ratio of true strain in the width direction, ε_w to that in the thickness direction ε_t as measured in a tensile test on a flat specimen:

$$r = \frac{\varepsilon_w}{\varepsilon_t} \quad (5.27)$$

5.6. Strain-rate Dependant Elasto-Plastic Models

Before examining any of the models discussed in this section, please make sure that you have read and understood section 5.5.1 which describes the effects of strain-rate-dependency.

There are many strain rate dependant elasto-plastic material models available. The Perzyna visco-plastic model models the effect of strain rate and can be combined with any of the rate-independent models to model a full elasto-plastic response with strain rate effects included. The Perzyna model is given by:

$$\sigma = \left[1 + \left(\frac{\dot{\varepsilon}^{pl}}{\gamma}\right)^{m}\right]\sigma_{y} \tag{5.28}$$

Where, $\dot{\varepsilon}^{pl}$ is the equivalent plastic strain rate, γ is the material viscosity parameter, m is the strain rate hardening parameter and σ_{y} is the uniaxial yield stress.

A similar model is the Pierce visco-plastic material model which is given by:

$$\sigma = \left[1 + \frac{\dot{\varepsilon}^{pl}}{\gamma}\right]^{m}\sigma_{y} \tag{5.29}$$

If the strain rate parameter, m, is very small then the Pierce model is usually more suitable than the Perzyna model.

The Cowper-Symonds model describes the relationship of dynamic stress to strain rate and is given by:

$$\sigma = \sigma_{y}\left[1 + \left(\frac{\dot{\varepsilon}}{C}\right)^{\frac{1}{P}}\right] \tag{5.30}$$

Where, $\dot{\varepsilon}$ is the strain rate, C and P are the Cowper-Symonds strain rate parameters and σ_{y} is the uniaxial yield stress. This model has been incorporated into various other models in order to account for strain rate, as an example a strain rate dependant power law plasticity model may be defined by:

$$\sigma = \left[1 + \left(\frac{\dot{\varepsilon}}{C}\right)^{\frac{1}{P}}\right]k\left(\varepsilon_{e} + \varepsilon_{p}^{eff}\right)^{n} \tag{5.31}$$

Where, ε_{e} is the elastic strain to yield, ε_{p}^{eff} is effective plastic strain, k is the power law strength coefficient and n is the power law hardening parameter.

The Ramburgh-Osgood model describes a rate sensitive power law plasticity model according to the following relationship:

$$\sigma = k\varepsilon^{m}\dot{\varepsilon}^{n} \tag{5.31}$$

Where, ε is strain, $\dot{\varepsilon}$ is the strain rate, k is the strength coefficient, m is the hardening parameter and n is the strain rate sensitivity parameter.

Other forms of these equations are available whereby instead of specifying constants (such as m, n, C, P, etc.) the user inputs load curves which may describe: tangent modulus versus strain rate, elastic modulus versus strain rate, failure stress versus strain rate, etc. Based on the data provided in the load curves the FE software will apply various forms of the above models.

5.7. Specialised Plasticity Models

Other specialised plasticity models for metallic materials include:
- Various Barlat Models, which are anisotropic plasticity models used to specifically model the behaviour of aluminium under various loading conditions. These models are particularly useful for analysing metal forming processes using sheet aluminium.
- The Johnson-Cook model is a strain-rate and temperature-dependent plasticity model. This model is used to analyse problems where strain rates vary over a large range, and temperature changes cause material softening.
- The Zerilli-Armstrong model can be used to analyse metal forming processes and high speed impact events where the stress depends on strain, strain rate, and temperature.
- The Steinberg model is used to analyse failure in problems with very high strain rates of approximately 10^5 sec^{-1}.

Concrete

Concrete plasticity models are commonly provided in FE software to facilitate the analysis and design of reinforced concrete structures. Concrete exhibits a non-linear stress-strain behaviour with strain <u>softening</u> occurring after yield. This means that, in contrast to most metals, plastic deformation of concrete results in stress <u>decreasing</u> with increasing strain, as shown in figure 5.19.

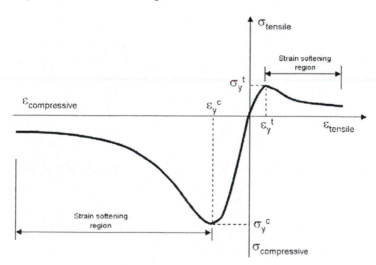

Figure 5.19: Typical Stress-Strain Response of Concrete

Concrete will generally fail by cracking in tension and by crushing under compressive loading, thus concrete material models allow for the specification of a number of experimentally derived material parameters in the material model. Typically, the uniaxial tensile cracking stress, uniaxial crushing stress, biaxial crushing stress, together with parameters which describe how quickly cracks will propagate are required as inputs to the material model.

Drucker-Prager

The Drucker-Prager material model is used to model the elasto-plastic response of granular type material such as soil, rocks or even concrete. This model works on the assumption that deformation prior to yielding is linearly elastic and deformation after yielding is perfectly plastic, with no strain hardening or softening. This model calculates yield and subsequent deformation based on three user defined constants: the cohesion value for the material, the angle of internal friction and the dilatancy angle.

The dilatancy angle controls the increase in material volume due to yielding. If the dilatancy angle is equal to the friction angle then the flow rule used is associative. If the dilatancy angle is zero or less than the friction angle then a non-associative flow rule is used as there will be no increase in material volume after yielding.

The yield criterion for the Drucker-Prager model is given by:

$$f = \alpha\sigma_m + \sqrt{\frac{1}{2}|\sigma_D|} - k = 0 \qquad (5.32)$$

Where, σ_m is the hydrostatic stress, σ_D is deviatoric stress, α is a constant related to the angle of friction and k is a constant related to the cohesion of the material.

In some cases an extended Drucker-Prager model is available, which allows the Drucker-Prager model to be combined with a model which incorporates strain hardening such as the bilinear isotropic or multi-linear isotropic.

Cast Iron

Cast iron behaves very differently in compression and in tension. In tension it is quite brittle, has low strength and cracks quite easily. In compression cracks do not easily form and the material has much higher strength and is more ductile, as shown in figure 5.20.

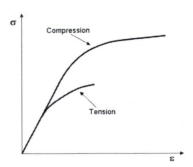

Figure 5.20: Typical Stress-Strain Response of Cast Iron

The cast iron material model uses different yield criterion for tension and compression: the von-Mises criterion is used in compression and the Rankine criterion is used in tension. Different yield strengths, flow curves and hardening laws are used in tension and compression. The elastic behaviour is assumed to be isotropic and identical in tension and compression.

5.8. Damage Models

5.8.1. Composite Damage Models

As already discussed, composite materials have many favourable features which makes their use attractive for high strength and high stiffness applications such as aerospace structures. A significant disadvantage of composite materials is that, once damaged, they lose much of their strength and stiffness. Laminated composite materials are particularly susceptible to impact damage which can cause de-lamination of the various layers, cracking and crushing of the matrix and eventual failure of the structure. The ability to predict the effects of damage is thus highly valuable to the designer or analyst. Composite damage models have been developed and provided as material models in some FE software.

One example is the Chang-Chang damage model which is used with fibre reinforced composite materials. This model considers damage due to four modes of failure: matrix cracking due to tension, matrix crushing due to compression, fibre breakage due to tension and fibre breakage due to compression. This model is often modified by using a Tsay-Wu damage model for failure of the matrix and the Chang-Chang model for the fibre failure.

These models work by determining the point at which failure in either the matrix or the fibres occur in each element. Each element can be in one of three states: intact, failed, or in the damage zone. If failure is detected then that element is labelled as having failed and can be treated accordingly (e.g. removed from the model and subsequent load steps). If the element is in the damage zone then it is in the processes of failing and is likely to be surrounded by other elements that are in the process of failing.

5.8.2. Concrete damage Models

Reinforced concrete is a very strong and adaptable material which generally fails due to crack formation in tension and by crushing under compressive loading. Several brittle failure models are available for simulation of damage and failure of concrete structures. Typically, the uniaxial tensile cracking stress, uniaxial crushing stress, biaxial crushing stress, together with parameters which describe how quickly cracks will propagate are required as inputs to these material models.

5.9. Specialised Material Models

5.9.1. Shape memory alloys

A shape memory alloy is a metal that "remembers" its original geometry and regains its original geometry automatically during heating or by unloading. These materials can undergo very large deformations during loading-unloading cycles without any permanent deformation. Such behaviour is often known as "super-elastic" behaviour. There are three main types of these alloys available: copper-zinc-aluminium, copper-aluminium-nickel and nickel-titanium. Nickel-titanium is the most popular and is generally known as Nitinol. A typical stress-strain curve for Nitinol is shown in figure 5.21 along with an idealised model used for FE material modelling.

The idealised model assumes that the stress-strain curve has three distinct phases: the austenite phase which is linear elastic, the transition phase and the martensite phase which is also linear elastic. The material model is defined by specifying the linear elastic portions of the idealised stress-strain curve and the points of change in the slope of the stress-strain curve.

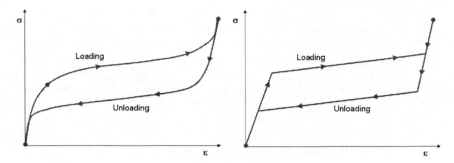

Figure 5.21: Typical Stress-Strain Curve for Nitinol (left) and idealised stress-strain curve (right)

5.9.2. Creep Models

We have already discussed visco-elasticity in section 5.4 and many of the concepts introduced there also apply to the analysis of creep. Creep is a rate dependent material nonlinearity where a material will continue to deform under a constant (i.e. non-changing) load. Materials that exhibit creep also exhibit stress relaxation, where if a deformation is imposed the stresses will diminish over time. These concepts have already been illustrated in section 5.4. Creep generally takes place in three distinct stages: primary, secondary and tertiary creep as shown in figure 5.22.

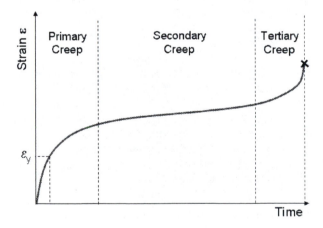

Figure 5.22: Creep Strain versus Time for a Constant Applied Load

Primary creep consists of the elastic response followed by decreasing strain rate. The strain rate eventually reaches a minimum and becomes almost constant and this stage is known as secondary creep or steady-state creep. In tertiary creep the strain rate begins to increase exponentially and failure occurs. Most creep material models implemented in FE software concentrate on modelling primary and secondary creep as tertiary creep is essentially the onset of failure and occurs more quickly than the other two stages.

There are a number of different models available for modelling creep material behaviour. Most are based on obtaining constants from material tests that are used to modify the equation for either primary or secondary creep, according to environmental factors (temperature etc.) and to take account of different creep mechanisms in different materials. A general form of these equations is given by equation 5.33.

$$\frac{d\varepsilon}{dt} = \frac{C\sigma^m}{d^b} e^{\frac{-Q}{kT}}$$

(5.33)

Where, C is a constant dependent on the material and the particular creep mechanism, σ is the applied stress, m and b are exponents dependent on the creep mechanism, d is the grain size of the material , Q is the activation energy of the creep mechanism, k is Boltzmann's constant, and T is the temperature. It is clear from equation 5.33 that increasing stress or temperature will change the shape of the creep curve shown in figure 5.22.

5.9.3. Foam models

Foam materials are often used as shock absorbers, to dissipate impact damage and as padding on many consumer products such as seats. These materials are also commonly used as core materials in sandwich structure composites (i.e. sandwich panels).

Closed cell foam models are used to model the behaviour of materials such as low density polyurethane foams, cross-linked polyethylene foams, EVA (Ethylene-vinyl acetate) foams, etc. These materials are commonly used as padding in sports equipment (e.g. boxing gloves, knee pads, exercise mats, martial arts etc.). Other applications include life preservers, floatation devices, acoustic insulation and impact limiters in automobiles.

Viscous foams are often used as energy absorbers to absorb impact during automobile crashes. These foams are also known as "memory foams" and are also used to make mattresses and applications in medicine where it is used as a pressure-relief material for patients needing to be free of pressure to sensitive body parts that require medical care. These foams were originally developed by NASA to protect astronauts from the tremendous pressures exerted on their bodies by G forces as they departed and returned through the Earth's Atmosphere. The material model for viscous foam requires the definition of load curves that describe how elastic stiffness and viscosity change due to applied deformation. These curves are normally input as power law curves.

Low density foams such as urethane are highly compressible and are used for many consumer products such as seat cushions etc. The material model for low density foams assumes hysteresis unloading behaviour under compression with possible energy dissipation. In tension the material model behaves linearly until tearing occurs. The material model uses a hysteresis unloading parameter, a decay constant and a shape factor to closely approximation the experimentally observed unloading behaviour of low density foams. A tearing stress and viscous coefficient are also required in the material model definition.

Crushable foams are used in impact absorption applications. As their name suggests they deform and absorb energy by crushing. The material models used for crushable foams require the definition of: a load curve which describes how yield stress varies with volumetric strain, a damping coefficient which describes the viscous behaviour of the material and the tearing stress.

Honeycomb foams are commonly used as in the core of sandwich panels and other similar applications. The material model for honeycomb foams is quite complex and is orthotropic in nature to take account of the fact that honeycomb structures behave differently along the cells and across the cells. Definition of the model requires specification of parameters related to the behaviour of the fully compacted honeycomb structure and load curves which describe the stress in each direction versus volumetric strain.

5.9.4. Gasket Material Models

A gasket is essentially a mechanical seal that is used to close the gap between two objects, generally to prevent leakage between the two objects under compressive loading. Gaskets are generally made from sheet materials such as gasket paper, rubber, silicone, felt, fibreglass, metals and polymers. A gasket material must be compressible so it will tightly fill the space it is designed for and taking account of any slight irregularities. The function of the gasket is to transfer force between two mating components. It is clear from this that gasket materials will be under high compression and will behave quite non-linearly. The majority of the gasket deformation will be in the through thickness direction and the stiffness contribution from the membrane and transverse shear directions will be much smaller, and thus is neglected in the material model.

Gasket material models require specification of two main sets of parameters. The first is the general gasket parameters such as initial stiffness, gasket gap etc. The second set of parameters relate to the pressure closure behaviour and includes gasket loading (compression) and unloading (tension) curves.

5.10 Summary of Chapter 5

This chapter describes each of the material models that are typically available for use in structural analysis in commercially available finite element software packages. This chapter should be used together with section 3.3 when building a finite element model. By working through table 3.01 and referencing the appropriate material in this chapter the most appropriate material model for the analysis being undertaken should be clear.

6

Modelling and Meshing

6.1. Introduction

As discussed in chapter 3, the most important step in a finite element analysis is turning the physical problem into a mathematical problem that is then solved using the finite element method. The procedure outlined in chapter 3 requires that we first decide which element type is appropriate, using chapter 4, then describe how those elements will behave (i.e. assign a material model), using chapter 5. Once the element type and material model have been decided, the next stage is to define the geometry of the mathematical model and then decide how it will be split into finite elements. This chapter is concerned with the process of creating the problem geometry (modelling) and then appropriately dividing this problem geometry into finite elements (which is known as "discretisation" or "meshing").

6.2. Modelling

6.2.1. Modelling Overview

The geometry of the finite element model will largely be dictated by the type of element chosen after consulting chapter 4. If a spar or beam element is chosen then the problem geometry will simply be a line or a set of lines. If a 2D element such as plane stress, plane strain or axisymmetric element has been deemed appropriate then a 2D planar geometry is required. If a shell element has been selected then a 3D thin shell type geometry is required. Finally, if a 3D solid element is deemed necessary then a representative 3D solid model of the problem will be required. In every case you should always try to use the simplest type of geometry and element that will get you the answer you require. There is seldom any advantage to using a more complicated element and, in fact, you will probably cause yourself more trouble by introducing more complexities into the problem! There is a common misconception among novice FEA users that 3Dsolid elements are somehow more "advanced" than beams or shells because you can use them to mesh a CAD solid model and because the mesh "looks like" the real structure. In fact the opposite is true: beams and shells can model bending behaviour and 3D solids cannot, so beams and shells are actually the more advanced element types!

Modelling the problem geometry is not, however, as straightforward as it may seem once the element is chosen. Frequently a problem will exhibit symmetry, which will allow reduction of the model size and convenient locations to introduce boundary conditions, and the geometrical model should be adjusted accordingly. In cases where a structure interacts with another structure it may be necessary to also model that structure or at least part of it. The structure may contain a myriad of detailed features related to its design or manufacture, which play no role in the type of analysis being carried out. These features are generally removed in a process called defeaturing which will aid the analysis and reduce model complexity. In some cases a complex structure may be better analysed by splitting it into a number of substructures and analysing these substructures independently. Each of these concepts will be discussed in detail in the sections below.

In order to aid your modelling process, I suggest that you try and answer each of the following questions and consult the appropriate section if you do not know the answer:

Q1:   Have you checked your model dimensions? (section 6.2.2)
Q2:   Is a truss or beam assumption valid for your model? (section 6.2.3)
Q3:   Is a planar assumption valid for your model? (section 6.2.4)
Q4:   Is a thin shell assumption valid for your model? (section 6.2.5)
Q5:   Is a combined beam/shell assumption valid for your model? (section 6.2.6)
Q5:   Does your model exhibit any symmetry? Can it be used to simplify the model? (section 6.2.7)
Q6:   Does your model require any defeaturing to simplify the model? (section 6.2.8)
Q7:   Does your model require cleaning up before analysis can take place? (section 6.2.9)
Q8:   Would using a cylindrical or spherical coordinate system simplify the model or make modelling easier? (section 6.2.10)
Q9:   Would the analysis be made easier by splitting it up into several more simple analyses, or by using sub-structuring or sub-modelling? (section 6.2.11)

6.2.2. Dimensions

The first step in building a representative geometry of the mathematical model is to ensure that you are using the correct dimensions. This is particularly important if you are using a CAD model generated in another piece of software or provided by a client or colleague. Assuming that a model is specified in metres, when in fact it is in millimetres, will result in an unrepresentative FE model and will cause great inaccuracy in results. Please refer to section 2.2 which discusses the importance of using the correct units and staying within a particular unit system.

6.2.3. Truss and Beam Assumptions

One of the most common errors made by novice users is to not recognise beam or truss problems. Whenever a problem consists of a simple beam, a combination of beams, a combination of long slender members, or any structure that has long constant cross sections, then a truss or beam model is appropriate.

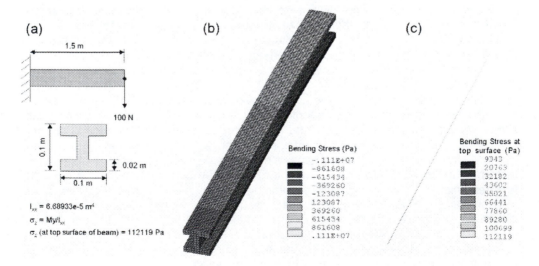

Figure 6.1: Recognising When a Beam Model is Appropriate

Consider the beam problem shown in figure 6.1(a). A novice user has incorrectly built a 3Dsolid model of the problem as shown in figure 6.1(b). In contrast an experienced FEA user has built the beam model shown in figure 6.1(c) and obtained a more accurate prediction for maximum stress with much less effort!

We can recall from chapter 3 that the essential difference between beams and trusses is that trusses can only experience axial tension and compression (i.e. they can only change in length), while beams can experience both axial deformation and bending. The main reason novices make the mistake of not using beams or trusses is that they assume that a 3Dsolid model will be more accurate. In fact beam and truss elements can have complex cross sections assigned to them and can more effectively model such structures, as seen in chapters 3 and 4 and figure 6.1. Beam elements are specifically designed to model bending and thus provide for better prediction of bending than solid elements, which in many cases can be too stiff in bending. Figure 6.2 shows a good example of a wire frame type structure that is particularly suited to modelling using beam elements. This type of structure would prove almost impossible to mesh using solid elements, yet is easily built and meshed using truss or beam elements.

Figure 6.2: A Wire-frame Structure Suited to Modelling Using Trusses or Beams

It is important to remember that beam and truss elements have section properties assigned to them, so even though the finite element model will just look like a collection of lines, the fact that a particular cross section is being used will be taken into account during the solution phase. A myriad of cross sectional shapes can be specified together with relevant area moments of inertia and torsional stiffness. In some cases tapered beams or trusses are allowed.

When using beam or truss elements, it should be noted that in most finite element codes the beam X axis is orientated along the length of the beam and the cross sectional axes, Y and Z are normal to the beam axis. The neutral axis of the beam is usually assumed to be on the line which defines the beam element, although some codes have a facility which allows specification of the distance of the neutral axis from this line.

6.2.4. Planar Assumptions

We have already discussed when and how plane stress and plane strain mathematical models are appropriate in chapter 3 and also in relation to the elements in chapter 4. For completeness, we will briefly revisit this material here.

All real world structures are three dimensional, however, if the geometry and loads of a problem can be completely described in one plane then the problem can be modelled as a two dimensional plane problem. It is important to restate that: for the planar assumption to be valid both the geometry <u>and</u> the loads must be constant across the thickness. Some examples of such problems are shown in figure 6.3.

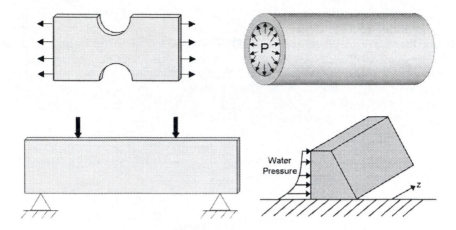

Figure 6.3: Examples of Problems Where a Planar Assumption is Valid

Plane stress is generally used for structures that are quite thin (i.e. the depth must be small compared to the cross section) like sheets or plates, however it is often used for problems where this is not the case. When using plane stress models it is possible to specify different thicknesses at various locations in the model as long as it is understood that the thickness will be applied equally on either side of the modelled cross section. This should be used with caution as the transition between the thicknesses will always be a sharp corner.

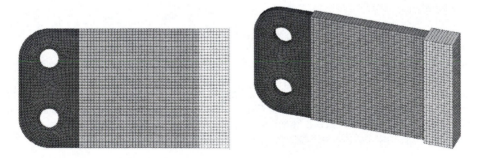

Figure 6.4: Use of Plane Stress Assumption to Model Multiple Cross Section Depths

Plane strain is generally used for structures that are very thick (i.e. the depth must be large compared to the cross section) where loading is constant across the thickness. It is normally assumed when using plane strain that the depth is infinite and thus effects from end conditions may be ignored. The specification of cross section depth in plane strain is thus not required.

6.2.5. Thin Shell Assumption

Many structures that are not suitable for a planar assumption and appear to require a 3D solid geometry can, in fact, be assumed to be thin shell structures and modelled accordingly. If the wall thickness of the structure is significantly smaller than the other dimensions of the problem then a thin shell assumption may be valid. A structure may be generally described as a thin-walled structure when its surface area to wall thickness ratio is 10:1 or greater. This assumption is valid for thin walled pressure vessels, aircraft structures, analysis of sheet metal forming, boat structures, submarines etc.

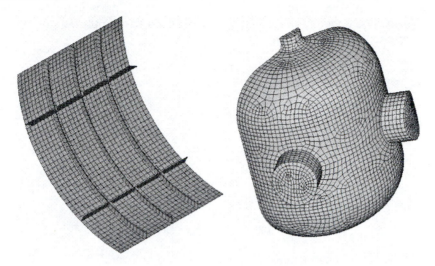

Figure 6.5: Use of Thin Shell Assumption to Model Thin Walled Structures; Aircraft Fuselage Panel (left) and Pressure Vessel (right)

When using the thin shell assumption it is important to be aware that using shell elements does introduce some extra complexities into the model. Shell elements, although appearing as just curved areas, do have a thickness associated with them which is usually distributed evenly on either side of the defined geometry. Hence, it is usual to generate a mid-surface geometry for a thin shell model. Many codes do, however, offer various options for offsetting the mid surface of particular regions in order to overcome modelling difficulties or allow for change in thickness to account for tapering geometries. Figure 6.6 illustrates the steps required to convert a H-beam into shell model.

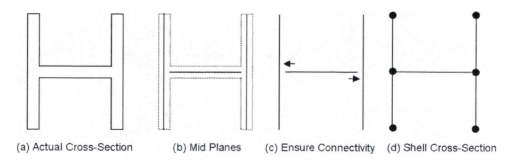

(a) Actual Cross-Section   (b) Mid Planes   (c) Ensure Connectivity   (d) Shell Cross-Section

Figure 6.6: Generating a Shell Model of a H Section Beam

The orientation of shell elements is also of utmost importance. Typically the x and y axes are orientated in the plane of the shell and the z axis is normal to this plane. A shell element has a top and a bottom side based on which direction the shell element normal, or positive z axis, is pointing. Is it is generally good practice to ensure that all shell element normal's are pointing in the same direction, in order to ensure that pressure loads are applied correctly and bending stresses are evaluated correctly.

Understanding shell element overlap and its likely effect on results is important for thin shell models that have shell elements meeting at corners etc. Figure 6.7 illustrates how such models will interpret the interacting shell element thickness overlapping each other. Obviously this will result in errors in stresses at these locations and it is generally suggested that such sharp corners should be transitioned via multiple elements incorporating an artificial fillet where appropriate. Alternatively, if the model is large enough, stresses at these locations can simply be ignored if they are not crucial to the design or analysis.

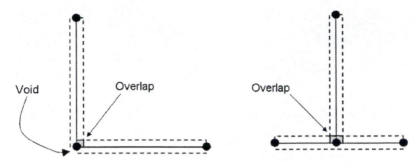

Figure 6.7: Shell Element Overlap

With the many recent advances in FE pre-processing software it has become all too common (and easy!) for the novice user to use 3D solid elements to mesh a thin walled structure which should have been correctly identified as a candidate for shell elements. A situation analogous to that shown for beams in figure 6.1 generally arises from this course of action, where a shell model would prove more accurate and would solve much quicker.

6.2.6. Combined Beam/Shell Assumptions

Several structures that may not seem suitable for a regular shell assumption may prove suitable for a combined beam and shell assumption. Stiffened or reinforced panels, such as that shown in figure 6.8, in particular, fall into this category.

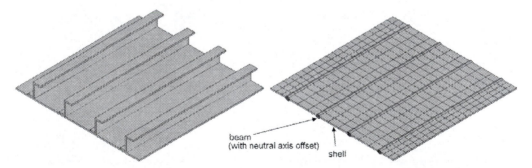

Figure 6.8: Panel Stiffened with Z-Section beams which is Suitable for a Combined Beam/Shell Modelling Approach.

When modelling a structure such as that shown in figure 6.8 in the manner shown on the right hand side of the figure, it is important to realise that you are artificially placing the centroid of the beam elements at the same location as the mid plane of the shell elements. It should be clear from the actual structure that, in fact, there is a significant distance between the mid-plane of the plate and the neutral axis of the Z-section beams. In order to compensate for this fact, most beam elements allow you to offset the beam neutral axis and shear centre so that they can be adjusted to act through a different location. By using this neutral axis offset the model will understand that the Z-section neutral axis is located at half the height of the beam and will thus correctly predict the stiffness of the combined structure. This type of modelling approach is used extensively in the analysis and design of aircraft structures.

6.2.7. Simplification Through Symmetry

When a problem cannot be simplified by a truss/beam assumption, a planar assumption or a thin shell assumption then a three dimensional model of the problem will be required. In most cases it will still be possible to reduce the model complexity by using symmetry. Indeed, regardless of whether the mathematical model uses truss elements or solid elements, is 2D or 3D etc. it should always be checked to see if it exhibits any symmetry in terms of geometry and loading.

There are four types of symmetry commonly encountered in structural analysis problems: reflective, axial, cyclic and repetitive, as shown in the figure below:

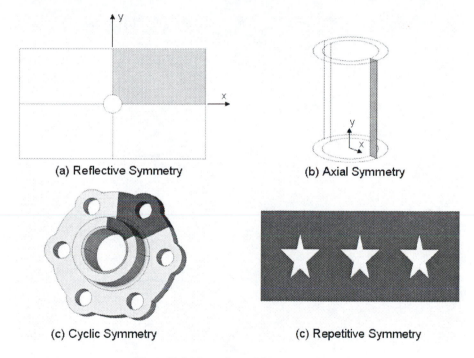

(a) Reflective Symmetry   (b) Axial Symmetry

(c) Cyclic Symmetry   (c) Repetitive Symmetry

Figure 6.9: Overview of Symmetry Types

If the geometry of the body <u>and</u> the external conditions (i.e. boundary conditions and loads) are also symmetrical then only one repeated part of the structure needs to be modelled. It is vital that the loading and constraint conditions are applied to the model in such a way that they truly reflect the symmetry of the problem. In order to illustrate this point we will consider a number of examples of each type of symmetry.

6.2.7.1. Reflective Symmetry 2D

Two-dimensional reflective symmetry is clearly evident in the case of the plane stress model of a flat plate with a central hole under uniform tension, as shown in figure 6.10. Provided that the correct boundary conditions are used, it is only necessary to consider one quarter of the full problem as indicated in the figure. In this case the correct boundary conditions require that displacements in the x-direction are constrained on the bottom edge of the quarter model and displacements in the y-direction are constrained on the left hand edge of the quarter model. Many commercial FE software packages have a symmetry edge option which can be used to specify symmetry edges to which the software will automatically apply the appropriate boundary conditions.

It is important to re-specify that <u>both</u> the geometry and loads must be symmetrical about the symmetry planes in order for reflective symmetry to be valid. Equally important is the fact that the correct constraints must be applied. If the constraints are not modelled correctly then the FE model will represent an entirely different problem from that intended.

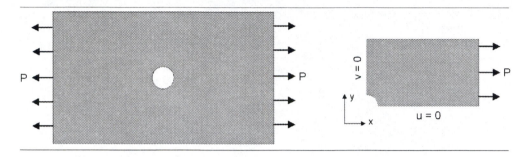

Figure 6.10: 2D Reflective Symmetry Used to Model a Plate with a Hole

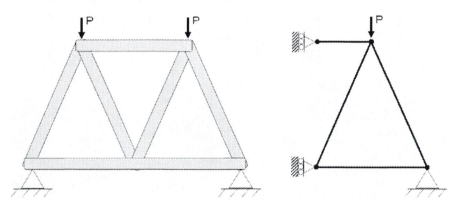

Figure 6.11: Framework (left) Modelled Using Beam Elements and 2D Reflective Symmetry (right)

6.2.7.2.   Reflective Symmetry 3D

Taking advantage of reflective symmetry in 3D models is important in order to reduce model size and hence reduce the number of elements required. Any reduction in element number means a corresponding reduction in solution time so this approach is highly valuable, particularly for complex 3D shell and beam problems and those requiring 3D solid elements. Figure 6.12 shows an example of a pipe cross joint structure subjected to a uniform internal pressure. It is clear that the structure has three orthogonal planes of symmetry where the geometry and loading are symmetrical across these planes. By using all three planes of symmetry the behaviour of the structure can be represented by one eighth of the actual problem geometry, as illustrated by the shaded section.

6.2.7.3. Axial Symmetry

We have already discussed axisymmetric elements in chapter 4 and have seen how axisymmetric problems are important in engineering analysis. Axial symmetry assumes that loads or boundary conditions due not vary in the circumferential direction and that the geometry is constant around the circumference. This assumption is applicable to any geometry that can be formed by sweeping a constant area around a defined axis, such as; straight pipes, pressure vessels, cones, circular plates, domes etc. Some relevant examples are shown in figure 6.13. Axisymmetric problems are similar to planar problems however in this case the distributions and loads are in the radial and axial directions.

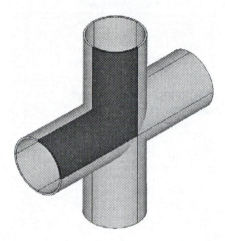

Figure 6.12: Reflective Symmetry in Three Dimensions

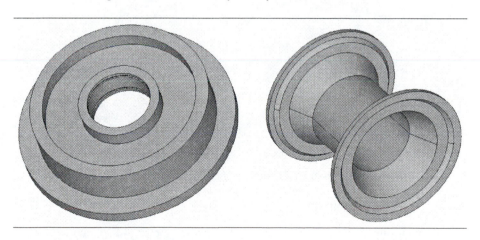

Figure 6.13: Examples of geometries formed by sweeping an area around a central axis, which are suitable candidates for axisymmetric analysis.

In general an axisymmetric model can only be subjected to axisymmetric loads, in some cases, however, it is possible to apply non-axisymmetric loads to axisymmetric models, but this process is quite difficult.

When setting up an axisymmetric model it is important to know which global axis your FE software requires you to use as the axis of symmetry. In most cases the axis of symmetry must coincide with either the global Y or Z axis. In either case negative x-coordinates are not possible as, due to the fact that the area being defined will be understood as being swept 360° around the axis, this would result in the model overlapping itself. If the structure being modelled contains a hole along the axis of symmetry it is imperative that proper spacing between the axis of symmetry and the 2D axisymmetric model is used, as shown in figure 4.42.

6.2.7.4. Cyclic Symmetry

Cyclic symmetry is used where geometrical features are repeated about an axis and the problem can be represented by only modelling one instance of the repeating feature with the appropriate boundary conditions. Good examples of cyclic symmetry are spline fittings, propellers, fans, turbines, flywheels and rotors.

Applicable types of loading for cyclic symmetry problems include centrifugal forces, radial displacements or uniform fluid resistance. Certain geometries which may appear suitable for cyclic symmetry are invalid due to the fact that the loading they experience does not fall into one of the above categories. Take for example a automobile wheel or a gear wheel, both experience a loading which is constantly orientated in one direction (vehicle weight or tooth loading) which cannot be satisfactorily included in a cyclical symmetrical model.

Figure 6.14 shows two examples of structures suitable for a cyclic symmetry assumption, with the portion required to represent the geometry shaded in each case.

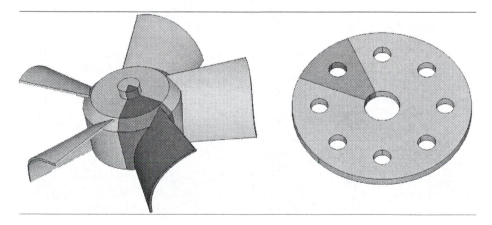

Figure 6.14: Examples of Cyclic Symmetry

6.2.7.5. Repetitive Symmetry

Instances of repetitive symmetry are not as common as the other types of symmetry mentioned above in structural analysis. Examples include evenly spaced cooling fins on a long pipe, repeating window holes in an aircraft fuselage or train carriage, and the geometry of some stent structures. For repetitive symmetry problems the common boundaries or the repeated segment are constrained in a perpendicular direction, as shown in figure 6.15.

6.2.8. Defeaturing

In many cases the structure to be analysed will contain small features or details that are important for the manufacture or assembly of the component under consideration or are important from an aesthetic point of view, but which will have no effect on the structural behavior of the component. Such features can make the finite element model very complex and can make meshing almost impossible, whereas removing them from the analysis may not effect the results significantly.

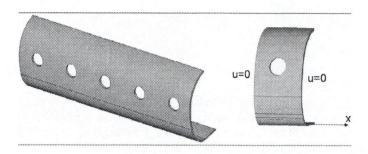

Figure 6.15: Example of Repetitive Symmetry

In general small fillets, screw threads, splines, small holes, small protrusions, small bosses, decorative features, etc. are candidates for removal in order to aid mesh generation and hence reduce the analysis complexity. This process is known as *defeaturing* or *model simplification*. We have already examined defeaturing briefly in section 3.3.3 and figure 6.16 shows an example of this process.

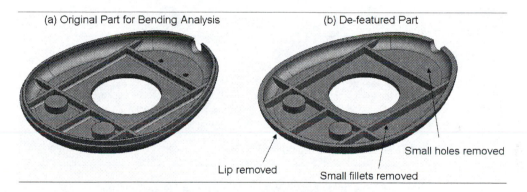

Figure 6.16: Defeaturing Example

It should be understand, however, that for some structures, such "small" details (such as fillets or holes) can be locations of maximum stress, and might be quite important, depending on your analysis objectives. You *must* have an adequate understanding of your structure's expected behaviour in order to make competent decisions concerning how much detail to include in your model. In general if a small feature is far away from expected stress concentrations or load paths then it can be safely removed. In some cases it may even be possible to remove large features from the model if initial simulations have shown that they make no contribution to the structures overall response to the given loading and their mass contribution is not required.

In some cases, only a few minor details will disrupt a structure's symmetry. You can sometimes ignore these details (or, conversely, treat them as being symmetric) in order to gain the benefits of using a smaller symmetric model.

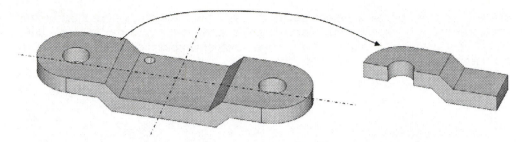

Figure 6.17: Removing Non-Symmetric Features to Allow for a Symmetrical Analysis

You must weigh the gain in model simplification against the cost in reduced accuracy when deciding whether or not to deliberately ignore un-symmetric features of an otherwise symmetric structure. In some cases it may be worth performing both a coarse symmetrical and non-symmetrical analysis in order to gauge the effect of removing non-symmetric features.

The most important thing to understand about de-featuring is that it is concerned with minor changes to the model to facilitate a finite element analysis. One shouldn't try to eliminate or modify significant geometric entities which will result in effective redesign of the model.

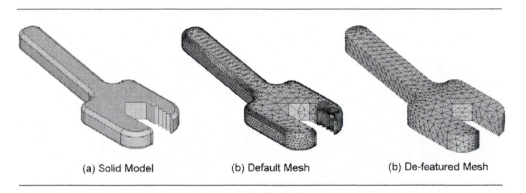

(a) Solid Model   (b) Default Mesh   (b) De-featured Mesh

Figure 6.18: Example of how De-featuring Effects Mesh Generation

Figure 6.18 shows an overview of the defeaturing process for a wrench solid model and its effect on mesh generation. The original solid model, shown in figure 6.18(a) contains rounded edges which are there for ergonomic reasons (i.e. to make a more comfortable grip) and to aid manufacturing. Their presence will have little or no effect on the load bearing capacity of the wrench during service, so removing these from the model to aid meshing is appropriate. The grooves machined into the grip of the wrench are another matter, clearly these are important for the operation of the wrench and, depending on what type of analysis is being carried out, would be expected to alter the in-service behaviour of the wrench. Having said this, for a basic analysis, say for example to determine bending stresses in the handle due to applied force by the user, the grooves could legitimately be removed from the model. Figures 6.18(b) shows the default mesh obtained when the original solid model was meshed in a commercial FE software package. The presence of the rounded edges and the groves has clearly caused many problems to the automatic mesher and the mesh obtained is not of high quality, as the transition between element sizes happens too quickly. In comparison, figure 6.18(c) shows the default mesh obtained from the de-featured solid model, with the rounded edges and grooves removed. It is clear that the de-featured mesh is far simpler and of better quality than the default mesh.

6.2.9. Model Clean-up

In many cases when a solid model is imported into a FE software package the transfer process may have resulted in some unwanted geometric effects being introduced into the model. In such cases model clean-up is imperative as it will reduce the probability of meshing problems later in the analysis procedure. While solid modelling is a powerful tool if models are generated using poor procedures then undesirable features such as short edges, short lines, small voids or gaps, sliver surfaces, small areas etc. may be present in the model. The interface between 3D CAD and FEA is seldom straightforward. A solid CAD geometry will almost always require significant modifications before it is suitable for meshing with finite elements.

A clean solid model is one which permits a mesh to be generated easily and contains surfaces of consistent size and shape ratios to avoid the generation of poorly shaped elements. The important thing to realise is that, although a solid model may look ok, this doesn't necessarily mean that it is geometrically intact. In such cases, the CAD file will contain physical features that are difficult to mesh, such as thin areas. The objective of model clean-up is to eliminate disproportionately small geometric entities (such as very small lines or loops), extraneous features, or areas with disproportionate size in one dimension (sliver areas). An important point to note is that removing these features will not significantly effect the analysis but will facilitate it. It should be remembered at all times that the finite element method is only an approximate theoretical method in any case, hence making these small changes will not significantly effect accuracy. In order to illustrate the importance of this step, figure 6.19 shows an example of a poor solid model and the mesh that results from using this model.

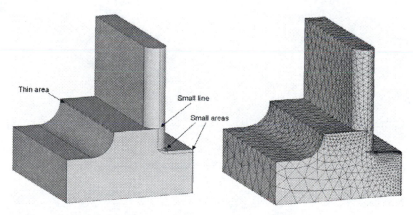

Figure 6.19: A poorly generated solid model (left) and the resultant mesh (right)

The right hand side of figure 6.19 shows the finite element mesh that would result from using the CAD model shown on the left. It can easily be seen that the poor CAD geometry results in far many more elements than are required in the areas with small lines/areas and also a large local transition in element size, which is unacceptable and will result in spurious results.

The model cleanup procedure begins once the model has been transferred from the CAD system into the FE software. The first step is to check for any gaps or voids in the model. Although the geometry may look perfect, very often careless geometry creation results in tiny gaps in a solid model. Take for example, the instance where an incorrect geometry is entered, so that a new surface is created 0.001mm away from the previous surface. The model will probably look ok but will prove impossible to mesh. Most FE software comes with in-built tools for finding and repairing such gaps. Similarly, small lines and areas (as shown in figure 6.19) must be detected and merged to larger lines/areas in order to simplify the model. Again, most FE software has facilities for detecting and merging these entities.

6.2.10. Coordinate Systems

Most analyses are carried out with models that are defined in a Cartesian coordinate system. This is the standard x,y,z coordinate system used for all the models described thus far in this book. In certain cases, however, the analysis can be made easier if an alternative coordinate system is used to define the model geometry, apply the loads and/or interpret the results. For example, a torsional analysis is often much easier to set up when defined in a cylindrical coordinate system, as shown in figure 6.20. The figure on the left of figure 6.20 shows an attempt to model torsion of a circular bar in a Cartesian coordinate system. It can be seen that the application of loads and boundary conditions is unsatisfactory as the loads or constraints cannot point in the correct direction. In comparison, the figure on the right of figure 6.25 shows the same loads applied in a cylindrical coordinate system. It can be clearly seen that the cylindrical coordinate system ensures that the torsional loads and constraints are accurately and easily modeled. The only change between the two models is the coordinate system used!

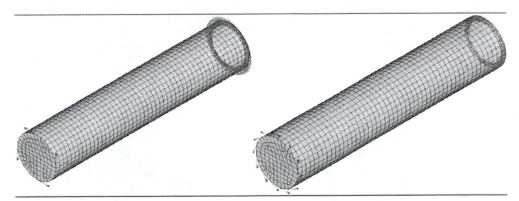

Figure 6.20: Definition of Loads in a Torsional Analysis, Cartesian coordinate system (left) and cylindrical coordinate system (right)

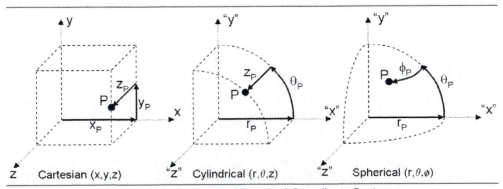

Figure 6.21: Overview of Standard Coordinate Systems

The standard coordinate systems available in most FE software are Cartesian, cylindrical and spherical as shown in figure 6.21. The Cartesian system is defined in terms of x, y and z. Their equivalents in the cylindrical system are r, θ and z, where r is radial distance, θ is circumferential distance and z is axial distance, as shown in figure 6.21. In the spherical system coordinates are defined in terms of r, θ and ϕ, where r is radial distance, θ is the major circumferential distance and ϕ is the minor circumferential distance. For a perfect sphere in a spherical coordinate system θ must be equal to ϕ at all points (refer to figure 6.21).

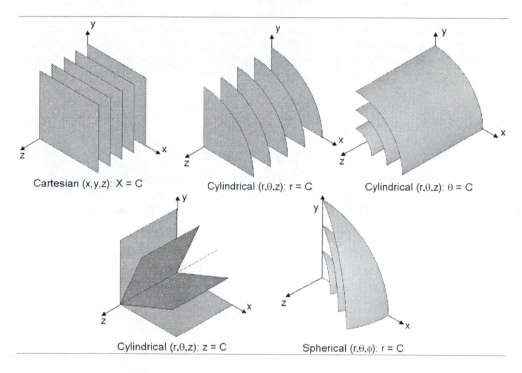

Cartesian (x,y,z): X = C

Cylindrical (r,θ,z): r = C

Cylindrical (r,θ,z): θ = C

Cylindrical (r,θ,z): z = C

Spherical (r,θ,ϕ): r = C

Figure 6.22: Surfaces of Constant Value in Various Coordinate Systems

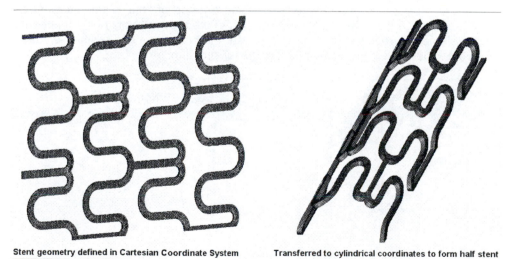

Stent geometry defined in Cartesian Coordinate System   **Transferred to cylindrical coordinates to form half stent**

Figure 6.23: "Wrapping" a complex geometry around a cylinder by transforming from a Cartesian to a cylindrical coordinate system.

In order to further understand the various coordinate systems it is useful to consider surfaces of constant value within each coordinate system. For example, in a Cartesian coordinate system specifying x = 5 represents the y-z plane at x = 5. These surfaces are often useful when generating models in various coordinate systems and an overview of the more common constant value surfaces are shown in figure 6.22.

In many cases even if a model is built in one coordinate system it is possible to later rotate the mesh into another coordinate system. This was done in order to generate figure 6.20. The model was built in a Cartesian coordinate system and later "rotated" into a cylindrical coordinate system in order to apply loads. The geometry in figure 6.20 is rather simple but for complex geometries this procedure is highly valuable. Consider, for example, a complex repeating pattern that one wished to generate as a cylinder or a sphere. The pattern can more easily be defined in a Cartesian system and later "wrapped" around a cylinder or sphere by rotating the mesh into the appropriate coordinate system. This type of procedure is very useful when modeling stent geometries as shown in figure 6.23

6.2.11. Sub-modelling and Sub-structuring

In many cases the problem to be analysed can be very complex and may be better analysed by splitting it up into a number of smaller models and then combining these smaller models later to obtain results for the full problem. This process is known as sub-structuring.

Similarly in certain problems there may be a specific part of a large model where detailed results are required while only general results are required for the remainder of the model. In such cases the region of interest can be extracted from the large problem, given a finer mesh and using specific mathematical techniques can be integrated with the larger model. This technique is known as sub-modelling.

We will discuss both techniques in the sections below.

Sub-structuring

Sub-structuring essentially works by condensing a group of finite elements into one "super-element". A separate analysis of the part of the model that is being made into a super-element is required in order to generate the properties of the super-element before it can be used in the large model of the full problem.

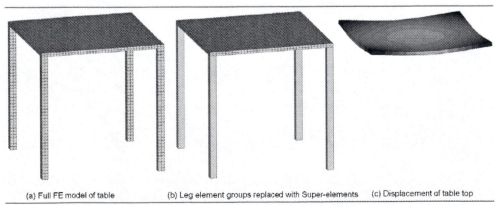

(a) Full FE model of table   (b) Leg element groups replaced with Super-elements   (c) Displacement of table top

Figure 6.24: Overview of Sub-structuring Analysis of a Table Using Super-elements

Figure 6.24 shows an elementary example of sub-structuring. The table shown in the figure consists of four identical legs which are subjected to the same load due to symmetry of the table and the fact that a uniform pressure load is applied to the table top. An appropriate analysis of a table leg was thus carried out and used to generate a super-element which represented the leg. The leg super-elements were subsequently incorporated in a model with the table top as shown in figure 6.24(b). The resultant displacement distribution in the table top is shown in figure 6.24(c) which is exactly the same distribution that would have been obtained had the full table model been solved.

Clearly, sub-structuring is very useful for reducing the computational time required to run very large or very complex models. This method is very useful for structures that have repeating geometrical entities and structures that have regions that exhibit non-linear behaviour. If a structure has repeated geometries (e.g. the four legs of a table as in the example above) it is possible to generate one super-element for each instance of the geometry and simply make copies of it at the required locations. In a non-linear analysis the linear portion of the model it is possible to sub-structure any linear portions of the model (e.g. a linear elastic body which is interacting with elasto-plastic bodies) so that the element matrices for the linear part of the model will not require recalculation at every equilibrium iteration.

This methodology is particularly useful where computer storage space or computational power is an issue. Using sub-structuring it is possible to analyse a model in several pieces, where each piece is a super-element that is small enough to be stored or solved on the available computer.

Sub-Modelling

Sub-modelling works by taking a specific portion of a large coarse model and making a detailed fine sub-model of the region of interest. The idea is that the coarse model captures the global problem behaviour while the sub-model captures local behaviour with a finer mesh at the area of interest. Take for example the 2D model of a three blade fan shown in figure 6.25(a). A fillet is placed between the fan blade and the fan disc but due to the coarse mesh used and the high stress gradient around the fillet; stresses around the fillet will not be predicted accurately. In order to overcome this a finely meshed sub-model of the region around the fillet is created in order to correctly model the local behaviour as shown in figure 6.25(b).

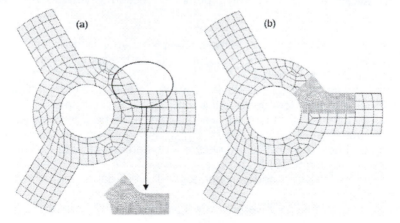

Figure 6.25: (a) Coarsely meshed model of a three blade fan (b) finely meshed sub-model superimposed over coarse model.

Sub-modelling may sometimes be referred to as the "cut-boundary displacement method" or as the "specified boundary displacement method". The cut boundary is the boundary of the sub-model which represents a cut through the coarse model. Displacements calculated on the cut boundary of the coarse model are specified as boundary conditions for the sub-model. Sub-modelling is based on St. Venant's principle, which states that if an actual distribution of forces is replaced by a statically equivalent system, the distribution of stress and strain is altered only near the regions of load application. The principle implies that stress concentration effects are localized around the concentration; therefore, if the boundaries of the sub-model are far enough away from the stress concentration, reasonably accurate results can be calculated in the sub-model.

Sub-modelling has several advantages over other modelling methods such as: eliminating the need for complicated meshing techniques to ensure smooth transitions in mesh size, and allowing one to experiment with different designs in the region of interest.

6.3. Meshing

6.3.1. Meshing Overview

After reading section 6.2 you should know what the geometry of your finite element model will look like. You should already have decided on an element type after consulting chapter 4. This section is concerned with putting these two decisions together and creating a mesh of finite elements based on the model geometry.

Even if you are already decided on the type of element you wish to use, it will generally be available in at least two variations: a linear variation which has straight edges and doesn't have mid-side nodes and a quadratic variation which has curved edges and extra mid-side nodes. We will begin by discussing the merits of each type. We will then discuss element distortion and identify what element shapes are allowable in a given mesh. Following this we shall discuss mesh transitions and refinement and illustrate examples where meshes should be refined and how this should be approached. Finally we will discuss mesh convergence and give practical examples of good and bad meshes with comments.

In order to aid meshing your model, I suggest that you try and answer each of the following questions and consult the appropriate section if you do not know the answer:

Q1: Are linear or quadratic elements more suitable for your model? (section 6.3.2)
Q2: Are you aware of areas in your mesh that will require a finer mesh due to stress concentrations? (section 6.3.3)
Q3: Are you aware of special considerations for meshing the particular element type you are using? (for beams, shells and solids see section 6.3.3)
Q4: Have you ensured that there are smooth transitions between element sizes or types? (section 6.3.4)
Q5: Have you checked your mesh to ensure that there are no excessively distorted elements? (section 6.3.5)
Q6: Do you know the difference between a good and a bad quality mesh? (section 6.3.6)
Q7: Have you performed a convergence test on your mesh? (section 6.3.7)

If you have answered each of these questions you should be well on the way to constructing a high quality mesh for your problem.

6.3.2. Choosing between linear and higher order elements

Regardless of which element type you have selected it will usually be available in at least two subtypes: a linear formulation and a quadratic formulation. Your software may offer higher orders such as cubic formulations. The difference between these formulations has been discussed in detail in chapter 4 for each element type, but will be revisited here in terms of consequences for mesh generation.

Linear Elements (no mid-side nodes)

For many types of structural analyses linear elements can give a reasonably accurate solution for a relatively small amount of computer time. Having said this, it is very important to realise that they are based on linear interpolation functions and thus cannot give accurate results for regions where stresses are changing rapidly. In particular, degenerate forms of linear solid elements (triangles in 2D and wedges or tetrahedrals in 3D) can be very inaccurate particularly in critical regions. Take a look at case study C to see an example. We will talk about element distortion later on, but it is important to note that linear elements should not be used in an excessively distorted shape. This means that quadrilaterals should be meshed so that they are as close to a square as possible and triangles as close to equilateral triangles as possible. This will not always be possible for complicated shapes, thus necessitating the use of quadratic elements.

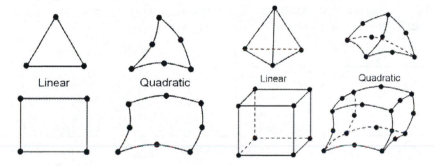

Figure 6.26: Overview of Linear and Quadratic 2D and 3D Elements

In the case of shell elements, linear formulations will obviously only give flat elements. When modelling a curved surface it is important to use enough linear elements to model the curved surface accurately and it is generally recommended that linear shell elements do not extend over more than a 15° arc. In general if a structure contains a significant amount of curvature and/or regions with rapidly changing stresses, then quadratic elements will always give superior results to linear elements.

Quadratic Elements (with Mid-side Nodes)

For structures that are highly curved quadratic elements usually give better results than with linear elements. This is particularly true when using triangles, wedges or tetrahedrons. In fact the warnings given in regard to these shapes for linear elements do not generally apply for their quadratic versions. Triangles with mid-side nodes can give quite accurate results as shown in case study C.

Quadratic elements do however have a number of peculiar traits that the user should be aware of. Firstly, distributed loads and pressures are allocated to the nodes in a non-straightforward manner, take a look at the theory of the 1D quadratic spar in chapter 4 to see why. You should see that rather than the load being distributed equally across the three nodes in thirds, that 2/3 is assigned to the mid-side node and 1/6 to the corner nodes. Similarly, reaction forces in quadratic elements follow this non-intuitive pattern. In most cases this won't be a problem, but in certain cases care must be taken. Secondly, the mass at mid-side nodes will be greater (for similar reasons) than the corner nodes – this can be important for dynamic analysis as there is a non uniform distribution of mass around the nodes.

When using quadratic elements it should be remembered at all times, that quadratic elements have significantly more nodes than linear elements and using even a few quadratic elements can thus increase the problem size to a great degree. If we take for example a mesh that requires the use of 1,000 brick elements. If a linear brick element is used then this will result in 8,000 nodes. If a quadratic brick is used then the resultant number of nodes will be 20,000!

In general, if computational resources are not an issue then quadratic elements are preferable to linear elements. In some types of highly non-linear problems quadratic elements can result in excessive solution times and linear elements are preferred.

6.3.3. General Meshing Guidelines

The general rule for building a finite element mesh for structural problems is that you should use a coarse mesh for areas where stress/strain is not changing rapidly and a fine mesh in areas where stress/strain is rapidly changing.

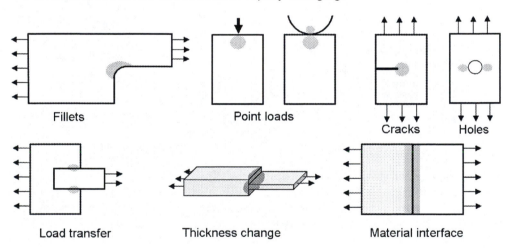

Figure 6.27: Examples of situations where a more refined mesh is needed in the shaded area

This raises the obvious question: "How do I know where stress is changing rapidly when I haven't done the analysis yet?" Well the answer is you don't, but you can do an initial analysis with a coarse mesh to find these regions and then re-mesh the problem with a fine mesh in the appropriate regions. Figures 3.17 and 3.18 in chapter 3 show an example of this process. Regions in which you can expect to find rapidly changing stresses are illustrated in figure 6.27 and include areas:
- Close to fillets or sharply curved edges.
- Close to concentrated point loads or concentrated reactions.
- Close to cracks or holes in the structure.
- Close to regions of load transfer: joints, welds etc.

- Close to regions where there is an abrupt change in thickness, material properties or cross section.

Meshing Beam Elements

Beam elements are generated by meshing lines that represent the neutral axis of the beams to be analysed. In cases where the line does not represent the neutral axis, a neutral axis offset can be specified at the element setup stage. Meshing beam elements basically involves splitting lines up into smaller lines, which are the beam elements. The greater the dominance of bending in the problem, the greater the number of beam elements that are required to capture bending behaviour. Beam meshes require convergence to determine the exact number of elements required, see section 3.3.7.

Various properties of beam elements need to be set in the mesh. Cross sectional orientation defines which way is "up" for the beam cross section, e.g. is an I beam or a H beam required? This is normally achieved by specifying a point or node as an orientation reference. If a neutral axis offset is required this should also be set up prior to meshing. Some beam elements allow tapering of the cross section along their length and this must be set up and specified during the meshing process. Case study B shows an application of beam elements and case study C shows a mesh convergence process for beam elements.

Meshing Shell Elements

Shell element models require a mid-plane surface in order to define the mesh. In most cases a solid model of the problem will be available and some FE software or CAD software suites have an automatic mid-plane surface generator available. Where this is not the case then the shell geometry will have to be manually generated. Lines and curves can easily be extruded into shell geometries for most purposes. Most of the information in this chapter that applies to 2D elements applies equally well to shells: mapped meshing with quadrilaterals is generally preferable to free meshing and/or triangular elements.

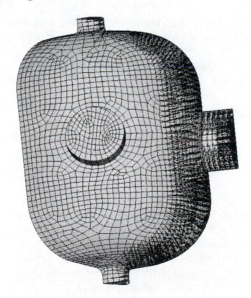

Figure 6.28: Shell normal's in a pressure vessel model point outward; meaning a positive pressure will act as a positive internal pressure.

A particular problem with shell elements is ensuring that shell element normal's are all facing in the correct direction (see figure 6.28). This becomes very important when pressure loads are applied as the normal specifies the direction in which the load will act. If all the normal's on a surface under pressure do not face in the same direction then artificial warping of the surface will be introduced. If any shell element normal's are found to be facing in the wrong direction then it is usually a matter of simply flipping them over so that they point in the other direction.

<u>Solid Elements</u>

Much of the discussion that follows is specific to 3D solid elements so it will not be repeated here. The main points are that mapped mesh bricks are preferable to tetrahedrons. If tetrahedrons must be used then they should at least be quadratic elements. In many cases where a mapped mesh at first may seem impossible it is usually possible to split the volume into a number of smaller volumes (known as volume segmentation) for the purposes of meshing (see figure 6.29). This may sound like quite a lot of work, and it is, but the quality of the results obtained will be worth it.

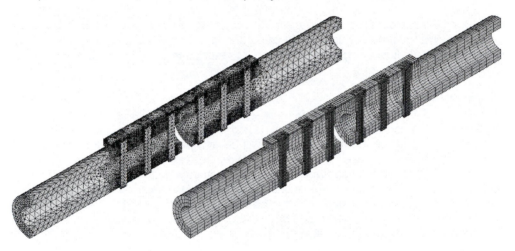

Figure 6.29: ½ symmetry FE models of two hollow cylinders bridged by a plate screwed into each cylinder. Model on the left is default tetrahedral mesh and the mapped mesh on the right was generated by volume segmentation.

6.3.4. Mesh Transitions

<u>Changes in Element Size</u>

In the section above we established that there will be certain regions in the model that will require a finer mesh (i.e. smaller elements) than the rest of the model. The next question is: how do we incorporate these smaller elements into the mesh?

The most important aspect of this is to ensure that there is a gradual change of element size from small to large. Placing large elements directly next to very small elements is asking for trouble. This would result in artificial stress concentrations being introduced as the FEM tries to interpolate displacements from the very small to the large element. In all cases, care should be taken to ensure that element size changes gradually from the fine part of the mesh to the coarser parts. Figure 6.30 shows an example of a good mesh transition and a poor mesh transition with the corresponding stress results. Figure 6.30(a) shows a plane stress model of beam bending with small elements at the built-in end of the beam where high stresses are expected. In this case the transition between small and large elements happens gradually along the length of the beam. In contrast figure 6.30(b) shows a similar model but with a sudden change of element size. The resulting un-averaged plot of element stresses is shown below each model. It can be seen that the stress contours are reasonably continuous across the element boundaries in (a) but this is not the case in (b), particularly in the region of the change in element size.

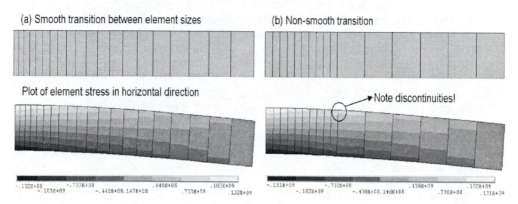

Figure 6.30: Smooth transitions in element size (left) produces better results.

Changes in Element Type

Frequently, it will be necessary to use more than one element type in a FE model. We have already given an example of combined beam and shell models in figure 6.8 and combined beam/solid or shell/solid models are also appropriate for various applications. Great care must be taken when joining different elements as there will be inconsistencies at the interface which could result in loads not being transferred between the different elements. In order to be consistent the two elements must have the same DOF: for example if a 3Dshell and 3Dsolid are to be joined they must both have U_x, U_y and U_z as displacement DOF and the same number and type of rotational DOF.

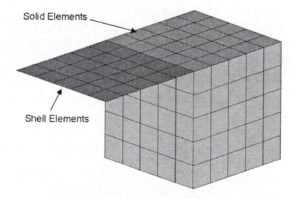

Figure 6.31: Connecting inconsistent shell elements to solids results in a hinge at the interface unless appropriate boundary conditions are used.

Let's consider an example of inconsistent elements in order to appreciate why this is an important issue. Suppose we have a 3Dbeam element with three displacement and three rotational DOF and a linear 3Dsolid brick element with three displacement and no rotational DOF. If the 3Dbeam element is joined to this 3Dsolid and a load is placed on the beam then the nodal forces corresponding to the displacement DOF will be transmitted to the solid element but the nodal moments corresponding to the rotational DOF of the 3Dbeam will not be transmitted.

Another example is shown in figure 6.31, where shell elements have been joined to solid elements along an edge of the solid part. The shell elements have rotational DOF whereas the solids do not. The unintended result here is that the interface will act as a hinge as the solid element cannot deal with the local rotations on the attached shells. This can be overcome through clever use of boundary conditions as we shall see in chapter 8, but it is important to recognise that undesirable results would be generated due to the hinge effect and the possibility of this happening should always be considered when using different element types.

Similar inconsistencies will exist between other element types with differing types or numbers of DOF and the user should be aware of these inconsistencies before attempting to combine different element types in a mesh. In some cases, the particular inconsistency may not be a problem as boundary conditions may be set that can overcome the inconsistency. In general, you should try to ensure that you use consistent elements.

6.3.5. Element Shape and Distortion

As we know, the finite element method essentially works by approximating the distribution of an unknown variable in a piecewise manner across the problem to be analysed. It should be obvious that this distribution will only be reliable if the shapes of the elements are not excessively distorted. If you consider that the element formulations are designed with the optimal shape of the element in mind (i.e. squares, equilateral triangles, cubes or equilateral pyramids) then it should be clear that the more the elements diverge from this optimal shape the more errors in the element formulations start to become increasingly important.

Therefore all elements should be as "regular" as possible with a finite element mesh. For 2D problems the fundamental shapes are triangles and rectangular and the best results are obtained when equilateral triangles and squares are used. Similarly for 3D the fundamental shapes are pyramids and bricks and the best results are obtained when equilateral pyramids and cubes are used. In practice, it will not be possible to build such regular meshes as complex boundaries, holes, fillets etc will require element shapes that deviate from these ideals. The allowable limits of distortion vary according to the problem but generally speaking if stress is changing rapidly throughout the problem then the results will be very sensitive to element shape.

Accordingly, most FE software have facilities for checking for poorly shaped elements and attempting to remedy them. Automatic meshing tools will always attempt to minimise the number of poorly shaped elements but this may prove impossible because of constraints placed on the mesh in terms of size or refinement etc.

One measure of the quality of an element's shape is aspect ratio, see figure 6.32. This is the ratio of the longest side of an element to the shortest side. As a general guide, aspect ratios of less than 3 give accurate results, while ratios of between 3 and 5 are generally acceptable, but aspect ratios greater than 5 should not be used unless they appear in non-critical areas of the model.

Other methods of quantifying element distortion can be made by measuring the skew and taper of the elements. A check on the internal angles of elements is a very useful indicator of element distortion. Allowable limits for the angles again depend on the problem under consideration, but general guidelines are shown in figure 6.33. It can clearly be seen from this figure that quadrilateral elements have a greater range of allowable angles than triangles and quadratic have a greater range than linear elements.

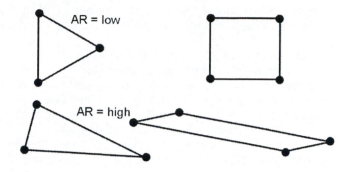

Figure 6.32: Illustration of low and high aspect ratio 2D elements

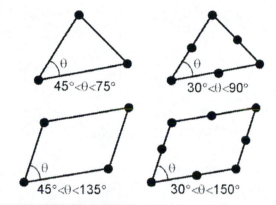

Figure 6.33: Allowable internal angles in linear and quadratic 2D elements

In general in 2D modelling given a choice between triangles and quadrilaterals for a similar nodal distribution you should pick quadrilaterals. Take a look at the results of case study C in chapter 10 to see why. Triangles are great for easy mesh generation and for meshing corners and fillets etc. but in general with a little planning a similar mesh can be made from quadrilaterals which will give much more accurate results.

In 3D modelling, bricks should be used over wedges and wedges used over tetrahedrals. Most modern FE software will automatically mesh in tetrahedrals as it is easier to generate a mesh in this manner. Tetrahedrals and wedges can produce acceptable displacement results but generally do not give accurate results for stress. This has a lot to do with the fact that these element shapes are very stiff. In recent years there have been some great improvements in element formulations for tetrahedral elements and there is no doubt that they are getting better, but they are still to be used with caution.

In some cases it will be impossible to avoid having some poorly shaped elements in the mesh. In such cases, it is considered acceptable for up to 5-10% of the total elements to be poorly shaped as long as they are not located in critical regions, such as stress concentrations (i.e. are far away from the region of interest).

6.3.6. What makes a good quality mesh?

In order to answer this question we will take a look at some examples and point out aspects of these meshes that affect the mesh quality. First, let's look at figure 6.34 which shows a ¼ symmetry plane stress model of a plate with a hole subjected to uniform tension. Figure 6.34(a) shows the default mesh produced by the automatic meshing tool in the available FE software. This mesh is unacceptable for a number of reasons:

1. Only three elements have been used around the curve of the hole. This is not enough to correctly model its highly curved geometry.
2. We know there will be a stress concentration around the hole, but the mesh around the hole is not fine enough to correctly model this.
3. The size ratio between the largest and smallest element in the mesh is too large, generally this should be no greater than 10:1
4. The mesh mixes quadrilaterals and triangles (near the hole) which is generally not good practice. A full mesh of quadrilaterals would be preferable.

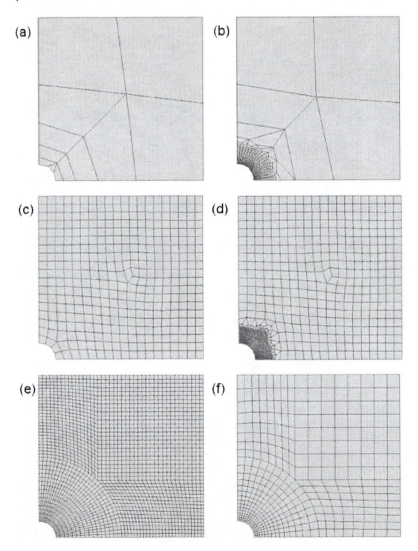

Figure 6.34: Various meshes for a ¼ symmetry plane stress analysis of a plate with a hole.

Figure 6.34(b) shows an attempt to correct the problems with mesh (a) by refining the mesh around the hole. In this case the previous mesh was simply corrected automatically by the software by specifying the locations where refinement was required. This mesh is still not satisfactory because:

1. The automatic refinement has introduced many poorly shaped and distorted elements into the mesh.
2. The size ratio between the largest and smallest element has gotten worse following the refinement.
3. The transition from tiny to very large elements happens too quickly, accordingly this mesh would not be expected to produce accurate results.

Figure 6.34(c) shows an attempt to make a more regular mesh by specifying a constant element edge length throughout the model. While this mesh is clearly much better than the previous two meshes it still has the following problems:

1. The mesh density in the region away from the hole is appropriate but there is no increase in mesh density near the hole.
2. Only four elements have been used around the curve of the hole, while this is better than in case (a), it is still not enough to correctly model its highly curved geometry.
3. There is one poorly shaped element in the centre of the mesh. While it is within acceptable limits it would be preferable to not have it there.

Figure 6.34(d) shows an attempt to refine the mesh in part (c) at the region around the hole using an automatic refinement, similar to part (b). In this case:

1. The automatic refinement has introduced many poorly shaped and distorted elements into the mesh.
2. The size ratio between the largest and smallest element has now become a problem.
3. The transition from small to large elements happens rather quickly and the refined region is quite localised.

Even with these problems this mesh is probably the best so-far, but there is room for much improvement.

Figure 6.34(e) shows a mapped mesh that was generated by splitting up the plate area into five separate areas and then meshing each area individually. In this case a constant element edge length has been used throughout the model:

1. This mesh is far more regular and superior to any of the meshes above.
2. However, there is no increase in mesh density near the hole and
3. The aspect ratio of the elements next to the hole is a little high.
4. The mesh density is probably a little too high in the upper left hand section of the plate, where stress will be relatively constant.

In order to overcome these problems with mesh (e) and arrive at the optimal mesh for this problem, mesh (f) shows an improvement of mesh (e) where biasing and set line divisions have been used to control the distribution of elements. This mesh has all the qualities required for this problem:

- A refined mesh near the hole.
- Smooth transitions between small and large elements.
- A regular, organised and smooth mesh made up from near square shaped elements.

Mapped Vs Free Meshing

A mapped mesh is one where the mesh has been forced into a pattern by specifying the number of nodes on all edges of the surface or volume being meshed. Basically the description of the number of nodes or elements on the edge will determine how the interior of the area or volume will be filled with elements. Mapped meshes will always provide a more uniform and better looking mesh, and hence better results.

In order to use mapped meshing an area must have either three or four sides. Areas with greater than four lines defining its boundary cannot be map meshed. In the case of a volume to use a mapped mesh (with bricks) the volume must be brick shaped (have six sides), wedge or prism shaped (five sides) or pyramid shaped (four sides). If the volume has more that six sides (including internal edges due to holes etc.) then mapped meshing will not be possible. Figure 6.35 shows the difference between mapped and free meshes for areas and volumes. In each of these meshes a standard global element edge was assigned. Note how a finer tetrahedral mesh is required to properly represent the curved edge of the volume.

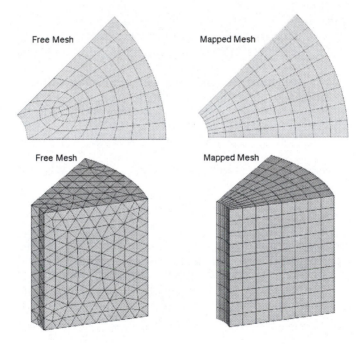

Figure 6.35: Mapped and Free Meshes for Areas and Volumes

6.3.7. Mesh Convergence

An important question often asked by new FE users is "how fine should the FE mesh be in order to obtain accurate results?" There is, unfortunately, no straightforward answer to this question. The method used to determine the optimum number of elements in a FE mesh is called "mesh convergence". The size and number of elements in a finite element model are inversely related: as the number of elements increases then the size of each element must decrease and consequently the accuracy of the model, usually, increases. This is the principle that mesh convergence works on.

Obviously in order to check the accuracy of your mesh you could compare your results with available experimental results or known accurate analytical results and then refine the mesh in regions where the difference between the FE and comparable results is too great. This process however, depends on comparable results being available which is not always the case and is rather crude. Another crude method of checking mesh quality is to run an initial analysis with a uniform medium density mesh then re-run the analysis with a mesh that is twice as dense in critical regions and compare the two sets of results. If there is little difference in the results then the first mesh is probably acceptable.

Having an acceptable mesh density is very important. If the mesh is too coarse then the results may contain serious errors and vital information like regions of high stress can be missed due to poor mesh resolution. If, however, you take a brute force approach and use an overly fine mesh throughout the model, your model will take a lot of time and computational resources to solve and may still be inaccurate as the mesh could be over-converged.

Let's consider an example in order to illustrate the principle of convergence: figure 6.36(a) shows a beam subjected to a bending load at its free end. We are required to build a FE model to determine the bending stress in the beam. Figure 6.36(b) shows the exact solution which shows that stress varies in a quadratic manner along the beam length. If linear 2D beam elements, that assume a linear variation in displacement within the element, are used then the effect of increasing the number of elements is obvious: the FE solutions gradually approach the true stress distribution as shown in figure 6.36 (c-f).

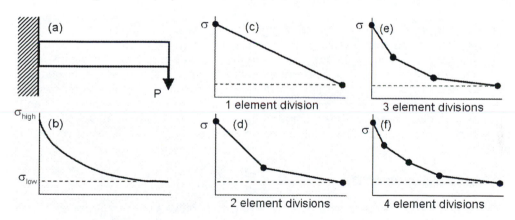

Figure 6.36: Effect of increasing number of elements on accuracy of results.

We can continue this process and keep on increasing the number of elements, but the question is: how do we know where to stop? If we select an important result from our analysis, such as the stress at a particular point in the middle of the beam and then record the result from each model (with increased number of elements) we can combine all these results in a convergence graph, as shown in figure 6.37.

The convergence curve shows how the accuracy of the analysis increases with increasing element numbers. As the number of elements gets very large so the FE predicted stress should approach the exact solution. Note that the advantage of using two elements rather than one is clear as it represents a 50% increase in accuracy, however going from 8 to 12 elements only results in an increase in accuracy of approximately 5%. In this case it is obviously worthwhile using the four extra elements, however, for a 3D solid model where the number of elements are in

10,000's it may not be worthwhile going from 80,000 to 120,000 elements for a 5% increase in accuracy.

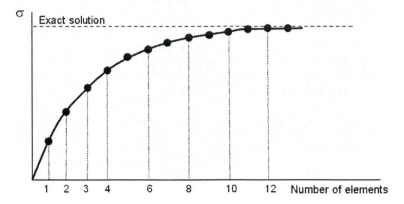

Figure 6.37: As the number of elements increases the predicted stress approaches the exact solution.

The above example is very elementary as we are dealing with equally spaced line elements with one node at either end. A better approach to the above problem would be to place more elements at the built-in end of the beam and gradually make the element size bigger as we move towards the free end of the beam (i.e. mesh biasing). Introducing biasing would complicate the convergence process slightly but it would still follow a similar trend to figure 6.37.

In reality our problems will not be so straight forward as we may wish to converge a number of different result items. For example, we may be interested in both the deflection and stress in the beam, and these may converge at different mesh sizes. Case study C in chapter 10 shows examples of this type of convergence. In general displacement results will converge faster than stress results and it could be possible that by the time that stress is converged, displacement has been over-converged and is starting to diverge again! This illustrates the point that in many analyses there will be competing factors and the analyst has to make an informed decision on mesh quality.

Philosophy of Convergence

Let's take a look at exactly what we are trying to achieve with convergence tests. When we build a FE model we make assumptions regarding geometry, material properties, loading and boundary conditions. There will be uncertainties in each of the assumptions that we make. For example we assume that the geometry is as shown on the engineering drawings, but in practice due to manufacturing errors the geometry will never approach this ideal. We make an assumption that the material is linear elastic when in fact no material really behaves this way. We assume that the loads or boundary conditions are constant when in nature this is never the case. Then when we mesh the FE model we introduce a further uncertainty: uncertainties about the quality of the mesh. Thus the total uncertainty of the FE model can be represented by:

$$U_{total} = U_{geom} + U_{mat\,model} + U_{load + BC} + U_{mesh}$$ (6.01)

If we have made informed choices in regard to selecting the most appropriate material model, boundary conditions and loads then we will have eliminated as much uncertainty as possible from these areas. This means that the only contributor to the total uncertainty in equation 6.01, that we can now significantly influence, is the uncertainties due to the mesh.

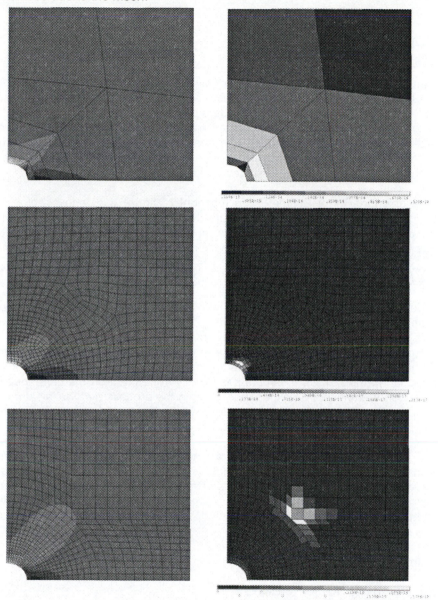

Figure 6.38: Plots of un-averaged von-Mises stress (left) and energy error estimates (right) for different meshes. Paler regions are regions of greatest error where further refinement could improve results.

As mentioned above, attempting to converge the model to just one result (e.g. a particular stress or displacement) is not a good idea as the global behaviour of the model should always be considered at all times. It is good practice to compare your model with other converged models, where these are available.

It is important to understand that convergence is not just related to the mesh. The applied loads and boundary conditions will have a considerable effect on the optimal mesh for each problem. For example a mesh converged to capture bending of a beam will not necessarily be the converged mesh to capture tension or compression of the beam. If you change the loads on a FE model then you must re-converge the mesh for this new loading scenario.

Error Estimates and Convergence

Most FE post-processors offer a facility to calculate element error estimates and plot these values on the FE mesh. These error estimates measure the relative difference between results (e.g. displacement or stress) across an element edge or at a node. The most commonly used error estimate is a structural energy error estimate. This energy error is similar in concept to strain energy. The structural energy error is essentially a measure of the discontinuity of the stress field from element to element. Figure 6.38 shows plots of structural energy error estimates for various meshes of the square plate with a central hole subjected to uniform tension problem.

The plots of energy error estimates on the right of figure 6.38 show regions with increasing error in increasingly paler colours. It can be seen that as a more refined mesh is used the error estimate becomes increasingly smaller and the number of elements with an error reduces. We will discuss error estimation again in chapter 9.

Adaptive Meshing

An obvious extension of being able to determine mesh error estimations is to use this information to automatically control the quality of the mesh. This process is known as "adaptive meshing". Adaptive meshing is a process whereby an automatic algorithm is used to solve the problem, check the error estimates, re-mesh the problem accordingly, solve again, re-check error estimates etc. Theoretically, this process should result in a converged mesh but it is usually dependant on the user specifying an initial mesh. If the user has experience of using the process and the initial mesh is well specified then the process usually works well, however, for new users adaptive meshing often causes more confusion than it is worth!

Adaptive meshing is a very useful tool for non-linear analysis such as in metal forming analysis where there are large amounts of plastic deformation. Elements can become excessively distorted due to this deformation and give spurious results. Adaptive meshing can be used during the simulation to automatically increase the mesh density in regions that are experiencing excessive strains and thus allow for a more robust and reliable solution.

6.4. Summary of Chapter 6

After completing this chapter you should:
- be able to simplify your model by using a beam assumption, planar assumption, thin shell assumption, or any combination of these, where appropriate.
- be able to recognise and simplify your model using the inherent symmetry in the problem.
- understand the concepts of model defeaturing and clean-up and be able to appropriately apply these concepts.
- understand different coordinate systems and how models can be built and/or analysed in these systems.
- understand the concepts of sub-modelling and sub-structuring and be able to appropriately apply them to your FE model.

- understand the various factors that contribute to the accuracy of a finite element mesh and understand the concept of mesh convergence and be able to carry out a mesh convergence test on your FE model.

7

Boundary Conditions and Loading

7.1. Introduction

In section 3.4 we outlined the general theory of the FEM using the minimum potential method. You will recall that we ended up with a statement of the global problem for a static structural analysis in equation 3.27: $[K]\{U\} = \{F\}$. We can recall from equation 3.28 that the global stiffness matrix is a function of the strain-displacement matrix [B] and the material property matrix for each element in the finite element mesh. The strain-displacement matrix in turn, is a function of the shape function matrix [S] for each element. Therefore, by selecting an element type (chapter 4), assigning a material model (chapter 5) and specifying a mesh layout and hence element connectivity (chapter 6) we have fully defined the matrix [K] in the above equation.

This chapter is concerned with the next step in the process which is to populate the displacement vector {U} with problem boundary conditions and the global load vector {F} with the problem loading conditions. If we expand equation 3.27 to represent a problem with n degrees of freedom then we have:

$$
\begin{bmatrix}
k_{11} & k_{12} & \cdots & k_{1N} \\
k_{21} & k_{22} & \cdots & k_{2N} \\
\vdots & \vdots & \ddots & \vdots \\
k_{N1} & k_{N2} & \cdots & k_{NN}
\end{bmatrix}
\begin{Bmatrix}
U_1 \\ U_2 \\ \vdots \\ U_N
\end{Bmatrix}
=
\begin{Bmatrix}
F_1 \\ F_2 \\ \vdots \\ F_N
\end{Bmatrix}
\tag{7.01}
$$

In order for the problem to be solved at least one of the rows in [U] and [F] must be populated. This means that for any problem there must be a boundary condition upon at least one point/node in the model and a load at another point/node. There is simply no way around this as the mathematics of the method depend on it. Where boundary conditions do not exist naturally then they must be invented, for example by using symmetric boundary conditions. So, by placing boundary conditions on your model you are populating various rows of the {U} vector with zeros (normally) and by placing loads on your model you are populating various rows of {F} with the magnitude of these loads.

If we return to chapter three again, we can recall that the global force vector is a combination of body forces, surface loads and concentrated loads:

$$
\{F\} = \{F\}^B + \{F\}^S + \{F\}^C
\tag{7.02}
$$

Where the individual components for each element, m, and node, i, are given by:

$$
\{F\}^B = \sum_m \int_{V_m} [S]_{(m)}^T \{f\}_{(m)}^B dV_{(m)} \qquad \{F\}^S = \sum_m \int_{S_m} [S^S]_{(m)}^T \{f\}_{(m)}^S dS_{(m)} \qquad \{F\}^C = \sum_i \{F\}^i
\tag{7.03}
$$

Notice from equation 7.03 that body forces $\{f\}^B$ and surface loads $\{f\}^S$ are applied to elements and are thus dependant on the element shape function [S] while point loads $\{F\}^i$ are applied directly to nodes and are thus unaffected by the element equations.

We will begin this chapter by discussing the three types of loads given by equation 7.03. We will then discuss how these loads are effected when a dynamic analysis is required. Following this we will discuss practical issues in relation to applying boundary conditions and loads to your model with relevant examples.

7.2. Types of Loads

The loads applied to any stress analysis problem can be broken down into three distinct categories that correspond to the three equations in 7.03: body loads, surface loads and point loads. It is important to realise that in all three cases the eventual result is a force being specified at particular nodes and thus populating the global force vector {F}.

Body Loads

Body loads as the name suggests act on the whole body under investigation. The most obvious example of a body force is the weight of the body itself. The weight of the body acts as a force per unit volume and thus is applied accordingly to every element that makes up the finite element mesh of the body. Other common body forces include: centrifugal and centripetal forces, and the acceleration felt when a body changes velocity (e.g. car braking suddenly). Basically any acceleration that a body experiences will cause a body force to be imparted on the body.

Body loads are applied directly to each element as a force per unit volume based on the percentage of the total volume that the particular element makes up. The body load is then distributed to the nodes that make up the element via the element shape function as shown in equation 7.03.

Surface Loads

Surface loads act on a surface or an edge of the body under investigation. Typical examples include pressure loads on the surface of a solid (e.g. an internal pressure on a pressure vessel) or a distributed load on a beam or the edge of a 2D plane stress model.

Surface loads are applied directly to the face of the selected elements as a force per unit length. The surface load is then distributed to the nodes along that face via the element surface shape function $[S^S]$ which is derived from the elements general shape function $[S]$, as shown in equation 7.03.

Point Loads

Point loads are concentrated loads that act at or through a particular point on the body. Typically these are forces or moments applied to a specific node in the finite element mesh. Since the force is applied directly to a node (or group of nodes) it can be directly placed in the global force vector {F}.

7.3. Dynamic Loads

As discussed in section 3.4.3, the equation [K]{U}={F} only applies to static analyses. In a dynamic analysis the applied forces will change over time and thus this equation is only valid for a specific point in time. In order to consider the inertia forces that result due to loads being applied over time, d'Alembert's principle is used to include the inertia forces as part of the body force equation. Similarly, as some energy is always dissipated during vibration in a dynamic analysis, velocity dependant damping forces are introduced as additional contributions to the body forces. Thus, for a dynamic analysis, the body load equation from 7.03 becomes:

$$\{F\}^B = \sum_m \int_{V(m)} \left[\{S\}_{(m)}^T \left[\{f\}_{(m)}^B - \rho_{(m)} [S]_{(m)} \{\ddot{U}\} - c_{(m)} [S]_{(m)} \{\dot{U}\} \right] dV_{(m)} \right. \qquad (7.04)$$

where $\{f\}_{(m)}^B$ is the body force applied to element m, $\{\ddot{U}\}$ is the nodal point acceleration vector, $\rho_{(m)}$ is the mass density of element m, $\{\dot{U}\}$ is the nodal point velocity vector and $c_{(m)}$ is the damping property parameter of element m.

In this case the global problem (i.e. equilibrium) equations become:

$$[M]\{\ddot{U}\} + [C]\{\dot{U}\} + [K]\{U\} = \{F\} \qquad (7.05)$$

where {F} and {U} are now time dependant, [M] is the global mass matrix and [C] is the global damping matrix. [M] and [C] are given by:

$$[M] = \sum_m \int_{V(m)} \rho_{(m)} [S]_{(m)}^T [S]_{(m)} dV_{(m)} \quad \text{and} \quad [C] = \sum_m \int_{V(m)} c_{(m)} [S]_{(m)}^T dV_{(m)} \qquad (7.06)$$

So the equilibrium equation for a dynamic problem still has the same requirements in order to obtain a solution. [M], [C] and [K] will be known from the element type selection, material model specification and mesh. Boundary conditions will result in certain rows of $\{\ddot{U}\}$, $\{\dot{U}\}$ and $\{U\}$ being set to zero or other specific values. Application of time-varying loads will result in certain rows of {F} being prescribed and thus the problem can be solved.

7.4. Practical Overview of Loads and Boundary Conditions

Many novice analysts spend much time building and meshing a FE model only to have the analysis rendered useless by careless application of B.C.'s (boundary conditions) and loads. Given today's modern and easy to use FE software it is very easy to create a flashy model of a particular problem apply some arbitrary loads to it and produce some kind of an answer. The problem is, if your loads and B.C.'s are poorly chosen then, as a consequence, your results will be meaningless: "garbage in equals garbage out". FE post-processors make the results of even a totally incorrect and dangerous FEA look colourful, convincing and nice to look at. The key point here is that no matter how good the results look, if you haven't modelled the problem loads and B.C.'s correctly then they are meaningless.

Consider, for example the turbine blade shown in figure 7.01. At first glance the results look fine, lot's of nice stress contours, values for high stress and low stress etc. However, let's ask some reasonable questions: What loads were applied to the blade? Was the loading static or dynamic? How was the blade mounting considered? Was a rotational velocity applied to model the in-service behaviour of the blade? Was the centre of rotation for the applied rotational velocity placed at the central axis of the turbine? Was turbine start-up and wind-down modelled by placing appropriate acceleration loads on the blade?

These are all reasonable questions to ask of any turbine blade analysis. In fact, the plot of von-Mises stress shown in figure 7.01 resulted from a static analysis where the bottom of the blade was constrained in all directions and a point load was applied to a node near the blade tip. This means that the result shown in figure 7.01 is meaningless as it doesn't tell us anything about the in-service behaviour of the blade!

Figure 7.01: FEA of a turbine blade; plot of von-Mises stress

You could place this result in front of a lot of engineers or designers and probably do a good job of convincing them that it was a meaningful result before someone started asking questions like those listed above. This is the danger of FEA! People see the nice colourful pictures and tend to forget their previous engineering training!

The point of the above is that selection of proper loads and B.C.'s is vital in order to generate meaningful results that can be used for design and problem solving. In the rest of this chapter we will discuss methods to help us make the proper choices.

7.5. Boundary Conditions

7.5.1 Overview

Boundary conditions, as the name implies, are required to represent everything in the problem environment that is not explicitly defined in the model. Take, for example, a cantilevered beam: it is usual not to model the wall/structure onto which the beam is attached but to constrain the movement of the beam nodes at that location to *represent* the wall.

Boundary conditions should never restrict deformations that would be allowed by the problem environment. They should also be specified so that they do restrict any deformations that are not allowed by the problem environment. For example, if your problem is a structure placed on a very slippery surface subjected to downward pressure, then it would be appropriate to constrain movements in the vertical direction at the base of the structure (to model the reaction force from the surface), but not appropriate to constrain horizontal directions as the structure is allowed to slide in reality.

Let's consider an everyday example in order to illustrate some of the above points. The table shown in figure 7.02 supports a central load due to an object being placed on it. We are required to build a FE model of this scenario and hence apply appropriate boundary conditions. As a first guess let's assume that the load is uniformly distributed over the table surface and that the legs of the table do not need to be modelled as their sole purpose is to provide a reaction from the floor. Thus we will provide constraints at the points where the legs join the table top. So, in effect, our constraints are modelling the legs.

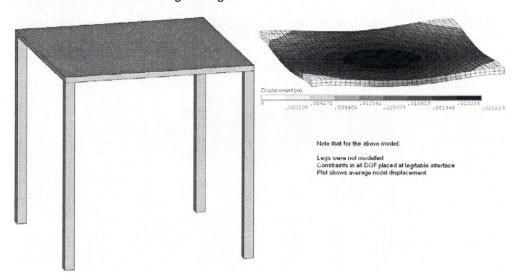

Note that for the above model:

Legs were not modelled
Constraints in all DOF placed at leg/table interface
Plot shows average nodal displacement

Figure 7.02: FEA of table top due to uniform loading with many assumptions about constraints (rigid constraints)

We have made what seem like simple and reasonable assumptions about the boundary conditions above, so let's examine the implications of these assumptions:

1. Assuming that the legs do not need to be modelled means we are assuming that they are rigid and hence they cannot compress or bend.
2. By neglecting the legs we have assumed that any possible sliding of the legs on the floor due to side load components resulting from bending of the table top are also neglected.
3. The imposed boundary conditions at the leg/table interface will also force the table bottom to remain horizontal at the region of the imposed constraint.

Clearly there are some serious problems with these assumptions and they do not come close to modelling what happens in reality.

In order to investigate how poor these assumptions were, a second model was solved in which the legs were included and constraints were placed at the leg/floor interface: so in effect the constraints in this case were modelling the floor. The results from this model are shown in figure 7.03. It can be seen from the figure that in this case that deformation of the legs is being considered and hence we can reasonably assume that the results from this model would be more accurate than the previous model. Note that in this case the maximum displacement of the table top is 0.05m whereas in the model shown in figure 7.02 the maximum displacement was 0.028m. So by including the table legs in the analysis the maximum displacement has almost doubled! This tells us that the assumptions made regarding constraints for the first model resulted in a model that was far too rigid because there was too much constraint in the model. So, to clarify, making poor assumptions about the constraints in a model will effect its rigidity which will in turn affect the predicted deformation which in turn will effect any stress results. This illustrates the importance of choosing correct boundary conditions.

This raises the question, "How do I choose the correct boundary conditions for the problem?" Essentially you have to make a decision as to where in the environment you stop your modelling and place boundary conditions so as to represent the environment beyond the model. If we reconsider the above example: making a decision to stop modelling at the top of the legs was a poor decision. Making a decision to model the legs and stop at the floor was better, but it is possible that going further and modelling the floor and placing contact elements between the legs and the floor could give even better results!

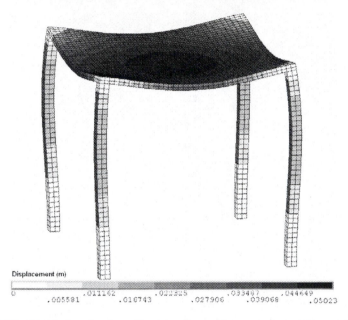

Figure 7.03: FEA of table top due to uniform loading with legs included (less rigid constraints)

Experience and common sense tell us, however, that although we would see a further increase in deformation if we modelled the floor with a frictional contact between the legs and floor, it would probably be a relatively small increase. So the best method we have for judging the accuracy of the model is by measuring how much it deforms and comparing this deformation to reality. By comparing the results from the first model (figure 7.02) to reality it is immediately apparent that they were not satisfactory. It is good practice to try several different "levels" of constraint and to compare the results, in a similar manner to mesh convergence, in order to determine the optimal constraint strategy for the problem under investigation.

We will now discuss the various types of boundary conditions available.

7.5.2. Fixed Supports

The most basic type of boundary condition available is a fixed support boundary condition. This boundary condition constrains regions of the model in all directions (i.e. all DOF). This type of boundary condition can generally be applied to individual nodes, groups of nodes and regions of geometry (such as points, lines, areas etc.).

Fixed supports should be used with care as they can induce coupled strain effects which may not actually be present in the structure in practice. See section 7.5.9 for details of coupled strain effects. Apart from cantilevered structures, the use of fixed supports in FE models is not as common as they may first appear. Remember that you are specifying that there is no deformation in *all* directions. This means that the structure is fixed rigidly in space at the location of the constraint and that thickening or thinning of the structure is also not allowed at this location. In practice there are very few situations where this is the case.

Placing a fixed support on a single node or point or on a line will create an artificial stress concentration at that point, as shown in figure 7.11(a). If you have to use fixed points or lines then you should ignore stress results in their vicinity. Remember that it is impossible to fix a point or a line in reality! A fixed support on a surface or an area shouldn't be a problem unless the surface is very small, in which case the same advice applies.

Figure 7.04 shows several examples of fixed supports used in FE models.

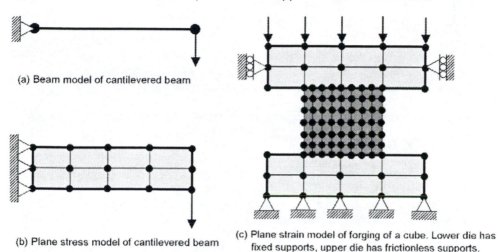

(a) Beam model of cantilevered beam

(b) Plane stress model of cantilevered beam

(c) Plane strain model of forging of a cube. Lower die has fixed supports, upper die has frictionless supports.

Figure 7.04: Examples of fixed supports in various FE models

7.5.3. Frictionless Supports

Frictionless supports differ from fixed supports in that they result in only certain DOF being constrained. For example a frictionless support may allow deformation in the x and y directions but will not allow movement/sliding in the z-direction. As the name implies, these constraints are generally used to model the effect of an adjoining structure that prevent the model from moving across the boundary between the two structures.

Frictionless supports are also used to allow a model (or part of a model) deform or move in specific directions. Consider the upper die in figure 7.04(c); we want the die to move down due to the applied load and hence to cause deformation of the cube. If we allowed the upper die to be free from constraint, then any small round-off errors during the solution would cause the upper die to slip off the cube and fly off into space and hence rigid body motion would ensue (see section 7.5.10). On the other hand, if we constrained the upper die in all directions (i.e. a fixed support) then it would not move at all. The frictionless support allows the upper die to move in the vertical direction but doesn't allow any horizontal movement of the die.

Frictionless supports are also used extensively to model symmetry as will be discussed in the next section. Other examples of frictionless supports are shown in Figure 7.05

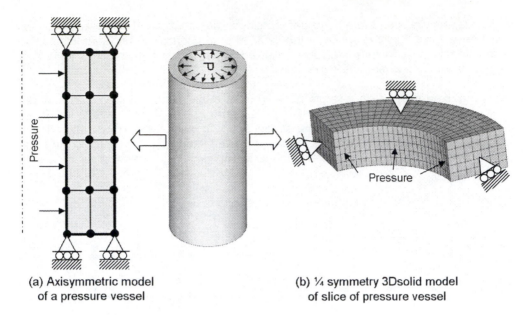

(a) Axisymmetric model
of a pressure vessel

(b) ¼ symmetry 3Dsolid model
of slice of pressure vessel

Figure 7.05: Examples of frictionless supports in FE models of a thick walled pressure vessel.

7.5.4. Symmetry Constraints

As already discussed in chapters 3 and 6, taking advantage of any symmetry that a problem exhibits is an excellent way of simplifying a model. Recall that in order to use symmetry to simplify a problem *both* the loads and the geometry must be symmetrical about the plane of symmetry.

In order to use symmetry you have to use constraints to account for the parts of the model that you are not modelling due to symmetry. This is generally achieved by providing frictionless supports on symmetry cut planes that constrain motion normal to the cut plane. Many modern FE software packages have a specialised symmetry boundary condition command that automatically determines the normal to the surface or line that defines the cut plane and thus places the appropriate frictionless support boundary condition. For examples see figures 6.10, 6.11, the right hand side of figure 7.05 and the left hand side of figure 7.11.

7.5.5. Multipoint Constraints and Coupled DOF

We have already mentioned that placing a boundary condition at a single point will cause an artificial stress concentration to be generated at the point of application. We will see later that the same is true for a force applied to a single node.

In reality, it is impossible to apply a point load or constraint as even the sharpest knife edge or the finest needle have a finite radius. In a finite element model, however, a point or a node has an infinitely small area. As we know stress is load/area so the local stress around the load or boundary condition is effectively infinity. This phenomenon is known as a "singularity".

Figure 7.06 illustrates the consequences of applying either a boundary condition or a load to a single node or group of nodes. Figure 7.06(a) shows a model of a flat plate where two nodes have been constrained on the upper surface and a pressure load has been applied to the line on the lower surface. It can easily be seen that two singularities have resulted, as expected. Note that the singularities dissipate quickly over the adjoining elements and thus the global behaviour of the structure is not affected. Similarly figure 7.06(b) shows a model where the upper surface is held via frictionless constraints and a force is applied to a node on the lower surface. A singularity at the load application point has resulted.

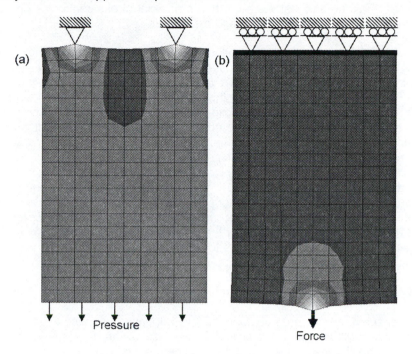

Figure 7.06: Applying boundary conditions to single nodes (a) or loads to single nodes (b) results in artificial stress concentrations.

Coupled degrees of freedom or multipoint constraints (MPC) are a method of avoiding singularities by coupling the DOF of various nodes within the model. In general a MPC or coupled set must have a master node and one or more slave nodes that "follow" the master node. Essentially the master node forces the slave nodes to take on the same magnitude of deformation in the coupled DOF. Coupled DOF are typically used to maintain symmetry on partial models, to model pin joints, hinges, universal joints and slide joints between two nodes and for forcing portions of the model to behave as a rigid body.

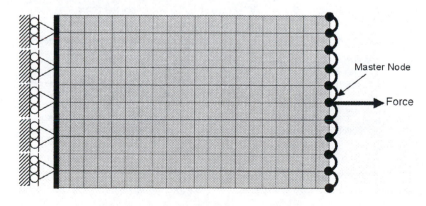

Figure 7.07: Using coupled DOF to apply a force to a surface

Figure 7.07 shows an example of how coupled nodes can be used to apply a distributed load to the edge of a plate. The nodes on the right hand edge of the plate have had their DOF in the horizontal direction coupled to the central master node. A point force is then applied to the master node and since the DOF of all the slave nodes are all coupled to this node, they will move to the right in parallel with the master node. This means that there will be no artificial stress concentration induced due to the point load. Compare this to figure 7.06(b)!

MPC's have many important applications in finite element modelling, such as modelling various types of joints and fasteners, distributing loads, and helping to tie dissimilar element types together. Let's revisit the shell to solid joint problem we examined in figure 6.31. Remember that the result of this model was that the interface acted as a hinge as the solid elements could not deal with the local rotations on the attached shells. MPC's offer a neat way of eliminating this problem: by coupling translational DOF of each shell node at the interface to two nodes on the solid elements then this unwanted rotation can be effectively constrained.

A more advanced form of MPC's and coupled DOF are constraint equations. By using equations a more complex relationship between the master and slave nodes can be defined. Simple coupling makes the slave node move in the coupled DOF a distance exactly equal to that moved by the master node. Using constraint equations it is possible to instruct the slave nodes to move a function of the distance travelled by the master (e.g. slave moves three times the amount the master moves).

7.5.6. Constraints as loads/ Forced Displacement

In certain cases it can be more useful to specify a location in space where a node or groups of nodes will end up at the end of the analysis, rather than specifying a force or pressure. This is known as a forced displacement constraint.

A forced displacement will make a node move by a specified distance and will then hold the node at that location. The forced displacement will always overcome and material resistance (i.e. stiffness) or other parameters (e.g. loads) and will always move the node to the specified location regardless of any other parameters in the model. Because of this it is important that reaction forces at forced displacement nodes are checked to ensure that the forces required for the displacement are realistic.

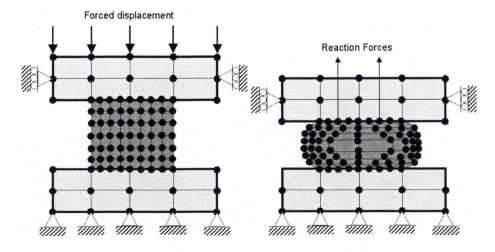

Figure 7.08: Example of forced displacement constraint used in forging of a rectangular block.

An example of the forced displacement technique is in the forging of a rectangular block model shown in figure 7.04(c). Suppose we wished to know the force required to reduce the height of the block by 30%. We can place an appropriate forced displacement constraint on the upper die and then check the reaction forces afterwards to determine if the force is possible with our available experimental apparatus.

7.5.7. Constraints and Coordinate systems

All loads and boundary conditions applied to an element are interpreted in that elements nodal coordinate system. By default these coordinate systems are parallel to the global coordinate system, however, if required the nodal coordinate systems can be "rotated" to another coordinate system in order to apply constraints. This is very useful to applying supports that are inclined at an angle to the global coordinate system (figure 7.09a) or for applying radial constraints via a cylindrical coordinate system (figure 7.09b).

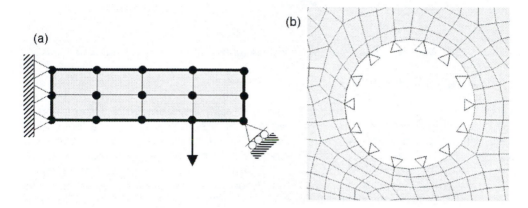

Figure 7.09: Rotated nodal coordinate system used to apply: (a) inclined frictionless support and (b) radial frictionless support to model rigid press-fitted pin

7.5.8. Non-linear and Time Varying Constraints

A non-linear constraint implies a constraint that changes in some manner over the course of the simulation. The most obvious example of this is contact. A model may be relatively unconstrained at the beginning of the simulation but as it deforms it may come into contact with other bodies and thus be more constrained in its further deformation. This applies to almost all metal forming processes, but let's consider a simple example for now.

Consider the beam subjected to bending in figure 7.10a. At the start of the simulation the beam is unconstrained at the free end, however after bending deflection reaches a certain threshold the beam will come into contact with the support at A, thus further deflection of the beam will be more constrained than the deflection up to this point. Any form of contact in your model can be considered a non-linear constraint.

Figure 7.10b shows an alternative type of non-linear contact: time varying constraints. All of the constraint types mentioned above can be set up to "switch on" and "switch off" at various points during the simulation. In this elementary example we have a plate which is initially supported with fixed constraints on both edges. At a specified time the right hand fixed constraint is removed, turning the problem into a cantilever problem.

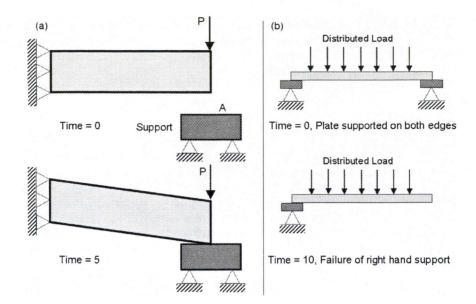

Figure 7.10: Non-linear boundary conditions: (a) contact and (b) time varying constraints.

7.5.9. Coupled Strain Effects

Recall from the definition of Poisson's ratio in chapter 2, that material deformation in one direction is dependant on deformation in other directions. Let's take a look at the plate with a hole problem again to investigate how boundary conditions influence coupled strain effects. Figure 7.11(a) shows a ¼ model of the plate with a hole. In order to properly model the symmetry of the structure and take account of the parts of the plates not modelled, the x DOF has been constrained on the left hand edge and the y DOF has been constrained on the bottom edge of the ¼ plate. The resultant stress distribution is what is expected for the plate with hole problem.

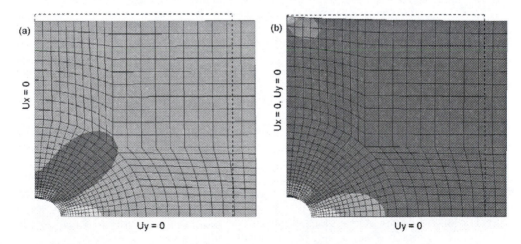

Figure 7.11: ¼ model of plate with a hole with different boundary conditions

In contrast figure 7.11(b) shows the same model but in this case both the x and y directions have been constrained on the left hand edge. Notice how, in this case, an artificial stress concentration has been introduced in the top left hand corner. It is also clear that the deformation of the plate in the x-direction has been reduced. It is important to realise that not only have we introduced coupled strain effects but we have modelled a totally different problem! Figure 7.11(a) models a plate with a hole in simple tension, figure 7.11(b) approaches a scenario that is almost like a form of biaxial loading!

7.5.10. Over-constrained and Under-constrained models

A model that is made artificially stiff due to poorly applied boundary conditions is referred to as being over-constrained. The example of the table shown in figures 7.02 and 7.03 illustrates this point. Over-constraining can be caused by: using excessive constraints (as in the case of the table), using redundant constraints or by introducing unwanted coupled strain effects into the model.

Redundant constraints are essentially constraints that are not needed due to a previous constraint not allowing that motion anyway. Take for example a shell element that has been constrained such that its nodes cannot move in the local z-direction. Further constraining rotations about the x or y axis would be redundant as rotation is not allowed by the first constraint. The problem with this is that, even though the rotational constraints have no effect on the movement of the element, they will likely introduce extra local stiffness into the model.

Under-constraining is generally caused by a model that has too few boundary conditions to prevent rigid body motion taking place. Imagine a 3D model of a cube that has been constrained in the x and y directions and then has a force applied in the z-direction. The cube will simply slide along the xz and yz planes and keep on moving because there is nothing to stop it. This is rigid body motion.

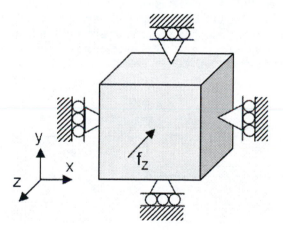

Figure 7.12: Rigid body motion results from under-constrained models. In this case the cube will move to infinity in the negative z-direction without any deformation of the cube.

7.6. Loads

7.6.1. Overview of Loads

Loads in FEA are used to model anything which causes the structure to deform or to displace. Remember that the output of a structural FEA is always a list of displacements of the nodes and that other quantities, such as stress, are calculated from this list of nodal displacements. Applied loads result in the right hand side of equation 7.01 (for a static analysis) or equation 7.05 (for a dynamic analysis) being fully specified. Either there is a finite value for the force at a node or it is zero.

Regardless of what type of load is applied, in a structural analysis, it always results in various forces being prescribed to particular nodes.

Structural loads can take the form of forces, moments, pressures, velocities, accelerations or periodic excitations. In most cases these will correspond directly to the load as it is applied in reality. We will discuss each of these in the following sections.

Using the correct units is vitally important when applying loads. If incorrect units are used at this stage, then obviously the results from the FE model will be worthless, so it is always good practice to re-check your dimensions and units before applying loads. See section 2.2 for more information on units and consistent unit systems.

7.6.2. Forces

Applied forces are the most basic type of load available in FEA. In general, forces can only be applied to points or nodes. As discussed in section 7.5.5, applying a force to a single node or point in a 2D or 3D model will result in a singularity and hence an artificial stress concentration. Point loads can generally be applied directly to line elements such as beams or trusses without any problems. If a force must be applied to a single point in a 2D or 3D model then either the stresses near the load application point must be disregarded or a number of nodes should be coupled using MPC's in order to avoid the stress concentration. Remember that in practice it is impossible to apply a force at a finite point!

Applied forces can be specified as a single value or as a graph or table. For a simple small deflection linear static analysis it is usual to input a single value for an applied force as it is assumed that the load is applied very slowly so as not to introduce dynamic effects. For large deflection static analysis and transient (i.e. time varying) analyses, it is more usual to specify a time versus force curve which defines how the force is applied over the course of the simulation.

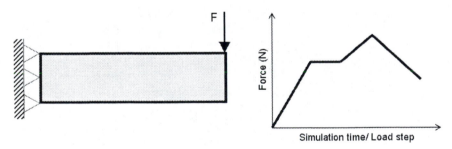

Figure 7.13: Point forces can also be defined as time varying

Axisymmetric elements require special treatment as in an axisymmetric model input values for force are based on the full 360° of the model. As an example, if a cylindrical shell of radius r has a distributed load of P N/m applied to its top edge, this would be applied as a force of $2\pi rP$ N to the top surface of the axisymmetric model, as shown in figure 7.14.

7.6.3. Moments

Moments can be directly applied to nodes belonging to elements that have rotational DOF such as beams and shells in much the same manner that point forces are applied. Moments cannot be directly applied to elements that do not have rotational

DOF such as trusses, plane stress, plane strain, axisymmetric and 3D solid elements.

With some creativity moments can be applied to solid element faces by overlaying the face with specialised surface effect elements or shell elements, or by using MPC's. Either way, this is quite messy and should be avoided if possible.

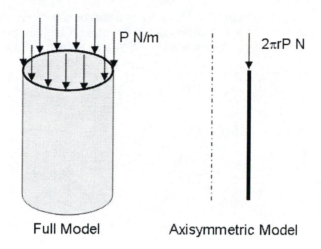

P N/m   2πrP N

Full Model   Axisymmetric Model

Figure 7.14: Forces applied to axisymmetric models are based on the full 360°

7.6.4. Pressures

Pressure loads can be applied to the edge of 2D elements or to a surface of a 3D element. Pressures are usually interpreted as being normal to the surface that they are applied. Applying pressure loads to 2D elements or 3Dsolids is relatively straightforward, you just choose the edge/surface you want the load to act on and then specify its magnitude. This is, unless, you require to specify a pressure load that is not normal to the surface. In such a case, a surface effect element is used to transmit the pressure load to the desired surface. See section 4.17 for more information on surface effect elements.

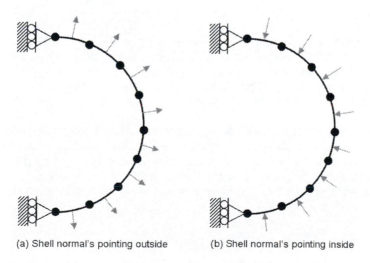

(a) Shell normal's pointing outside   (b) Shell normal's pointing inside

Figure 7.15: Section through a shell model of a spherical pressure vessel. An applied pressure load would model positive internal pressure in (a) and a vacuum in (b)

When applying pressure loads to shells it is important to determine in which direction the shell normal is facing. By definition, shell elements have a defined inside and outside surface and the normal points from the inside to the outside surface. If the shell normal's are orientated in the wrong direction then an expected positive pressure could be interpreted as a vacuum as shown in figure 7.15. It is normally a simple matter to flip the shell element normal's to point in the other direction.

Pressure loads may also be applied as gradients whereby the pressure at one end of the surface is a function of the pressure load at the other. This can be very useful for simulating the presence of load transfer pins in holes as shown in figure 7.16.

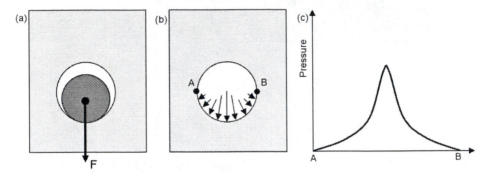

Figure 7.16: Pressure gradient used to simulate load transfer from pin to hole.

In order to apply a pressure load uniformly to a surface of a volume it is usual to ensure that a mapped mesh has been used to mesh the volume. This means that the brick elements will all have the same face pointing in the same direction. Remember that a brick has six sides: automatic meshing tools do their best to ensure that the bricks are all aligned in the same way, e.g. all bricks have surface 2 facing in the z-direction. This is much more easily achieved, and easier to check with a mapped mesh, than with a tetrahedral mesh. Due to the fact that tetrahedrals have to be orientated in many different directions to make up a volume it is highly unlikely that all element sides making up a surface of the volume will be of the same numbered edge.

7.6.5. Velocities

In a similar manner to forces, velocities can be specified for nodes or groups of nodes in a FE model. Linear velocities (i.e. in the x, y or z-directions) are usually only used for non-linear analysis such as impact analysis, where the velocity of the projectile may be specified. Certain forms of metal forming analysis are often defined by specifying the punch velocity in the finite element model.

Structural analysis of rotating structures (such as turbine blades, propellers, engine parts etc.) requires specification of rotational velocities. Rotational velocity can be specified in two ways: a global velocity can be specified for the entire model about the global Cartesian origin or a specified velocity can be assigned to groups of nodes. In the second case it is generally possible to define a new origin for the centre of rotation, if this is required.

7.6.6. Accelerations

The most common acceleration load used in FE models is gravity. In order to take account of a structure's self weight a global specification of gravitational acceleration is generally used. Linear acceleration can be specified to act in any of the global Cartesian directions. In order for self weight effects to be experienced by the model equation 7.06 requires that a value for density must be specified in the material model definition.

In a similar manner to rotational velocities, rotational accelerations can also be applied to the model either globally or locally to groups of nodes and the origin of the rotation, if different from the global origin, can also usually be specified.

7.6.7. Non-Linear Loads

The above discussion about loads and load types has, for the most part, assumed that the load magnitude, orientation and distribution will remain unchanged throughout the simulation. Three types of non-linearity effect loads: large deformations, contact and follower forces.

If during the simulation, the surface or edge on which a load is applied is caused to deform to a great extent then a large deformation analysis is required. This is because of the fact that an update of the load orientation is required as the surface will probably have significantly changed its orientation due to the large deformation.

We have already discussed contact in the boundary conditions section of this chapter but what is important to understand in terms of loads is that contact reactions do not develop as the load is applied but only when the contacting bodies touch and deform against each other.

Follower forces are loads that are locally defined in relation to nodes or elements and not a global coordinate system. When a follower force is used, as the structure deforms the follower force adjusts its orientation so that it is always acting in the same direction relative to the geometry. Figure 7.17 shows an example of a follower force and compares with a linear force.

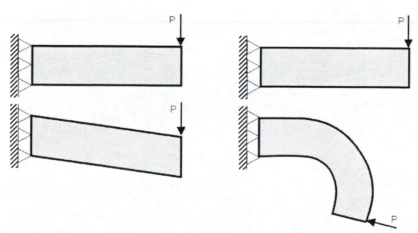

Figure 7.17: Comparison of linear (left) and follower (right) loads on a cantilevered beam.

7.6.8. Multiple Loads and Load Steps

If your structure has multiple loads then you have two choices available for dealing with the multiple loads. Obviously, you can just perform a single analysis (i.e. single load step) in which all loads are considered. Alternatively, you can use multiple load steps in which each loading condition is applied separately and the results are combined later to obtain the solution to the full loading scenario.

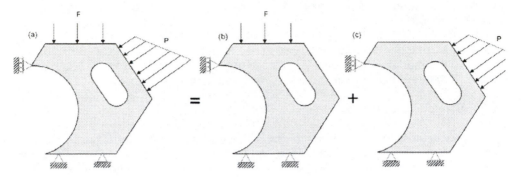

Figure 7.18: Multiple loading scenarios (a) can be solved in separate load steps (b) and (c).

A load step is basically one set of loading conditions for which you obtain a solution. By using multiple load steps it is possible to isolate the structures response to each separate loading condition and combine these responses in any desired combination during post-processing. Being able to combine different load steps during post-processing allows you to study many "what if" scenarios.

7.7. Comparison of Loads and Boundary Conditions

In order to illustrate many of the concepts discussed in the previous sections we will consider the example of a steel tensile test specimen as shown in figure 7.19 below.

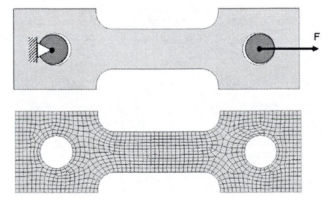

Figure 7.19: Plane stress finite element model of a tensile test specimen

The upper part of figure 7.19 shows an overview of the actual tensile test. The specimen is loaded via two pins. The pin on the left holds the specimen in place while the pin on the right moves towards the right and hence exerts a force on the right hand hole. We will now consider a number of different loading and boundary conditions that may be used to model the behaviour of the tensile test specimen during the linear phase of the tensile test.

Figure 7.20 shows the first attempt at modelling the problem. In this case, it was decided to build as representative a model as possible in order to use it as a benchmark to compare all subsequent models. Accordingly, the pins were included in this model and 2D surface to surface contact elements were used between the pins and the specimen at the appropriate points. It was assumed that the pins had a very small clearance fit with the specimen at each end and that the contact was frictionless. A fixed support constraint was applied to the left hand pin which fixed it in all DOF. The DOF of the nodes in the right hand pin were coupled and the master node was subsequently constrained in the vertical direction and had a point force of 100 N applied in the horizontal direction. This meant that the right hand pin behaved as a rigid body (i.e. didn't experience any deformation) which was acceptable given that we are not interested in the deformation of the pin in this study.

The bottom half of figure 7.20 shows the deformed shape of the specimen with contours of von-Mises stress. Note that the deformation of the specimen has been exaggerated for clarity, by a factor of 500. The maximum displacement in this case was 3.47×10^{-6} m and the maximum stress was 43 MPa, which indicates that the linear elastic assumption is valid as displacement is low and stress is below the yield stress for the material.

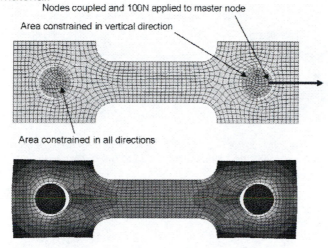

Figure 7.20: Representative finite element model of specimen and pins with contact elements (top) and results for von-Mises stress with lighter colours indicating higher stress (bottom)

In order to try and simplify the model, the pins were then removed from the model and the nodes on the right hand circle were constrained in all directions, while the nodes on the right hand circle have been coupled and the horizontal force was applied to the master node. Figure 7.21 shows an overview of this model and the subsequent results. Clearly, in this case, the loads have been perfectly distributed on the circles rather than concentrated on a smaller contact region, as would happen in reality. This model is a poor choice as it assumes that the pins can pull on the inside portion of the curves which is physically impossible in reality. In this case the maximum displacement and maximum stress were less than before: 1.95×10^{-6} m and 30 MPa. Notice that the stress distribution is totally different to that experienced in our benchmark as shown in figure 7.19.

An obvious way to improve the above model would be to apply the load and constraints to only the outside semi-circle at each pin hole. Figure 7.22 shows an overview and results from this type of model. Clearly this model has produced a deformed shape that is much closer to the benchmark than the previous model, as the holes are now free to nearly deform into the shape they would assume in reality. There are still some problems with this model however. The fact that the load and constraint are applied to nodes just on the semi-circles means that at the top and bottom of each circle there is a stress concentration around the last node to be loaded. The maximum displacement and stress in this case are 2.93×10^{-6} m and

137 MPa. So the displacement is less than in the benchmark but the stress is higher! This tells us that this method is introducing artificial stress concentrations, as we already know, at the top and bottom of the circles where the assumed contact with the pins is "lost".

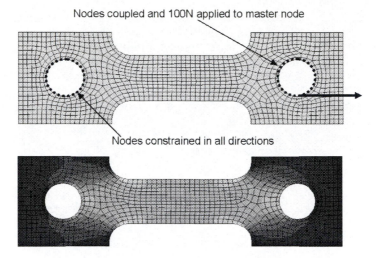

Figure 7.21: A much simplified model of the tensile test the pins have been removed and the BC's and loads are applied to the circles

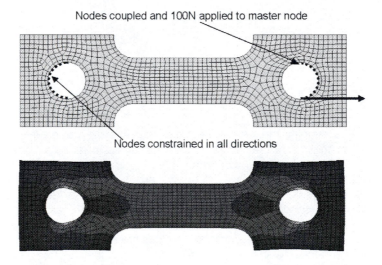

Figure 7.22: In this case BC's and loads are applied to the outside semi circles

Figure 7.23 shows another attempt at improving the results by changing the loading strategy. In this case the constraints were kept as above (left outside semi circle fixed) but a pressure load was applied to the right hand outside semi circle. The top of figure 7.23 shows the results from a uniform pressure load over the semi circle while the bottom shows the results obtained when a pressure gradient was applied to the semi circle. In this case a gradient similar to that shown in figure 7.16 was used with the pressure at the top and bottom on the semi circle being 10% of that on the centre-line of the specimen. It is clear from both of these results that the distribution of von-Mises stress at the loaded end is approaching that experienced in the benchmark.

Practical Stress Analysis with Finite Elements

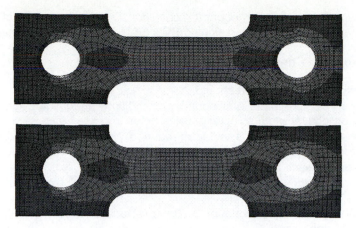

Figure 7.23: Pressure loads applied to right hand outside semi circle: uniform pressure (top) and pressure gradient (bottom)

Finally, figure 7.24 shows a complex method of modelling the load transfer from the pins. In this case a much smaller section of each of the circles has been selected to constrain and load. This is based on the rationale that at any time (during the elastic deformation) only a very small section of the pin will be in contact with the circles on the specimen. Hence a group of nodes around the centreline of the specimen were selected for constraint in the left hand hole. In order to further improve the results, these nodes were rotated into a local cylindrical coordinate system so that they became radial constraints (rather than constraints in the x-direction). Similarly at the right hand hole a tapering pressure load was applied in much the same manner as above, but to a smaller line in this case. It can be seen from figure 7.24 that the resultant von-Mises stress distribution is very similar to that obtained using the model with the pins and contact elements shown in figure 7.19.

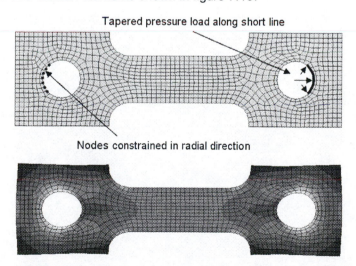

Figure 7.24: Radial constraints are used on a shorter line of nodes and a tapered pressure load is applied to a shorter line – gives results similar to the benchmark.

The point of all this is not to show a highly complex method of getting the same results as obtained with the contact model, but to illustrate the effects of various different assumptions regarding boundary conditions and loads and to stimulate you to begin to think creatively when assigning BC's and loads to your model.

7.8   Summary of Chapter 7

After completing this chapter you should:
- Understand how loads and boundary conditions modify the global problem equation and facilitate solution of the problem.
- Understand each of the different boundary condition types available for structural analysis.
- Understand each of the different load types available for structural analysis.
- Understand the effects of loading assumptions on the outcome of your analysis.
- Be able to logically select appropriate loads and boundary conditions for your finite element model.

8

Solution

8.1. Introduction

The previous chapters have discussed selection of a finite element type, assigning a material model to describe fully the element behaviour, dividing the problem geometry into a number of these elements and assigning loads and boundary conditions to specific nodes. At this stage the problem has been fully defined and now requires solution. This chapter discusses the various options available for solution of structural problems, illustrates how these options are practically implemented using commercial FE software, outlines the methods used in the solution of finite element problems and details why solutions often break down.

8.2 Preparation for Solution

Before attempting to solve your model it is important that you re-check all the inputs to your analysis. Identifying a problem at this stage could save a lot of time later on and will not result in a waste of computational resources. The following checks are recommended for all models as a matter of course:

(i)   Dimensions and Units:

It has already been stated earlier in this book but it is worth repeating: the most common error made by new users of FEA is not using the correct units. With this in mind before setting up the solution is a good time to re-check your dimensions and units.
-Get a list of nodal coordinates for key point on the model and compare the nodal distances to the dimensions of the problem. If they are not the same then it is likely that a dimension has been entered incorrectly.
-Check the magnitude of the applied loads. Have you used the correct units?
-Check the material model definition. Were the correct units used in the specification of the material model?
-Are the units used for dimensions, loads and material properties consistent? See chapter 2 for more information.

(ii)   Mass and Volume Checks:

You will know the volume of your problem geometry from the problem specification. Most FE software calculates the volume of the defined elements before solving the problem. Upon comparing the volume of the defined elements with the required volume there should be very little difference. If there is a big difference then it is likely that the wrong dimensions were used and/or the mesh is too coarse.

If you have defined density as part of your material model then the FE software will calculate the mass of the FE assembly. If you have information about the mass of the actual structure then these can be compared directly. Even if you don't have this information then a mass check can be useful as you should have some idea of the "ballpark" on the required model mass. If the mass of the FE model is significantly different from the actual mass of the structure then either the wrong dimensions have been used and/or the wrong material properties or material model has been used.

Mass checks are particularly important if you are carrying out a dynamic analysis. The response of the structure will be directly related to its mass and its distribution of mass, so it is important that the model mass is representative of the actual structure.

(iii)   Mesh Checks

These days most FE software perform automatic mesh checks to detect poorly shaped elements and this has been covered in chapter 7. A manual re-check can be useful at this point in order to identify mesh problems which the automatic check may have missed.

Viewing the mesh from several angles is useful to ensure that there are no extra unwanted geometries or features. Most FE software allows you to shrink the elements which is useful to highlight any holes in the mesh where elements are missing. This is unlikely with today's modern automatic meshing algorithms but should be checked nevertheless. Boundary line plots are useful to show up parts of the model that are not connected.

It is good practice at this stage to issue a "delete all nodes" command. Since most FE software will not allow you to delete nodes that form part of an element, this command will only delete stray nodes that are not connected to the current model. Conversely, it is important to make sure that you have selected or enabled all of the elements and nodes that make up your model. Most FE software allows you to select portions of the model (and hence disable other parts) in order to apply boundary conditions etc. It is all too easy to forget to enable the full model before issuing the solve command.

(iv)   BC and Load Checks

You should plot the loads and constraints applied to the model and compare them with the problem specification. Ensure that they are applied at the correct location and that they act in the proper direction. Be especially careful of local coordinate systems and ensure that the display is set up to show different coordinate systems correctly.

Once you have completed these checks you should be ready to begin your solution. The following sections describe each of the various solution types available for structural FEA and offer tips on how to ensure that the solution runs smoothly.

8.3   Static Analysis

A static analysis is the most basic type of analysis available in most commercial FE software and is often the default analysis. As the name implies, a static analysis assumes that all the previously defined loads and boundary conditions are applied so slowly that the structure does not behave dynamically.

Realistically this is never the case as if you move any structure from a rest position to motion (i.e. deformation) then there will always be some vibration effects present. In many cases, however, these effects can be so small to be considered negligible and hence the popularity of static analyses in FEA. Even though static analyses are just that, they can contain steady-state inertia loads such as gravity or even constant rotational velocity.

8.3.1 Linear Static Analysis

By default many FE software packages use a linear static analysis. A linear static analysis assumes that material properties are linear, deflections are small and that loads and boundary conditions do not change with time. This type of analysis is useful for sub-yielding loading of metallic structures etc.

In terms of a structural static analysis a linear analysis means that you are using a linear elastic material model. Any other material model type makes it a non-linear static analysis, which is covered in section 8.3.2. If you are sure that a linear static analysis is valid for your model then you will need to know at what point the linear elastic assumption breaks down for the material type that you are using. Normally this means knowing the yield strength of the material. All stress results should be checked against this yield strength to ensure that stress results are still within the elastic region. If this is found to not be the case then a non-linear static analysis will be required.

The loads in a linear static analysis cannot vary over the course of the simulation: a single value must be entered for each load type. This rules out time varying loads which will require, at least, a non-linear static analysis if not a transient analysis.

The expected deformations in a linear static analysis should be very small. Remember we are working within the elastic region of the material's stress-strain curve. This normally accounts for much less that 1% strain and so you should expect deformations of the order of mm or below unless your geometry is very large.

In summary, a linear static analysis is suitable if all of these conditions are satisfied:
    1. Dynamic effects are not important in your analysis
    2. You are using a linear elastic material model
    3. Your loads and boundary conditions remain constant over the course of the solution.
    4. There is no contact in your model
    5. Deflections are very small

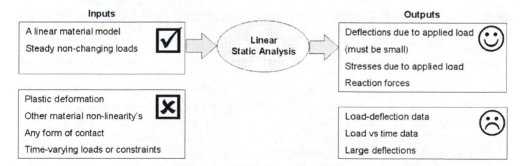

Figure 8.01: Allowable inputs and obtainable outputs from a linear static analysis.

8.3.2 Non-linear Static Analysis

Static analyses, however, can also be non-linear and can include: large deformations, contact, plasticity, creep, stress stiffening, hyper-elasticity etc. In a non-linear static analysis it is possible to define time varying loads etc. but it is still assumed that there will be no dynamic effects when these loads are changed during the analysis.

In a non-linear static structural analysis any material model which doesn't include time effects can be used. This pretty much allows for all of the material models in chapter 5, except for those that take account of strain rate.

Load steps and sub-steps

There is a paradox in many FE software packages in that a non-linear static analysis does not consider time but you can setup the solution so that "time steps" are used. Time steps may also be referred to as load steps or sub-steps and their function is to divide the analysis up into a number of sub-steps to ensure that the solution runs smoothly. Using sub-steps ensures that the applied load is not completely applied in one go but rather a number of smaller cumulative solutions are run where: a portion of the load is applied, a solution for the piecewise deformation due to this load is obtained and then the next load increment is applied. This method is very important for large deformation problems as if all the load was applied in one step then the equations would most likely become highly unstable. Using sub-steps allows a large deformation to occur over a number of steps and thus ensures that the system equations remain stable and hence allow for a solution for the full increment to be obtained. For more information on load steps see section 8.9.2.

Most FE solvers have an automatic time-stepping/sub-stepping procedure which once activated will automatically divide the problem into sub-steps if a large deflection is detected. Other solvers require you to specify the number of sub-steps in advance or at least take a guess that the automatic sub-stepping algorithm can use as a starting point. Using sub-steps has the advantage that you can later examine the results at any of the sub-steps, thus you can view results for 50% of load application, as well as the full load application. This is very useful, for example, if the structure exceeds the material UTS at some point during the simulation as you can simply go back to the sub-step previous to the UTS being exceeded, determine the applied load at this point, and thus make a judgement as to the maximum load that the structure can withstand before failure.

The main point of this is that if your analysis has large deformations (e.g. metal forming, non-linear elasticity etc.), large amounts of contact and non-linear or time varying loads, or any combination of these then splitting the solution up into sub-steps will be required.

In summary, a non-linear static analysis is suitable if all of these conditions are satisfied:
    1. Dynamic effects are not important in your analysis.
    2. You are using a material model which doesn't include strain rate effects.
    3. Your loads and boundary conditions remain constant over the course of the solution or change relatively slowly.
    4. Contact is allowed in your model.
    5. Deflections can be large provided they are accounted for in the solution set up, i.e. you must tell the solver you wish a large deflection analysis.

The main differences between the setup of a non-linear static solution in comparison to a linear solution are:
    1. You must tell the solver that large deflections are expected.
    2. In most cases you must use solution sub-steps.

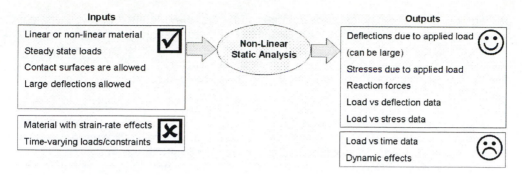

Figure 8.02: Allowable inputs and obtainable outputs from a non-linear static analysis.

8.4   Modal Analysis

A modal analysis is used to determine the natural frequencies and associated mode shapes of a structure. Almost all structures will undergo some type of vibration or time-varying loading during their service life and an understanding of the structure's natural frequencies is useful in order to avoid exciting these frequencies during service. The typical output from a FE modal analysis is a list of frequencies and a plot of model deformation at these frequencies. Animations of deformation can be very useful here in order to visualise the mode of vibration.

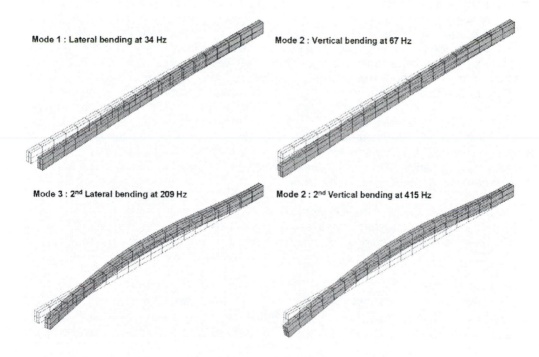

Figure 8.03: Results from modal analysis of a cantilevered beam

Obviously this is a dynamic analysis, so dynamic effects are included and the density of the material must be included in the material model. The mass of the model and the distribution of the mass throughout the model are of prime importance in a modal analysis. It is important to note also, that symmetry assumptions cannot be used for a modal analysis. If you use reflective symmetry (for example) you will eliminate a percentage of the mass of the structure and thus the results from your modal analysis will be meaningless. When performing a modal analysis it is important to always model the full structure. The exception to this rule is cyclic symmetry: because modal analysis of cyclically symmetrical structures such as turbine blade discs etc. is so important most FE software has specialised routines for modal cyclic symmetry.

In general, modal analysis is considered a linear analysis, thus non-linear material behaviour is not allowed. Contact regions are also not generally allowed in a modal analysis. Applied loads are not required in a modal analysis as you are simply trying to obtain the natural frequencies of the structure. If you are seeking to obtain the frequency response of a structure due to a specific excitation force then a harmonic analysis is required, see section 8.5. An exception to this rule is pre-stressed analysis: it is possible to pre-stress a structure using a static analysis and then perform a subsequent modal analysis to determine the modal response of the pre-stressed structure. Constraints, however, are allowed in a modal analysis and should be as representative as possible. Over-constraining can be a serious problem in modal analyses as an over-constrained model will behave too stiffly and will thus over-predict the modal frequencies.

Various types of damping can be applied to the structure in a modal analysis. Damping can be inputted as part of the material model or as a global damping constant applied to the whole system. You should consult your software's documentation to see what type of damping is available and how it is implemented.

A number of different mathematical methods for formulating the problem and obtaining a solution are usually available in order to solve many different types of problem. The most popular methods available are: Lanczos methods, Subspace iteration methods and Householder methods. There may be other specialised methods available in your software which are specific to particular problems. The best method to use will vary from problem to problem and will depend on the size and form of the equations being solved (i.e. number of DOF in the mesh and whether the equations are symmetrical or not). Modal analysis is a specialised field and further discussion of the method is beyond the scope of this book. For further information get any of the excellent texts available on Modal analysis or see the help documentation from your FE software.

In summary a modal analysis is suitable if:
1. You are interested in the natural frequencies of vibration of your structure.
2. You want to see how the structure behaves at each of these frequencies.
3. You can set up a linear version of your model in order to determine these.

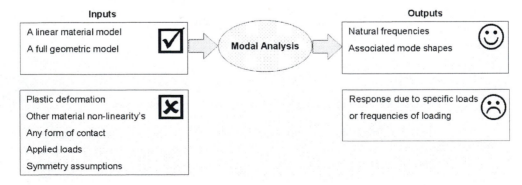

Figure 8.04: Allowable inputs and obtainable outputs from a modal analysis.

8.5   Harmonic Analysis

A harmonic analysis is used to determine the steady state response of a structure that is exposed to a sustained harmonic load or set of harmonic loads. A harmonic load is a load that varies sinusoidally at a single frequency. All the loads applied in a harmonic analysis must be at the same frequency, however, the load amplitude and its phase may be specified as being different at different locations in the modal. Harmonic loads are normally imparted on structures due to rotating or reciprocating machinery. Figure 8.05 illustrates two harmonic loads of equal magnitude and frequency but have different phase.

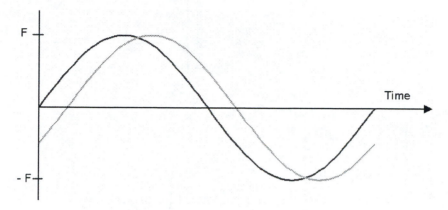

Figure 8.05: Two harmonic loads out of phase.

The response of the structure in a harmonic analysis will depend on its natural frequencies and, obviously, on the frequency and amplitude of the applied harmonic loads. The harmonic solution will give the magnitude and phase of the nodal displacements, velocities and accelerations. Post-processing allows the calculating of element stresses and strains from the displacement results. The response of the structure is typically examined at several frequencies and generally a graph of some response quantity (usually displacements) versus frequency is produced. Peak responses on the graph are then identified and stresses reviewed at those peak frequencies, see figure 8.06. This can be useful in order to determine harmonic loads that should be avoided in the structure's environment during service. Clearly if a certain harmonic frequency causes the structure to experience large displacements or stresses then that frequency is capable of causing failure of the structure. The frequencies marked f_1 and f_2 in figure 8.06(c) are frequencies that should be investigated in this manner.

In a similar manner to modal analysis, harmonic analyses are normally considered linear analyses, thus non-linear materials or contact elements are usually not allowed. All applied loads have to be sinusoidal in nature and must be of the same frequency. Both point forces and pressure loads are allowed. Loads with different amplitudes or phase differences can, however, be applied to different parts of the model. A harmonic analysis can also be performed on a pre-stressed structure provided that the harmonic stresses are much smaller than the pretension stresses, a typical example is a violin string. Constraints are also allowed in a similar fashion to modal analysis.

There are normally a number of different harmonic solution methods available which are individually applicable to particular problems. The most common methods used are: direct methods, reduced methods and mode superposition methods.

The direct method uses the full system matrices to calculate the harmonic response by directly solving the system equations. This method is the easiest to set up as it doesn't require any further specification of master degrees of freedom or other parameters in the model. For large problems this method can be costly in terms of computational time.

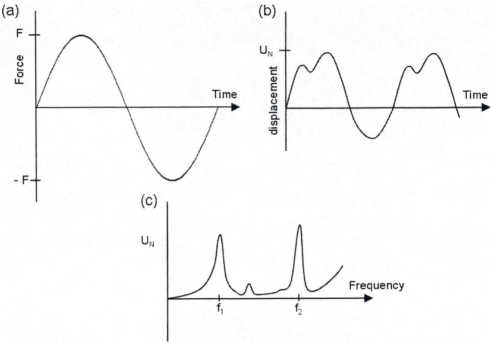

Figure 8.06: Typical results from a Harmonic analysis: (a) the applied load (b) vertical displacement of a specific node vs. time (c) Max vertical displacement of a specific node vs. loading frequency

The reduced method lets you condense the problem by specifying master degrees of freedom and using reduced system matrices. The solution then only solves for the specified master DOF and calculates the displacement of these points. After these have been determined the solution can subsequently be expanded to determine the displacement of all the nodes/DOF in the problem using a second solution step known as an expansion pass. This method is faster to solve and less computationally costly than the full method, however it has the disadvantage that pressure loads cannot be applied and that all forces must be applied at the assigned master DOF.

The mode superposition method works by first determining the natural frequencies of the problem using a modal analysis and then combining different ratios of the mode shapes to calculate the structure's response. It is much faster than the two above methods for many problems and has the advantage that it allows solutions clustered around the structure's natural frequencies. A significant disadvantage of this method however, is that it does not allow the use of constraints.

It is important to note that when a steady state harmonic load, such as that shown in figure 8.06(a) is applied to a structure there will be an initially unsteady response from the structure as it adapts to the newly applied load and moves from rest, after which the structure will have a steady state response. This "settling in" period is not predicted by a harmonic analysis, so if you are interested in this part of the analysis you should perform a transient analysis.

In summary a harmonic analysis is suitable if:

1. The structure to be analysed is expected to be subjected to steady state harmonic loads due to the presence of rotating or reciprocating machinery.
2. You wish to determine the structure's response to these harmonic loads and identify frequencies that cause unacceptable deformation of the structure.

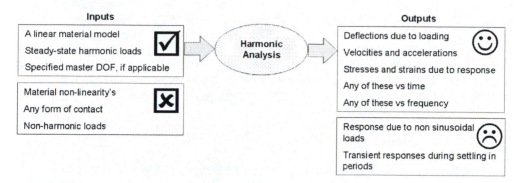

Figure 8.07: Allowable inputs and obtainable outputs from a harmonic analysis.

8.6   Transient Dynamic Analysis

8.6.1 Overview

A transient dynamic analysis (sometimes called time-history analysis) is a method used to determine the dynamic response of a structure due to the action of any type of loading that varies with time. This type of analysis is used to determine time-varying displacements, velocities and accelerations at each time step which can be used to calculate strains, stresses, and forces in the structure at each time step in the analysis. A transient analysis can consider any combination of static, transient, and harmonic loads. Obviously, inertia and damping effects are considered to be very important in a transient analysis.

8.6.2. Solution, Time-steps, and time-integration

Recall, that the basic equation solved by a transient dynamic analysis is given by:

$$[M]\{\ddot{U}\}+[C]\{\dot{U}\}+[K]\{U\}=\{F(t)\}$$
(8.01)

where, [M] is the global mass matrix, [C] is the global damping matrix, [K] is the global stiffness matrix, {U} is the nodal displacement vector and {F(t)} is the time-dependant load vector. At any time, t, during the solution these equations can be considered as a set of static equilibrium equations that also take account of inertia forces and damping forces. FEA solvers use time-integration methods to solve these equations at discrete time points and the time increment between successive time points is called an "integration time step", or more normally a "time step".

Most FE solvers use an implicit integration method where the basic approach is to assume that the solution for the discrete time t is known and that the solution for a discrete time t+Δt is required, where Δt is a suitably chosen time increment (i.e. the time step). During the implicit time integration solution the solver will increment a time step and then iterate to determine if the external forces, or loads applied to the structure are in equilibrium with the internal forces. The equilibrium check is performed by iteration since the displacements at the new time are a function of the displacements and accelerations at the previous step. The calculation also requires formation and inversion of global model mass, damping and stiffness matrices before equilibrium calculations can commence which is not a trivial task! The mass, damping and stiffness matrices have to be re-evaluated and constructed at each time step as the nodes will have moved their position during the previous time step. This means that the system matrices have to be evaluated, inverted then re-evaluated several times during the solution. As an example, suppose you have 500 nodes in your two dimensional model, using the implicit method will require forming and inverting the 1,000 row by 1,000 column mass, damping and stiffness matrices as many times as there are time steps in the solution! This illustrates the main reason why model size has a direct relationship to solution time. In fact, it has been established that the speed of the analysis is roughly proportional to the numbers of degrees of freedom squared!

8.6.3. Types of Transient Analysis

Again, there are normally a number of different transient solution methods available depending on the type of structure and problem being analysed. Similar to harmonic analysis there are three methods which are extensively used: direct methods, mode superposition methods and reduced methods.

For small models, or if high excitation frequencies are used, or if there are only a small number of time steps, then the direct method is most appropriate. The direct method is also recommended if highly accurate results are required. The direct method uses the full system matrices to calculate the transient response of the system. It is the most general and computationally expensive of the three methods and all types of non-linearities (plasticity, large deflections, large strain, contact, etc.) and all types of loads may be included. This method is the easiest to set up as it doesn't require any further specification of master degrees of freedom, calculation of mode shapes or setting of other parameters in the model.

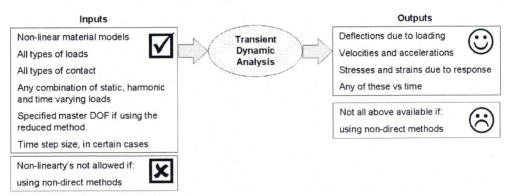

Figure 8.08: Allowable inputs and obtainable outputs from a transient dynamic analysis.

If the model is large or subjected to many excitation time steps then the mode superposition method is often more appropriate. As in a harmonic analysis; the mode superposition method works by first determining the natural frequencies of the problem using a modal analysis and then combining different ratios of the mode shapes to calculate the structure's transient response. The mode superposition method is much faster at obtaining a solution than the direct method but has significant disadvantages since the time step used must be constant (meaning that automatic time stepping, if available, cannot be used), it does not generally allow non-linearities in the model and not all load types are allowed.

The reduced method lets you condense the problem by specifying master degrees of freedom and using reduced system matrices. The solution then only solves for the specified master DOF and calculates the displacement of these points. Obviously this means that the reduced method will always obtain a solution much faster than the direct method but, again, this is offset by some significant disadvantages: an expansion pass solution is required after the initial solution; all loads must be applied at the master DOF; not all load types are allowed and the time step used must be constant (meaning that automatic time stepping, if available, cannot be used).

8.6.4. Breakdown of Transient Analyses

Although, from figure 8.08, it might appear as though a transient analysis is the answer to all your simulation problems it is important to note that there are several factors which can cause the solution to become unstable and breakdown or prevent the solution from converging.

If the applied loads are very fast loads (i.e. are applied over a very short period of time) then the implicit time integration scheme will often be unable to solve the equations or the system of equation can become unstable. This effectively rules out solution of most type of impact problems. If the model has large regions of contact (e.g. some metal forming processes) or contact conditions that change very rapidly (e.g. impact events) then the implicit scheme may also become unstable. In general, if the problem is highly non-linear, with non-linear materials, contact, large deflections etc. and subjected to dynamic or short time or impulse loads then an implicit transient analysis will most likely not be capable of providing a solution. In such cases an explicit dynamic analysis may be used, see section 8.7.

8.6.5. Summary

In summary a transient dynamic analysis is suitable if:
1. The structure to be analysed is subjected to time-varying loads that introduce dynamic effects in the structure.
2. The structure is subjected to any combination of static, harmonic and transient loads.
3. You require information about the initial response of the structure to a suddenly applied static or harmonic load before the structures response becomes steady state.

8.7   Explicit Dynamic Analysis

In section 8.6.2 above we described the implicit time integration method which requires the formation and inversion of the global system equations at each time step. Obviously this is very costly in terms of computational resources, particularly when the model is very large. When the number of DOF becomes large the CPU time per increment becomes very long. For problems with few time increments this is not so much of a problem but when a large number of time increments are required then it is likely that computational resources will be exhausted.

Explicit time integration offers an alternative method which can be used to more easily and quickly solve such problems. In explicit integration nodal accelerations are calculated directly from Newton's second law ($F = ma$) using a mass matrix that is diagonal. This eliminates the need to generate and invert large matrices as the response of each node is now independent. Nodal velocities and displacements are then calculated directly from accelerations and the time increment that the solution has progressed through. The term "explicit" is used since the calculation only refers to values of force, displacement, velocity and acceleration at the start of each new time step. The solution is obtained on an element by element basis thus large 3D models with thousands of degrees of freedom can be solved with comparatively modest computer storage requirements.

In order to fully understand the difference between implicit and explicit methods consider equation 8.02 which shows the general form of the implicit method.

$$U^{n+1} = f\left(\ddot{U}^{n+1}, \dot{U}^{n+1}, U^n, \ldots\right)$$
(8.02)

So the current nodal displacement U^{n+1} is a function of time derivatives of U^{n+1} (i.e. nodal acceleration and velocity) which are also unknown. Consequently the implicit method iterates (i.e. guesses) a solution for the nodal displacements and its time derivatives in order to solve the equation. This also, as we know, requires solving simultaneous equations and hence inverting the large system matrices.

In contrast the general form of the explicit method is shown in equation 8.03.

$$U^{n+1} = f\left(U^n, \dot{U}^n, \ddot{U}^n, U^{n-1} \ldots\right)$$
(8.03)

In this case the current nodal displacement is a function of the nodal displacement, velocity and accelerations from the previous time step. Since this information is already known a diagonal mass matrix can be used and the global equation is a system of linear algebraic equations which can be solved without using simultaneous equations.

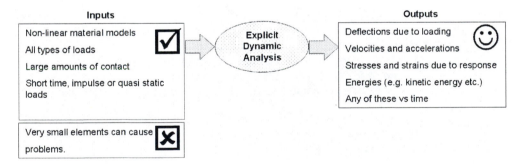

Figure 8.09: Allowable inputs and obtainable outputs from an explicit dynamic analysis.

The main disadvantage of this method is that time step size must be small for a stable solution. If too large, significant numerical errors will develop. For stable computations, the time step is normally selected by the solver code such that the time step is less than the time required for a stress wave to travel through the smallest element. This means that the solution time is essentially based on the size of the smallest element in the mesh and this must be taken into account when designating the mesh.

Explicit dynamic analyses are used to obtain fast solutions for short-time, large deformation dynamic, quasi-static problems with large deformations and multiple non-linearities, and complex contact/impact problems. Explicit dynamic analysis is essentially used for complex problems that are too difficult to solve using a transient dynamic analysis.

In summary an explicit dynamic analysis is suitable if:
1. Your model is highly non-linear with non-linear materials, large deflections and lots of contact regions or advanced contact types.
2. The applied loading is dynamic, ranging from short time impulse loads to quasi-static loads. (e.g. impact events to metal forming)
3. You have had difficulty solving the problem with an implicit transient analysis.

8.8 Specialised Types of Analysis

There are many types of specialised FE software available for solving particular problems. We will not consider all of these here, however, the two most common and thus important to be aware of are spectrum analysis and buckling analysis.

8.8.1 Spectrum Analysis

In a spectrum analysis the results of a modal analysis are used with a known spectrum to calculate displacements and stresses in the model. A spectrum is simply a graph of a response (usually displacement) versus frequency. Other responses that can be used in a spectrum include velocity, acceleration or force. A single point response analysis is used to determine the effect of the same response spectrum curve at various points in a model (e.g. all the support points). In contrast, a multipoint response spectrum analysis is used to determine the effects of applying a different response spectrum at different points in the model.

This type of analysis is generally used in place of a time-history analysis to determine the response of structures to random or time-dependent loading conditions such as earthquakes, wind loads, jet engine thrust, rocket motor vibrations etc.

8.8.2 Buckling Analysis

Buckling occurs due to a lack of stability in a structure subjected to a compressive load. It is important to realise that this behaviour is largely independent of the material strength and is dependant on the shape of the structure. A buckling FE analysis is used to determine the critical loads which will cause the structure to buckle and the buckled mode shapes (i.e. the characteristic shape associated with the structures buckled response). Two types of buckling analysis are typically available: linear buckling analysis (also known as eigenvalue buckling analysis) and non-linear buckling analysis.

A linear buckling analysis is used to predict the theoretical buckling strength of an idealised linear elastic structure. This method is similar to classical or "textbook" approaches to elastic buckling and has the same limitations. In reality most structures do not behave linearly and will have small imperfections (casting imperfections, voids, cuts etc.) that will prevent the structure from achieving its theoretical elastic buckling strength. This essentially means that linear buckling analysis will often give dangerous results and should not generally be used for design work.

Non-linear buckling analysis is usually much more accurate than linear buckling and essentially works by using a nonlinear static analysis to determine the effect of gradually increasing loads and to determine the point at which the structure becomes unstable. In a non-linear analysis the model can include details such as plastic deformation, large deflections, initial imperfections and gaps. In some cases it is also possible to analyse the post-buckled performance of the structure which is very useful where the structure buckles into a stable configuration.

8.9   Overview of Solution of the Problem

8.9.1   Basic Methods of Solution for Linear Problems

For most structural finite element analyses there are a number of solution methods commonly used by commercial FE software suites: Gaussian elimination methods, wavefront methods and iterative methods.

Gaussian Elimination

Gaussian elimination is an elementary mathematical method of solving systems of simultaneous equations by successively eliminating unknowns. A description of this method can be found in almost any relevant mathematical textbook and thus it is not proposed to repeat this information here. Since the governing equation of any FEA is essentially a set of simultaneous equations in matrix form, i.e. [K]{U}={F}, it is obvious that Gaussian elimination is a useful method for solving FEA problems.

The basic concept of the method is to eliminate a variable at a time by performing basic mathematical operations on the problem equations. This process continues until only one variable is left, which can then easily be determined. Since all the other variables will now be written in terms of the last variable it is a simple matter to back substitute to determine all the other variables.

The speed at which Gaussian elimination can solve a problem is directly dependant on the bandwidth of the problem. If you re-examine some of the examples from chapters 1 and 4 you will see that, in general the stiffness matrix in a FEA is banded. This means that non-zero terms are grouped in a "band" around the diagonal of the matrix, see figure 8.10

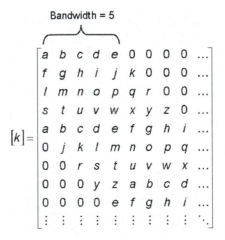

Figure 8.10: Bandwidth of a stiffness matrix

The smaller the bandwidth of the stiffness matrix, the faster the problem can be solved by Gaussian elimination. A problem with a bandwidth of 1 will yield an instant solution whereas a problem with a bandwidth of 10 will require a variable to be eliminated and the equations rewritten nine times before the first variable is solved, after which nine back substitutions will be required to fully solve the problem.

Bandwidth is greatly affected by the method used to number the nodes in a finite element model. If we consider the two plane stress model of a cantilevered beam shown in figure 8.11, it can be seen that both models have elements numbered across the height of the beam, the model on the top of the figure has nodes numbered along the length of the beam and the model at the bottom of the figure has nodes numbered across the height of the beam.

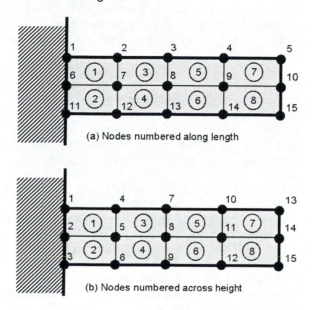

Figure 8.11: Two identical plane stress models of a beam with different node numbering methods.

If we examine the model with the nodes numbered in the horizontal direction (i.e. along the length) we can easily see that the degrees of freedom associated with nodes 1 and 7 will be linked as they are both part of element 1. Since this is a 2D model each node will have two degrees of freedom, thus the first two rows and columns in the stiffness matrix will correspond to node 1 while the 13th and 14th rows and columns will correspond to node 7. In the model with vertical numbering scheme (i.e. across the height) the nodes 1 and 5 will be similarly linked, corresponding to the first two and the 9th and 10th rows and columns respectively. This means that the bandwidth of the model with the nodes numbered in the vertical direction will be 10 while the bandwidth of the horizontally numbered model will be 14.

This shows that the bandwidth of the model is directly related to the manner in which the nodes are numbered. The bandwidth of a model can be calculated from equation 8.04.

$$B_{width} = \frac{(d+1)}{nDOF}$$

8.04

Where d is the greatest difference between node numbers in a particular element and nDOF is the number of degrees of freedom at each node. In order to keep the problem bandwidth as low as possible nodes should be numbered across the shortest dimension in the model.

The Wavefront Method

The wavefront method, sometimes known as the *frontal method*, is a direct elimination method that avoids constructing the full global stiffness matrix by eliminating degrees of freedom as soon as it can as it works through the model. The wavefront method formulates each of the individual element matrices, then reads in the DOF for the first element. Any DOF that has a known value (due to a boundary condition) or that can be expressed in terms of other DOF's is then eliminated and an equation is written to a storage file. The DOF's remaining in the solution constitute what is known as the wavefront. This process is repeated until all DOF's have been eliminated and the storage file contains a triangularised matrix. The individual nodal DOF can then be calculated by back substitution, thus solving the problem.

The wavefront is the number of DOF retained by the solver at any time that cannot yet be eliminated. The wavefront will expand and contract as the solution proceeds and will finally become zero when all DOF have been eliminated. Obviously the value of the wavefront directly effects solution time and the higher the number of DOF in the wavefront the longer the solution time. Choosing an element numbering scheme that specifies the order in which elements are processed by the solver can reduce the number of DOF in the wavefront and hence speed up solution.

Figure 8.12 shows the progression of the wavefront along the length of a plane stress cantilevered beam model. In figure 8.12a the stiffness matrix of element 1 is calculated and placed in a temporary matrix. Since node 1 only occurs in this element its DOF can now be removed from the matrix. Thus the DOF of node 1 are rewritten as functions of the DOF of nodes 2,4 and 5 and the temporary matrix will only contain information on these nodes. In figure 8.12b the stiffness matrix of element 2 is calculated and added to the temporary matrix. Nodes 2 and 3 are not used by any other element and thus its DOF can be removed at this stage and rewritten in terms of the other nodes. This process continues through all the elements in the model until only node 15 remains and its DOF are solved. A backward substitution then occurs and all the other DOF are evaluated.

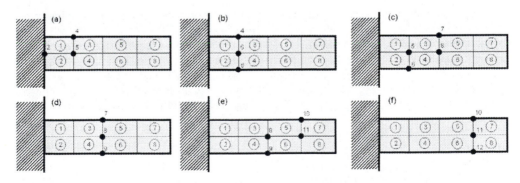

Figure 8.12: Progression of the wavefront in a 2D plane stress model

Notice that, in this case, if the elements had been labelled along the length of the beam rather than across the height then at any time during the solution there would have been more nodes in the wavefront and consequently more nodes in the temporary matrix and the solution would have been considerably slower. Clearly, for the wavefront method, the numbering of the elements is crucial and they should be ordered across the shortest dimension of the model. Most FE automatic meshing tools have algorithms that attempt to ensure that this is always the case. Some FE software also comes equipped with commands for checking the maximum wavefront size, identifying where it will occur and offering you options for element re-numbering.

8.9.2 Iterative Methods for Non-Linear Problems

Non-linear models cannot be solved using the direct methods described in the previous section. Instead an iterative method is required. Essentially this involves the solver making a guess (i.e. an iteration) at the value of one of the unknowns and thus solving the problem. Obviously, it is highly unlikely that this iteration will be correct so an error estimate is made on the original iteration. The error estimate is then used to adjust the original iteration and thus make a second iteration and the solution is repeated. The cycle of iteration, solution, error estimate, iteration continues until the error estimate is below a predetermined threshold.

A non-linear solution is obtained via a series of incremental solutions. In order to ensure stability the total load is applied in steps, known as load steps, which we discussed briefly in chapter 7. At each load step an intermediate solution is obtained using the incremental load. The final load orientations and magnitudes, deformations and stress levels from the previous load step are used as initial conditions for the current load step. The error estimate as described above is calculated as usual at each load step. If the error is too high then the load step may be automatically divided up into smaller steps. This process continues for each load step until the total load has been applied to the model or the solver identifies a diverging solution.

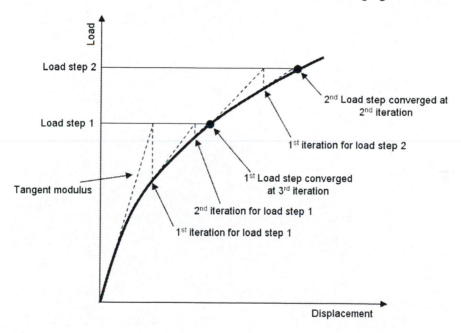

Figure 8.13: Overview of the Newton-Raphson method

Most finite element solvers use Newton-Raphson iteration methods to solve non-linear problems. This method uses the tangent modulus of the previous iteration to calculate the next deformed position using a linear solution. The calculated displacement is projected back onto the load-displacement curve parallel to the load axis. In most cases the point to which it projects will not be equivalent to the load required at the end of the load step so a second iteration is required. In this case the difference between the required load-step load and the projected position load is subtracted from the load step and the next displacement is iterated for by taking the tangent modulus of the projected position. This continues until the difference between the required load and the end of the load step and the projected position is within a specified tolerance – known as convergence. Figure 8.13 shows an overview of this procedure.

A number of convergence enhancement methods are usually available in FE solvers and these can be set up to aid convergence for models where convergence is a problem. Automatic load stepping is the most common of these, which attempts to optimise the iteration process by assigning load step increments that will help the solution to converge. Bisection is another convergence aid which is used to specify that if the error estimate is above a certain threshold (i.e. too high) then the current load step should be automatically divided in two.

For certain non-linear analyses the standard Newton-Raphson method may prove unsuitable as the tangent stiffness matrix may become singular thus causing convergence to become almost impossible. For such situations an alternative Newton-Raphson iteration method, known as the arc-length method can be used. As the name suggests the arc-length method forces the Newton-Raphson equilibrium iterations to converge along an arc as shown in figure 8.14. The arc-length method often prevents divergence even when the slope of the load vs. deflection curve becomes zero or negative.

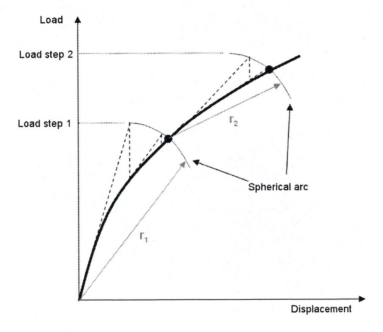

Figure 8.14: Overview of the Newton-Raphson arc-length method

Load steps, sub-steps and iterations

In the above discussion we have used a number of terms which require clarification at this point. A non-linear analysis can be considered to operate on three levels. At the top level are the load steps that the user defines when assigning loads. For example you may specify that pressure rises linearly to a value of 100 MPa over one second. You may then specify that pressure stays constant for three seconds. These are two load steps.

At the middle level, between each load step you can instruct the software to perform several solutions (sub-steps or time steps) in order to apply the load gradually and help the solution converge.

At the bottom level, the solution routine will use a number of Newton-Raphson equilibrium iterations in order to obtain a converged solution.

Most modern FE software has automatic time stepping and sub-stepping routines available to aid in the selection of appropriate steps, however, in general the automatic stepping algorithms require you to provide some sort of intelligent guess of the initial step size or number of steps in order to have a basis to work from. If the steps are too large then convergence may fail, however if they are too small then the solution may take a great deal of time. So how do you know how big/small to set the initial step size? Let's take a look at some guidelines:

- Gradually changing or moderately non-linear problems can usually be solved using relatively large sub-steps. A static analysis is so stable that it can be solved in a single load step so a problem that is only "slightly" non-linear shouldn't require many more steps. It is usually a good idea to chance specifying a large step size with such problems and letting automatic stepping take over if needed.
- Small sub-steps should be used at behaviour transition regions. A large step should not span a large change in the model state, for example crossing the yield point. Other examples where the model will experience an abrupt change in behaviour include the onset of buckling, contact due to high forces, contact due to high velocities, sudden changes in stress-strain behaviour (e.g. bilinear or multi-linear models), unloading of a plastically deformed model etc.
- Coming from the previous point, it should be obvious that the first sub-step should not cause yielding to occur.

By correctly specifying multiple load steps it should be possible to ensure that loads can be applied in large sub-steps up to the point of transitions in model behaviour. The load step that spans the transition point can then be broken into smaller sub-steps to help convergence through the difficult transition period.

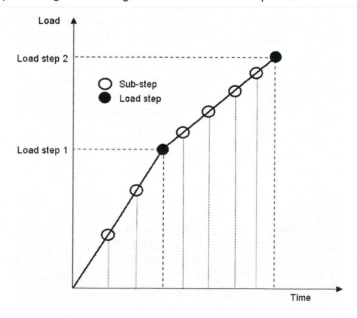

Figure 8.15: Overview of load steps and sub-steps

8.10 Fixing a Failed Solution

This section is concerned with how to deal with a solution that fails. When a solution fails most commercial FE solvers provide one or more, often cryptic, error messages which are supposed to help you diagnose the problem and make appropriate changes to the model or solution set up. In reality these errors messages are rarely of any hep, particularly to the novice user, hence in this section we will outline the reasons why solutions fail and how to overcome these difficulties. We will also discuss the more common error messages that are obtained, describe what exactly they mean and, again, describe how to change the model to eliminate these errors.

8.10.1 Problems with Computer Resources

A common reason for solutions to fail is because the solver runs out of computational resources needed to solve the problem, namely insufficient disk space or RAM. Insufficient processing power will rarely cause a solution to stop but it will cause the problem to take a very long time to solve. If your solution stops because of insufficient disk space or lack of RAM, the easiest course of action is to use another disk as the working directory for the solution or another computer, if available. If this is not possible then you will need to simplify your model.

The best way to simplify your model and free up computer resources is to reduce the number of DOF in the model. The most obvious way to do this is to use a symmetry assumption. If you haven't used symmetry up to this point then use it now. This is the time to ignore that one feature that stopped you from chopping the model in half! If this is not an option then you should look at other ways of reducing the number of nodes in the mesh. Further clean-up of the geometry and re-meshing could significantly reduce the number of nodes used. Local refinement of the mesh may also prove useful. If you have used a very fine mesh, then clear the mesh, re-mesh with a coarser mesh and refine the new mesh only at regions of interest. Manual meshing with volume segmentation, where you manually split the structure up into smaller regular volumes for meshing purposes, is a lot of work but will significantly reduce the number of nodes. One eight node brick element can be used to replace up to five ten node tetrahedrals using this method, thereby potentially reducing the number of nodes in the model by up to 80%, see figure 6.29 for an example.

8.10.2 Problems with Linear Static Analyses

A linear solution is inherently stable due to the static nature of the loads and the fact that large displacements are not allowed. As previously mentioned, a linear static analysis will solve in a single load step or sub-step, thus convergence is not usually an issue. As such, any problems due to load steps, sub-steps, convergence, excessive deformations, etc, do not apply to linear static analysis and will be discussed in the next section.

Insufficiently Constrained Models

The main reason for failure of a linear static analysis (and even many other types) is that the model has not been constrained adequately. We have already discussed this in section 7.5.10, but it is worth repeating here. Rigid body motion will result if a model is allowed to move in any direction without encountering resistance. Figure 8.16 shows some examples of rigid body motion.

Generally the FE solver will give the node numbers or locations and DOF associated with the error. A typical error would read something like: "The solver has detected that the value of UY at node 1256 is 20e6 which is greater than the limit allowed by the current analysis.......please verify your constraints." In such cases you should focus your debugging efforts on that part of the model.

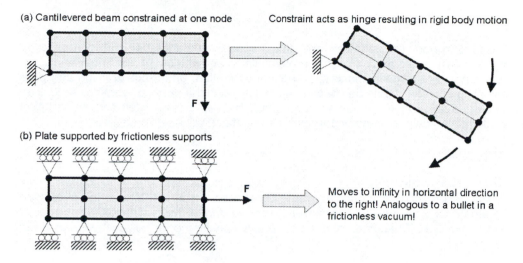

(a) Cantilevered beam constrained at one node   Constraint acts as hinge resulting in rigid body motion

F

(b) Plate supported by frictionless supports

F

Moves to infinity in horizontal direction to the right! Analogous to a bullet in a frictionless vacuum!

Figure 8.16: Examples of rotational and linear rigid body motion resulting from insufficiently constrained models.

Excessive Deformations

The other major error that can occur with a linear static analysis is when the solver detects large displacements without rigid body motion. In such cases the model is adequately constrained but certain nodes, or even the entire model has deformed excessively. Remember that a linear elastic material model must be used in a linear static analysis and this model only allows for infinitesimal strains; usually of the order of less than 1%. So if the solver detects strains in excess of this it may generate an error message and/or stop the solution.

In such cases it is likely that there is one of four things wrong with your model: you have entered an incorrect high value for the applied loads, you have entered a value that is too low in the material model definition, you have used dimensions that are too small in your model geometry or you have used a non consistent system of units. The first three are easy to check and it should be obvious where the problem is. If inconsistent units have been used it is sometimes more difficult to determine and you should reread the relevant section in chapter two to ensure that your units are consistent.

Poor Modelling Choices or Solution Setup

The only other reason why a linear static solution may fail to run is because information is missing from the model or material models or element types that are incompatible or not suitable for the analysis have been picked. For example if you specify a plane stress with thickness element behaviour or shell elements and then forget to specify the element thickness then the model will not be able to solve as it will assume a zero thickness of each of the elements. Specifying an acceleration or gravity load and then forgetting to define the density of the material is another common mistake. Incorrectly selecting thermal or fluid elements, for example, meshing your problem and then asking the solver to conduct a structural analysis is another common mistake. Similarly, applying non-suitable loads such as time varying, thermal or fluid type loads in the model will result in an error when attempting a solution. If you have followed the advice in this book you should have avoided such mistakes.

8.10.3 Problems with Non-Linear Analyses

Problems in non-linear analysis are almost exclusively concerned with convergence issues. Having said that all of the problems listed above, concerning linear analyses, also apply to non-linear analysis, so rigid body motion, under-constraining and poor modelling choices should be checked as explained in the previous section. As well as convergence problems, contact problems and excessive element deformation problems can plague a non-linear analysis and these will be discussed below.

<u>Excessive Element Distortion</u>

A non-linear analysis will always use a non-linear material model which will allow plastic deformation or non-linear elastic deformation of the elements. In either case the elements can become excessively distorted during the simulation if the applied load is too high, if the material model is poorly specified or if the mesh design is not adequate.

The most extreme case of element distortion is when the solver detects that the element boundaries have crossed each other as shown in figure 8.17. In this case the solver, when recalculating the element area or volume for the next load step will determine that the element has a negative area or volume. In such cases the solver will report one of the following:
-negative volume in element 2314
-negative Jacobian in element 2756
Recall, from chapter 4 that the Jacobian is used to determine element area so a negative value will imply a negative element area.

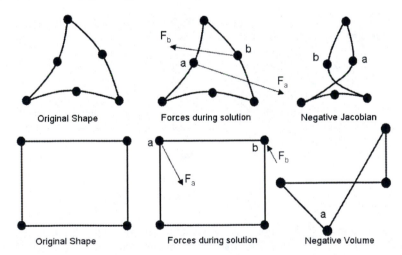

Figure 8.17: Excessive element distortion results in negative volume or negative Jacobian messages

If after receiving a negative volume or Jacobian message you have checked the material model definition, the applied loads and the unit system and they are all correct then a re-mesh is required, particularly at the region of the elements indicated by the error message. This can be achieved using adaptive meshing if available otherwise the mesh should be redesigned based on the expected deformation. For example if the model is to be compressed significantly in the vertical direction during the simulation and you are using quadrilaterals then the mesh should be generated so that the elements are initially elongated in the vertical direction. In this manner when the elements are compressed during the solution they will assume the ideal square shape, see figure 8.18.

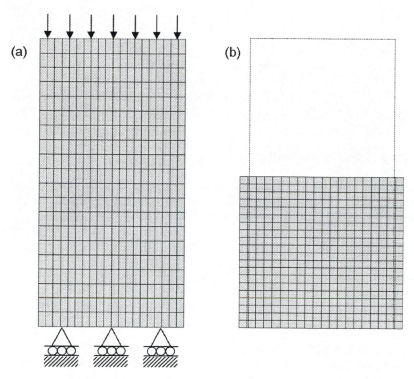

Figure 8.18: The mesh should be designed based on the expected results in order to avoid excessive distortions.

Contact Problems

Contact will usually be present in some form in a non-linear analysis. Contact is frequently the cause of problems in non-linear models via convergence difficulties or otherwise. Before using contact for a complex non-linear problem you should be sure that you know how to use contact and how to adjust the various parameters that define a contact pair or contact surface. If you can't run a simple analysis of contact between two spheres or a sphere and a plate and understand the effects of varying contact stiffness, friction coefficients, penetration tolerances, etc. then it is highly unlikely that you will get a complex non-linear model to solve!

You should be sure that contact regions will actually be in use when the solution is run. Often the analyst thinks that an area may be subjected to contact, when in fact it is not, and the solver becomes confused when trying to establish contact in these regions. Equally, unexpected regions can contact instead of the regions you predicted! If in doubt use a regular boundary condition/constraint instead.

If you are relying on contact being established to constrain certain parts of model be warned that sometimes, initially open contact regions can displace significantly before the contact is established and "miss" the target surface, resulting in rigid body motions. Most solvers will offer contact checking commands which will provide the initial status of any contact pairs (i.e. open or closed) which will allow you to identify such problem areas.

The bottom line is: spend a lot of time ensuring that you know what you are doing when you define contact, practice on simpler models and check the status of contact pairs throughout the available results or load steps to see how they are behaving.

<u>Convergence Problems</u>

Non-linear analysis is complex and it is not uncommon for a non-linear analysis to fail to converge on the first few attempts. In most cases some simple adjustments to the model or the solution parameters will aid convergence and ensure that your model solves. If you have a convergence problem then you should try one or more of the following suggestions

1.   Apply and remove the load gradually

For a non-linear analysis to converge it is necessary that the load is applied in increments small enough to ensure that the analysis will closely follow the structure's load-response curve. If you apply the load gradually you will normally reduced the number of Newton-Raphson equilibrium iterations required for each step and thus have a better chance of the solution converging.

Apply loads gradually is also common sense. Any system moved from rest abruptly will react with severe dynamic effects and will immediately stiffen. Understand that the only difference between a forceful push and an impact is how quickly the load is applied!

Even if it is required to have a load increase quickly it is good practice to have a "lead-in" period where the loading rate is slower in order to gently get the structure moving before faster loads are applied, see figure 8.19. It is equally important to not suddenly stop the model from moving and to provide appropriate "lead-out" periods where the loading rate is reduced.

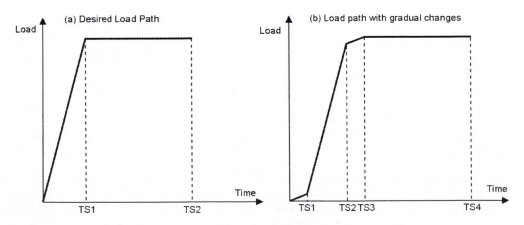

Figure 8.19: Applying and removing a load gradually using lead in and lead out periods often aids convergence.

Figure 8.19(b) also illustrates the importance of using a different load step or time step when you are making any change to the load. Trying to change the load rate during a load step is asking for convergence difficulties – reread section 8.9.2 if you don't understand why.

If you are running a non-linear static analysis another technique that may be used in overcoming initial instability due to applied loads is to run a "slow" transient analysis instead. Introducing time integration often prevents such problems from diverging.

2.   Change the way the loads are applied

Sometimes point loads or constraints will cause highly localised stresses which will make convergence difficult. If you run into difficulties then try to replace point loads with distributed loads. In certain cases it will be possible to use forced displacements instead of applied forces, see section 7.5.6. This approach is often useful to start a non-linear static solution closer to the equilibrium position or to control displacements through periods of instability, such as post buckling.

3.   Change the material model

The type of material model used may have a detrimental effect on convergence for the particular problem being studied. Try using an alternative material model if available, for example replace a bilinear model with an equivalent power law model.

Temporary artificial stiffness can be applied to unstable regions of the mesh using specialised techniques in order to artificially restrain the system during intermediate load steps and prevent unrealistically large deformations being predicted. Once the system has displaced into a stable configuration the artificial stiffness can be removed. This method is most often applied via specialised elements which are set to "switch on" and "switch off" at specialised times.

4.   Change the mesh

All of the advice given on mesh generation particularly applies to non-linear models. Non-linear solutions are particularly sensitive to the quality of the mesh. Ensure that your mesh is fine enough in areas of expected high stress, has smooth transitions and is as regular as possible. If you have been using triangles or tetrahedrals try and replace them with quadrilaterals or bricks. If you have been using linear elements try and replace them with quadratic elements.

5.   Check preliminary results

If the solution completed a few load steps before diverging then look at the results from these steps and try to identify areas of strange stress or displacement behaviour.

Some solvers allow you to plot the Newton-Raphson residual forces from the completed equilibrium iterations. Obtaining a contour plot of the distribution of the residual forces in the model will quickly identify regions that are causing convergence difficulties. This method can be very useful where convergence difficulties occur in the middle of a load step and particularly where the model has large amounts of contact and other non-linearities. By identifying a portion of the model where large residuals persist, such as a particular contact surface, you can focus on the non-linearities in that area instead of dealing with the entire model.

Contact pairs often cause difficulties in convergence and examining the completed load steps often highlights problem contact surfaces or entities. You should attempt to determine how and when contact occurs during the completed load steps. Some solvers have specialised contact analysis tools that help to determine regions where contact is unstable. Changing appropriate contact parameters can then help the solution to converge.

Generating a reaction force versus deflection curve for particular nodes is often useful to indicate areas of the model where instabilities may be occurring such as possible buckling.

6.   Use your solvers advanced solution control methods

Since convergence is such a problem in non-linear analysis, most commercial FE solvers offer a myriad of options and methods that can be switched on and off in order to try and improve convergence. The arc-length method discussed in section 8.9.2 is one such method. You should consult your software's documentation in order to determine what tools it offers to aid convergence.

It is also possible to relax the convergence tolerance in order to force a solution. Obviously this solution should not be used for design work or in analysis reports as the accuracy has been compromised, but rather, can be used to determine regions of stress or displacement anomalies which can then be treated accordingly. Once the required changes have been made to the model the convergence tolerance should be reset and a further analysis run to check if convergence can now be achieved.

7.   Don't despair!

Non-linear analysis is difficult. Analysis of non-linear problems is currently at the forefront of research. Several types of highly non-linear problems such as high energy impact events, simulation of certain metal forming processes and simulation of in-service behaviour of medical implants have still not been simulated to a satisfactory degree of accuracy by the research community. A great deal of experience is required to successfully solve many non-linear problems and obtaining convergence is something of a black art.

With this in mind it is important to realise when you are out of your depth. In such cases you should get advice from someone more experienced. If you don't immediately have access to a mentor then contact your software support team or join one of the many excellent user groups on the internet. A quick search on the internet will identify user groups for all of the popular commercial FE solvers.

8.11.   Summary of Chapter 8

After completing this chapter you should:
- Know how to perform the required final checks on your model to ensure that it is ready for solution.
- Understand the capabilities and limitations of each type of analysis type available in structural finite element analysis.
- Be able to determine which analysis type is suitable for your particular problem.
- Understand and explain the basic methods used to solve linear and non-linear structural problems.
- Understand the common problems that occur during solution of linear and non-linear problems and be able to apply common fixes in order to obtain a solution.

9

Post Processing

9.1.   Introduction

In many ways post-processing could be considered the most important part of the FE process. This is the part of the process where you examine the results generated by the FE model, manipulate them to show the information you want, and ultimately make decisions about the quality of your model and the related results. Only now do you get to understand how the applied loads affect your design and how good your finite element mesh is. There are many different methods of displaying and interpreting results and we shall discuss each of them in this chapter.

9.2.   Overview of Post-processing

The reason you have spent all that valuable time up to this stage in building a finite element model, applying relevant loads and boundary conditions and specifying a particular solution type is that you want an answer to some critical questions about your problem. These questions could be: What is the level of stress in a particular region? How much does the structure move when the load is applied? Will my design work when put into use? The post-processing phase of a finite element analysis is where you find out the answers to these and other questions.

It is important to understand that the FE solver will only produce the displacement of each node (at each time step if applicable) in any structural finite element analysis. Recall that the global problem equation is [K] {U} = {F}, and the solution returns the global displacement vector {U}. Return to chapters 3 and 4 and take a look at some of the worked examples if you don't understand why this is. In order to determine derived quantities such as strain and stress the results must be further manipulated, hence the term post-processing of results. An overview of most post processing procedures is shown in figure 9.01.

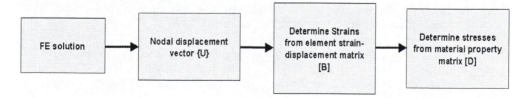

Figure 9.01: General procedure for post-processing

As strain is the change in element length divided by its original length, in a particular direction, then the element strain-displacement matrix [B] (see chapter 4) is used to transform the nodal displacement vector {U} for each element into the strain for each element. Once strain has been calculated, the element material property matrix [D] is used to relate element strain to element stress. Recall that the strain-displacement matrix for each element [B] is derived from the element shape functions [S], so by choosing an element type we specify [B] and hence how strain will be evaluated and by specifying a material model [D] we specify how stress will be calculated. This illustrates an important point of post-processing and shows how if an incorrect material model is specified, the predicted displacement from a FE solution may be correct but the predicted stress results will be in error.

The above illustrates how the shape functions [S] and their derivates [B] play an important role in post-processing. Consequently it is important to realise the limitations of various element formulations in terms of the results they can supply. For example, a linear spar element can provide a linear change of displacement in the element but stress within the element will be constant. In comparison, a quadratic spar element can have a quadratic variation of displacement within the element and a linear variation of stress. So, with the linear element stress is the same at all points in the element whereas with the quadratic element stress is different at each node. Figure 9.02 summarises this information for each of the common element types available in commercial FE software.

Element Type	No of Nodes	Displacement Distribution	Stress / Strain Distribution
Linear spar	2	Linear	Constant
Quadratic spar	3	Quadratic	Linear
Linear beam	2	Linear	Constant
Quadratic beam	3	Quadratic	Linear
2D Linear Triangles (CST)	3	Linear	Constant
2D Quadratic Triangle	6	Quadratic	Linear
2D Linear Quadrilateral (Q4)	4	Bi-Linear	Linear *
2D Quadratic Quadrilateral (Q8)	8	Quadratic	Linear
3D Linear Shell	4	Linear	Constant
3D Quadratic Shell	8	Quadratic	Linear
3D Linear tetrahedral	4	Linear	Constant
3D Quadratic tetrahedral	10	Quadratic	Linear
3D Linear hexahedral	8	Linear	Constant
3D Quadratic hexahedral	20	Quadratic	Linear

* not in all cases, depends on element distortion

Figure 9.02: Overview of element displacement and stress-strain capabilities

In reality, it is not as straightforward as this, as the element matrices of more complex elements are calculated using numerical integration and, after solution, stresses are usually only calculated at the integration points in the element. The integration points are usually not at the same location as the elements nodes. In general FE users are more interested in the results at the nodes so the stresses at the nodes are interpolated from the integration points to the nodes.

We will now introduce a general procedure for post-processing of structural finite element models which we will then discuss in detail in the subsequent sections of this chapter. Figure 9.03 shows the recommended general procedure which should be used for post-processing.

The first step is always to check the deformed shape of the structure. This tells us straight away if the loads have been applied correctly and if the boundary conditions are behaving as intended. Once the general deformation behaviour has been confirmed, the next step is to check the magnitude of displacement. If the magnitudes are too high or too low then this tells us that there is a problem with our applied load magnitudes, our model dimensions or our material model definition. In many cases this is also indicative of an inconsistent unit system being used. The next step in the process is to examine stress results. By checking contour plots of stress across the elements we can quickly determine how good the mesh we used was. If unacceptable discontinuities in stress contours are found then the model requires re-meshing with a more suitably refined mesh in order to capture the problem behaviour. If stress contours look ok then the displacement and stress results should be compared to the previous run to investigate if the model has converged. If the model has converged then the results can be deemed viable and can be further analysed.

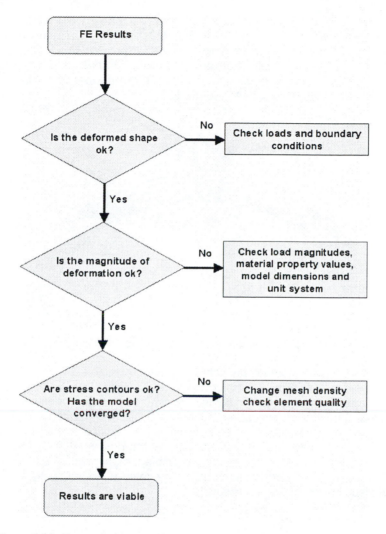

Figure 9.03: General procedure for post-processing of structural FEA results

Note that just because the results are deemed viable does not mean that they are accurate. A viable set of results is one in which: the displacement of the model is as expected, the magnitude of deformation is as expected and the stress results indicate that the model has converged. Further investigation of the result may later reveal that a particular assumption was invalid or that the model is unsuitable for the analysis being undertaken.

9.3. Displacement Results

The amount of textual output from even a relatively small finite element model will be very large. Long lists of numbers are produced detailing the displacement at each node in the model. In most cases trawling through a long list of numbers does not tell you very much about the behaviour of your model so graphical methods are used to present the information in a more concise and useful manner. Generating plots of deformed shape and contour plots of displacement provides a quick and easy way of checking that the model is deforming as expected/required.

<u>Deformed Shape Plots</u>

Checking the deformed shape of the mesh after solution is the first step in any model verification process. All FE software post-processors provide this facility and it will generally be at the top of the list of options in the post-processor. The ability to plot the deformed shape over the original shape is vitally important as it allows a direct representation of how the model has deformed.

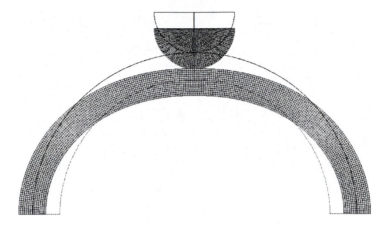

Figure 9.04: A typical plot of deformed shape overlaid on original model shape.

Some FE software will automatically scale the deformation of the model in order to aid the user's appreciation of the method of deformation. Consequently the magnitude of maximum deformation should always be checked and, if necessary, the scaling of deformed shape should be reduced/increased or even turned off. Typically infinitesimal deformations (i.e. linear elastic results) are scaled up as the deformation will usually be too small to see. Figure 9.04 shows the deformed shape result from a linear static analysis in which the maximum displacement was of the order of 10^{-5} metres. It is clear that the deformation has been scaled up significantly in this plot in order to aid visualisation of the deformation.

<u>Deformed Shape Animations</u>

Most modern post-processors offer the facility of animating the model deformation. In a linear static analysis, the post-processor takes advantage of the linear assumption to interpolate the load-deflection response into a number of steps and produces a frame for each interpolation. These frames are then combined into a short animation. This should be used with caution, as it should be understood that a linear static analysis always applies the load in <u>one</u> step: new users frequently misunderstand this when they produce an animation from the results of a linear static analysis. In a modal analysis the mode shapes at each frequency can be similarly animated.

For a non-linear static analysis or a transient analysis animated results have more meaning as a frame will be produced at each time step or load sub-step. In these cases the deformation at each frame is directly related to the deformation at a particular point during load application. It is also possible to get the post-processor to interpolate in between time steps or sub-steps if a smoother animation is required. This should be used with extreme caution as in this case the response in between steps is non-linear and the interpolation will generally assume a linear response.

Analysis of Deformed Shape

Once you have generated your deformed shape plot or animation the first thing you must check is if the model is deforming as you expected. If the model is deforming strangely then it is most likely that the applied loads or boundary conditions have been applied incorrectly. Remember that we are still only checking the shape of deformation at this step. Load magnitudes, dimensions, units and material properties will affect the magnitude of deformation. The only things that effect the <u>shape</u> of deformation are the loads and boundary conditions. Typical problems include: a boundary condition constraining the wrong DOF, a load pointing in the wrong direction, pressure loads pointing in the wrong direction due to shell normal's pointing in the wrong direction, pressure loads acting on the wrong face of 3D solid elements and contact elements not set up correctly.

If you can't figure out what is causing the problem, the best strategy is to change some of the boundary conditions and re-run the problem. Each re-run will tell you more about the model behaviour and should eventually identify the problem boundary condition or load.

Deformation Magnitudes

If the deformed shape looks as expected then the next step in our results verification process is to produce a contour plot of deformation magnitudes in order to check the magnitude of the predicted deformation. Figure 9.05 shows a typical plot of displacement magnitude for the deformed shape shown in figure 9.04. Remember that displacement is a vector quantity and the plot shown in figure 9.05 is a sum of displacement in the x and y directions. It is often more useful to generate contour plots of displacement specifically in the global problem directions. Figures 9.06 and 9.07 show corresponding plots for displacement in the horizontal and vertical directions for the same model. In each of these plots the maximum displacement is of the order of 10^{-5} meters which is as expected as these are results from a linear static analysis.

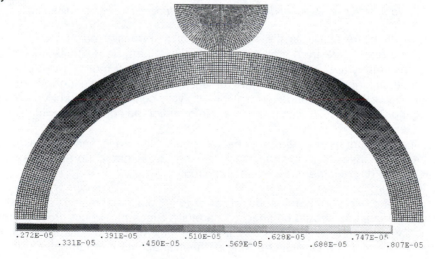

.272E-05		.391E-05		.510E-05		.628E-05		.747E-05	
	.331E-05		.450E-05		.569E-05		.688E-05		.807E-05

Figure 9.05: Contour plot of displacement magnitude

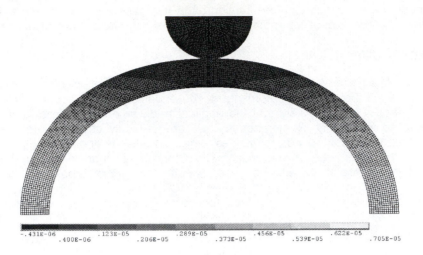

Figure 9.06: Contour plot of displacement magnitude in the horizontal direction

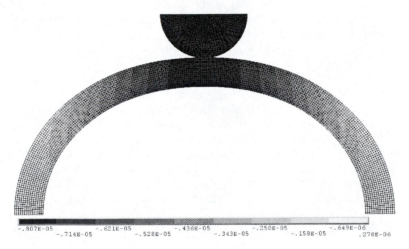

Figure 9.07: Contour plot of displacement magnitude in the vertical direction

Once these plots are available the magnitudes of deformation can be checked for viability. If the results are from a linear static analysis then the displacement should be very small. If they are not then there is a problem with your model. Most likely you have entered an incorrectly high value for an applied load, an incorrect material model parameter, incorrect model dimensions or, very commonly, you have not used a consistent unit system. Check chapter 2 for information on consistent units.

If everything appears ok then the assumption of a linear static analysis being valid should be questioned. Perhaps the loads are so large that a non-linear material model and large displacement analysis is required.

Even if you are viewing the results of a large displacement analysis then the deformation magnitude should be checked carefully. Since most post-processors will automatically scale results, excessively large displacements could appear ok in a deformed shape plot, but careful checking of the displacement magnitudes will show up the error.

If the displacements seem reasonable then they should be compared with analytical or experimental results where available. If the FE displacements do not compare well with the other results then it is most likely that the finite element mesh has not converged. In general even very basic analytical predictions should not be more than an order of magnitude different from the FE results. A convergence process, as described in chapter 7, should be undertaken in order to converge the displacement results. Even in cases where the only purpose of the FEA is to determine stresses, displacements should be converged before even looking at stresses. This is due to the fact that stress results are derived from displacement results, so stress will not converge before displacements, hence a converged model based on displacements forms the starting point for a model converged on stresses.

9.4.   Stress and Strain Results

Nodal (averaged) versus Element (un-averaged) Results

Since displacement of each node is the primary data produced by a structural analysis it is always presented in terms of nodes. When you produce a contour plot of displacement, as in the previous section, the displacement results are always averaged at the nodes in order to produce as smooth a contour as possible. This manner of presenting results is known as "nodal" or "averaged" results.

Stresses and strains, on the other hand can be presented in two forms: as nodal (averaged) or element (un-averaged) contours. Due to the nature of FEA which requires splitting the problem up into elements there will always be discontinuities in stress and strain across element boundaries. In reality this is not the case and stress/strain will be continuous throughout the structure. In a well converged finite element model the discontinuities across element boundaries will be minimal. In poorly converged FE models, especially near regions of stress concentration, these discontinuities can be large and appear as a jagged contour. As mentioned in chapter 7, examining these un-averaged results is crucial during mesh convergence and reducing the "jaggedness" is the key to mesh convergence in regions of rapidly changing stresses.

The un-averaged results can also be displayed as averaged (nodal) contours where, regardless of the quality of the un-averaged results, the contours will always appear smooth. This is because the averaged method takes the values for stress or strain across the element edges and averages them in order to produce a transition value that will make the contour continuous across the element edges. If the mesh has been properly converged and the un-averaged contours have been checked for reasonable continuity then the averaged contour plot will usually produce accurate results. If the mesh has not been converged and/or there are jagged un-averaged contours then the averaged contour plots are quite dangerous as they can mistakenly make the user think the results are smooth and that everything is ok.

The averaging method used in averaged results can generally be controlled by the user. This facility is important as in certain cases it is not recommend that the results be averaged across element boundaries. Examples of such cases include: between elements with different material properties, between shell elements of different thickness, between beam or truss elements of different cross section or between different type elements (e.g. at shell/solid interfaces). Figure 9.08 shows an example of averaged and un-averaged contour plots of stress for an un-converged model and figure 9.09 shows a similar example for a converged model. Notice how there is little difference between the averaged and un-averaged contours in the converged model, whereas there are significant differences in the un-converged model.

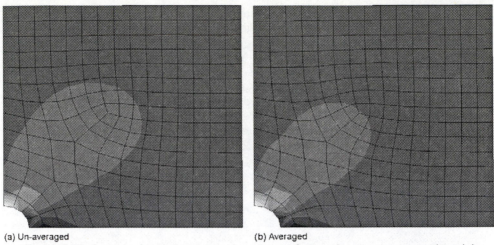

(a) Un-averaged   (b) Averaged

Figure 9.08: Averaged and un-averaged contours of stress in an un-converged model

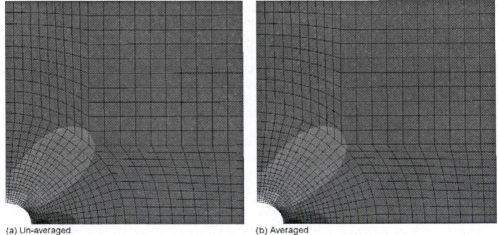

(a) Un-averaged   (b) Averaged

Figure 9.09: Averaged and un-averaged contours of stress in a converged model

<u>Types of Stress and Strain</u>

A wide variety of strain and stress types are usually available for plotting in commercial FE software. We will discuss the more commonly used strain and stress quantities here. In most structural fields the stresses in the model are of more interest than the strains. You will already have checked deformation magnitudes by this stage so checking of strain magnitude is usually not necessary. Exceptions to this rule include disciplines where strain values are of more interest such as metal forming, where failure is often based on failure strains. Also, if the FE results are being compared to an experimental test where strain gauges have been used to take measurements then plotting strains often allows direct correlation of experimental and FE results.

- Directional Strains and Stresses

The most obvious method of displaying strains and stresses is to plot the distribution of the stress or strain in a particular global or local direction. This can be done at each solution point in the model (i.e. integration point or node). This type of result is very useful when the problem has been set up so that a global or local direction corresponds to the orientation of a strain gauge on a prototype or if an analytical model produced results in a particular direction. Many types of problems will also have dominant directions in which the results are of more interest, e.g. cylindrical problems are usually split into the radial, axial and circumferential directions. By setting up appropriate coordinate systems results can be aligned with any desired directions. In general, positive values indicate tensile stresses/strains and negative values indicate compressive stresses/strains.

- Principal Strains and Stresses

The strains in each direction at each solution point can also be resolved into their principal components. See chapter 2 for a description of principal strains and stresses. The maximum principal stress/strain will always be the most positive of these (regardless of magnitude) and the minimum will be the least positive. In a 3D problem a 3rd principal stress/strain known as the median or mid principal stress/strain will be numerically in between the maximum and minimum.

- Stress and Strain Intensity

Stress intensity or strain intensity is often used as a measure of average stress or combined stress at a particular location. Stress intensity is the largest absolute difference between any two of the principal stresses, i.e.

$$\sigma_{int} = \max\left(|\sigma_1 - \sigma_2|, |\sigma_2 - \sigma_3|, |\sigma_3 - \sigma_1|\right) \tag{9.01}$$

Similarly, strain intensity is given by:

$$\varepsilon_{int} = \max\left(|\varepsilon_1 - \varepsilon_2|, |\varepsilon_2 - \varepsilon_3|, |\varepsilon_3 - \varepsilon_1|\right) \tag{9.02}$$

- Von Mises Stress and Strain (Equivalent Stress and Strain)

We have already defined von-Mises stress in chapter 2, but for convenience we will restate it here:

$$\sigma_{VM} = \frac{1}{\sqrt{2}}\sqrt{(\sigma_1 - \sigma_2)^2 + (\sigma_2 - \sigma_3)^2 + (\sigma_3 - \sigma_1)^2} \tag{9.03}$$

Similarly, von-Mises strain is derived from the principal strains and is given by:

$$\varepsilon_{VM} = \frac{1}{1+v}\left(\frac{1}{\sqrt{2}}\sqrt{(\varepsilon_1 - \varepsilon_2)^2 + (\varepsilon_2 - \varepsilon_3)^2 + (\varepsilon_3 - \varepsilon_1)^2}\right) \tag{9.04}$$

The von-Mises stress and strain are also sometimes known as "equivalent" stress and strain. It should be noted however that an equivalent stress is not necessarily a von-Mises stress. An equivalent stress is basically any stress that takes the three principal stresses or six directional stresses (3 normal and 3 shear) and turns them into a single number which can then be compared to a uniaxial measure of yield. This means that Tresca stress is also an equivalent stress. Similarly Hill's yield criterion given in equation 5.25 can be arranged into an equivalent stress form. Having said that, 90% of the time when equivalent stress is referred to it can be taken that von-Mises stress is implied.

Equivalent stresses are very useful for illustrating the average stress at a particular location, as well as predicting yield and failure, however they should be used with caution as an equivalent stress will always be positive and so will not differentiate between tensile and compressive stress states.

- Hydrostatic Stress

Hydrostatic stress is sometimes used as a method of visualising average stress at a particular location in the model. At any location the hydrostatic stress is the mean of the three principal stresses, i.e.

$$\sigma_H = \frac{1}{3}(\sigma_1 + \sigma_2 + \sigma_3)$$
(9.05)

- Specialised Stresses and Strains for Beams and Shells

Various specialised element types have specific directional stresses associated with them, for example beam elements will always have the capability to allow plotting of axial stresses, torsional stresses and bending stresses. In many cases you can pick the location in the beam cross section for which you wish results to be reported, for example you may be particularly interested in results at the top or bottom surface of the beam.

Similarly shell elements generally allow for shell membrane stresses and transverse shear stresses to be plotted. In order to produce stress plots the post-processor must be informed which location through the shell thickness the user is interested in. Typically the available choices are the top, middle or bottom surface of the shell. The direction of the shell normal will determine which side of the element the post-processor considers to be the top or bottom and this should be checked carefully before plotting results.

9.5. Reaction Force and Resultant Force Results

Determining the reaction forces and moments at the model constraints will often be an important part of a FEA. Reaction forces can be used as input loads to other FE models of adjoining structures which have been modelled as constraints in the current analysis. Checking that the sum of the reaction forces equal the applied loads is a fundamental method of checking model validity.

Similarly the resultant forces that occur in each element in the model during the solution can be plotted. Resultant forces are often useful for determining the strength of internal joints when assemblies are being modelled and can be used to determine suitable load magnitudes for sub-modelling procedures.

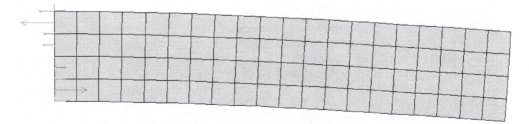

Figure 9.10: Typical reaction force display

9.6.   Methods for Display of Results

Contour plots

Contour plots are used to show how a result item varies over the geometry of the model. As mentioned above these plots can be either averaged across the nodes or un-averaged. Contour plots can be produced directly for 2D and 3D elements to show how the result changes across the geometry. For line elements (i.e. beams and trusses) contour plots can often be difficult to read so results can be mapped onto the line as shown in figure 9.11.

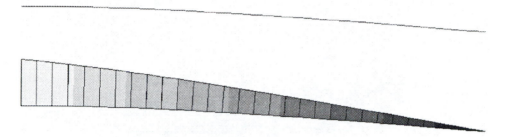

Figure 9.11: Contour plot mapped onto beam elements.

The top of figure 9.11 shows the deformed shape of the beam elements and the bottom shows a contour plot of maximum stress mapped onto this deformed shape.

Vector plots

Any result quantity that is a vector (i.e. that has both a magnitude and direction) can be plotted in a vector plot. Such quantities include displacement, rotation, velocity, acceleration and principal stresses. Vector plots of displacement are very useful in order to determine material flow when modelling metal forming processes. Principal stress vector plots show the direction of the principal stresses throughout the model and are useful for aiding strain gauge positioning in verification testing.

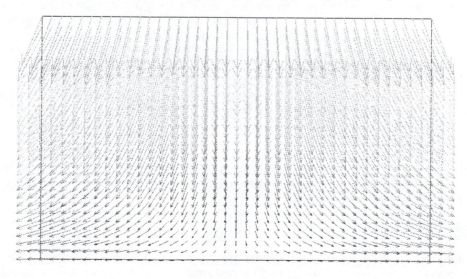

Figure 9.12: Vector plot of displacement showing material flow during forging of a rectangular block.

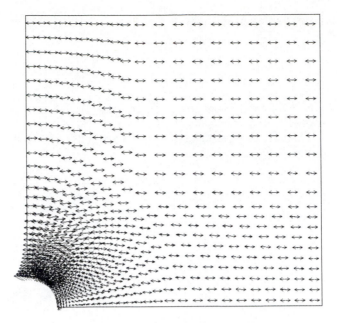

Figure 9.13: Vector plot of maximum principal stress in a ¼ model of a plate with a hole..

Animations

We have already discussed animation of deformations in section 9.3. For a non-linear static analysis or a transient analysis animated results are very useful to illustrate how a result quantity such as stress or strain develops and changes during a simulation. In a non-linear static analysis a frame is generally produced at each load step or sub-step thus allowing direct correlation of applied load with displacement, stress or strain. In a dynamic analysis results are produced at particular times, usually the time steps, again allowing direct correlation with applied load via the applied load versus time input.

As mentioned previously it is sometimes possible to get the post-processor to interpolate in between time steps or sub-steps if a smoother animation is required, however, this method should be used with extreme caution.

Path Plots

Path plots are essentially graphs that are used to illustrate how a particular result quantity varies along a particular predefined path through the model. Typically the path will consist of a line of nodes. Figure 9.14 shows an example of path plots for the plate with a hole problem.

Path plots are very useful for check results continuity and may be used as part of the mesh convergence process. As the mesh reaches convergence then the path plot should become more smooth.

Graphs

Graphs are typically used in non-linear and dynamic analyses to show how a particular result quantity varies during the simulation. In a non-linear static analysis the x axis is normally load step or sub-step, while in a dynamic analysis the x-axis will be time step or simulation time. Such graphs are useful for determining the point during the simulation where a particular threshold was exceeded (e.g. yield stress) and hence the applied load associated with this point. Figure 9.15 shows a typical graph from a dynamic analysis which shows how principal stresses and von-Mises stress developed during the simulation

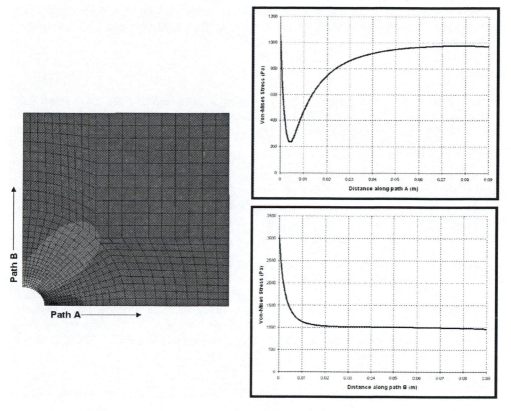

Figure 9.14: Path plots of Von-Mises stress along the side and bottom edges of a ¼ model of a plate with a hole.

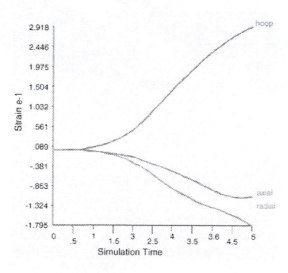

Figure 9.15: Graph of the development of principal strains in a particular node in a model during a FE solution.

Damage Plots

For material models such as concrete models and composite models that allow for prediction of damage, special plots are available which show which elements have failed, cracked or crushed due to the applied loading. Different FE software use different methods to illustrate the damaged elements. Some use symbols on the elements to indicate what type of damage has occurred, others use colours or contour plots.

Lists

Any result item can be listed or produced in tabular form during post-processing. List of stresses at each node in the model, displacements of each node, derived data at each element, etc, can be useful to determine specific points in the model that exhibit high stresses etc.

Lists of applied loads and reaction loads are a useful way to allow for equilibrium checks. As previously mentioned, it is good practice to always check that the sum of the applied loads equals the sum of the reactions in any particular direction.

Iso-surfaces and Slicing Techniques

An iso-surface is a surface of constant value through a 3D model. For example, you may be interested in seeing what portion of the model is above a certain stress threshold, by defining iso-surfaces you can clearly see the internal surface which divides the part of the model above that stress threshold to that below the threshold. Figure 9.16(a) shows a regular plot of von-Mises stress in a 3Dsolid model of a cantilevered beam subjected to an end load on one corner. Figure 9.16(b) shows the same result as an iso-surface plot.

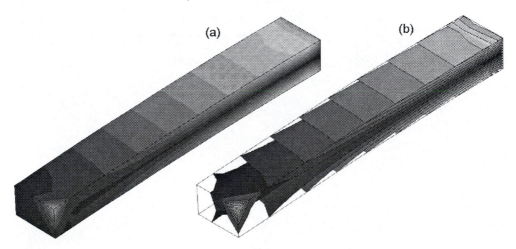

Figure 9.16: (a) Regular contour plot of von-Mises stress (b) Iso-surface plot of von-Mises stress.

Slicing techniques are also used to see what is happening in the interior of a 3D model. It is possible to view results as a series of slices through the model and to subsequently animate these results to show how the result quantity varies through the interior of the model as you move in a particular direction along the model. Figure 9.17 shows an example of two different slices through the model shown in figure 9.16(a).

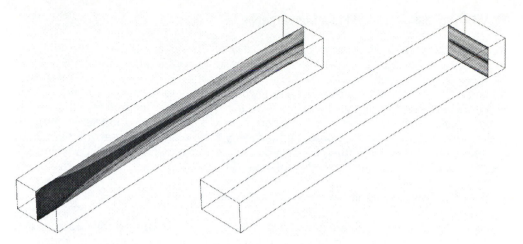

Figure 9.17: Longitudinal and transverse slices through a cantilevered beam with unbalanced tip load.

Results coordinate systems

As we know result data is calculated for each element in terms of either the local nodal coordinate system (for nodal results such as displacements) or the local element coordinate system (for derived data such as stresses or strains). In general results in local coordinate systems are not of any interest to the user so these results are automatically transformed into the global problem coordinate system before being displayed or listed. In most cases this will be the global Cartesian coordinate system.

If required, the results display coordinate system can be changed to another global coordinate, a user defined local coordinate system or back to the local element coordinate system. Results can then be displayed or listed in terms of this coordinate system. This procedure is very useful for problems that exhibit particular behaviour due to their geometry and where aligning results with a particular direction in the geometry may provide more useful results. A typical example is any tubular problem (e.g. expansion of a tubular pressure vessel) where the dominant problem directions will be the radial, axial and circumferential directions. By setting up an appropriate local cylindrical coordinate system results can automatically be displayed in the dominant directions.

See section 6.2.10 for more information on coordinate systems.

9.7. Results Verification

After you have displayed your results, checked the deformed shape, checked the deformation magnitudes and checked stress levels you will be at the point where you have to take an objective look at the results obtained and make some decisions. These decisions will be dependant on the reasons for conducting the analysis in the first place: if the analysis is part of a design process then you will have to make design decisions based on the results or, if the analysis is part of a wider analysis you may wish to compare FEA results with experimental results or results from other models. Regardless of why you want the results there are a number of verification steps which you should undertake for each FE model that you build.

The verification process will contain all or some of the following steps:

9.7.1. Quality Assurance – Model Convergence

The most basic method of ensuring the quality of FE results is to ensure that the model has converged. We have already discussed this at length in section 6.3.7 but the most important point is that the model should not contain very rapidly changing stresses. If the variation in stress between two adjoining elements is large then more elements are needed at this location. Similarly, path plots of results through the model should be relatively smooth (see figure 9.14 for a good example). A convergence curve should be produced for every finite element model you build and it is reasonable to assume that convergence has occurred when there is a less than 5% change in results between models of increasing mesh density.

9.7.2. Comparison of Results with Expected Results

Before completing the analysis you will most likely have an idea how the structure will behave. If the predicted FE results are very different to what you expected then you should examine the model carefully. It is possible that the FE results may be correct but in general when a model does something you don't expect it is a sign of trouble. Check your assumptions carefully, particularly the B.C.'s, loading and material model. Make sure that displacements are not scaled up and that you are using the correct coordinate system. Basically, assume that you are right and the FE solution is wrong and perform every check you can think of before agreeing with the results that the FEA has predicted.

9.7.3. Comparison of Results with Hand Calculations

Before the advent of FEA paper based engineering analysis and modelling still allowed for the design of many impressive structures and products which still work today. Just because you are now competent in carrying out a basic FEA doesn't mean that you should throw away all of these analytical tools. Many problems and systems can be simplified and their behaviour approximated with equations which allow paper based analysis or "text book" methods. You should always attempt to find a relevant paper based calculation to compare with your FE model results. There are several reference books available which list hundreds of equations for calculating stress and strain results in many different problems. Typical problems which are easily analysed with hand calculations included beam bending, thin and thick walled pressure vessels, plates with holes or notches, cylinders with holes etc.

In most cases the hand calculation will be much simpler than your model and you shouldn't expect a perfect match between the two, however, if the results are more that an order of magnitude in difference then it is likely that there is a problem with your FE model. Remember that these equations were used to design bridges and towers that have stood for hundreds of years, so they are reasonably accurate!

9.7.4. Using a Model Hierarchy

We briefly discussed the concept of using model hierarchies in chapter 1 and chapter 3. One of the most obvious ways to check the validity of your base model is to compare the results to those obtained using simpler and more complex models. If a problem is particularly difficult then setting up and running a much simplified test model is good practice in order to determine what "ballpark" the expected results will be in. Once this is known the base model can be run with an expected set of results already known from the simplified model. The basic methodology of constructing a simplified model is to make even more assumptions than made previously, e.g.

ignoring any features that don't allow symmetry, assuming plane stress even if this is not strictly the case, etc.

Similarly if you are reasonably confident that your base FE model is correct but you want to be more certain, then setting up and running an even more complex model can show you how confident you should be in the base model results. The methodology for generating the more complex model involves reducing the number of assumptions, e.g. introducing contact, modelling supports, using non-linear material models, creating a 3D model, etc.

Obviously using a more complex or simpler model will change the results however the magnitude of deformation should not change significantly.

9.7.5. Comparison to Experimental Results

By far the best method of verifying and validating FE results is to compare them with results obtain from an experimental testing programme. Many FE software companies market FE as a method to end experimental testing forever and this is what many customers expect when they commission an analyst to perform a FEA. While it is true that FEA has reduced the need for experimental testing, it most definitely has not eliminated the need for testing. Would you really trust the results of a FEA that has not been validated against the results of at least a few experimental tests? What about if the FEA was used to design the aeroplane you are about to travel on? I doubt it.

Always use experimental test results when they are available. Take all the information you can get: deformations, strain gauge results, strain rates, velocities, accelerations, photo-elasticity results etc. Compare these results against the equivalent results predicted by the FE model. If they are different, find out why and fix the FE model. Remember nature isn't wrong the FE model is! Given all the assumptions inherent in the FE process and all the assumptions you have made, if the FE and experimental results do not correlate then it is most likely that the FE model has problems, provided that the experimental results are reliable.

If possible you should try to be involved in the experimental testing process so that you can be confident that the methods used are reliable. Again, if possible, you should try a few FE simulations before experimental testing so that you have an idea how the structure will behave. Involvement in the test process will provide valuable information about the assumptions in your FEA. Similarly you should always try to carry out a few material tests on the material used in the design. Often material property data supplied by manufacturers is in error or doesn't take account of differences between batches. Performing your own material tests allows you to have confidence in the material model used in the FE model.

If and when your FE model results correlate well with the test results then you have a very powerful model on your hands. Now that you are sure that the model is accurately modelling nature you can confidently change model parameters and obtain an optimal design for the product under investigation.

9.7.6. Comparison to the Work of Others

In cases where it is not possible to carry out your own experimental tests and the model is so complex to render a hierarchy study redundant, it is good practice to perform a literature search for similar experimental or FE work reported by others. If a researcher has published a research paper on a similar model then you should begin your analysis by recreating those results. Once you have recreated the work of others in the field you can have reasonable confidence that the results from your model will be valid and accurate. You should preferably try to obtain experimental test results in this manner but when this is not possible it is satisfactory to compare your model to other FE models produced by other analysts. If your model behaves in a similar manner with similar magnitudes of deformation and/or stress then it is likely that your model is valid. In such cases, however, you have to consider the possibility that both you and the other analyst made the same mistakes!

9.8. Statistical Methods and FE Results

Once you have examined your results, validated your model and are confident that your model is satisfactorily modelling the real situation, the next step is often to vary different model parameters to see how they effect the results. Rather than simply changing parameters in an unplanned manner it is good practice to use statistical methods in a similar manner to "design of experiments" methodology. This section presents methods that can be used with validated finite element models: it is assumed that you have already obtained a solution validated your model results against experiments or other models.

9.8.1. Sensitivity Analysis

It is often required during a finite element analysis to determine the effects of changing various model parameters on the resultant performance of the model. For example, it may be required to explore the effects of varying material parameters or dimensions on the resultant maximum von-Mises stress in the model. Many users approach such a task in a haphazard manner and change parameters arbitrarily without any plan or overall strategy. There are statistical methods available for planning such analyses known as factorial analysis which provide a structured method where many variables can be changed simultaneously and results obtained much quicker than if an unplanned approach was used.

There are many methods available for performing a factorial analysis. In this section we shall discuss the Taguchi method which is the most popular method in engineering analysis. This method is used to identify input parameters and their interactions that have the largest effect on the FE results. Traditionally analysts have used the "change one input at a time" approach in order to determine the effect of each input on the FE results. This method is costly in terms of analyst time, as a new model needs to be built each time an input is changed, and in terms of computational resources. The Taguchi method allows determination of the sensitivity of the FE model to variation of a large number of input parameters using much fewer models than traditional methods.

Using the Taguchi method, the different input parameters (such as material properties, dimensions, loads etc) are known as factors. Each factor is assigned levels which equally divide the range of each factor. For example if Young's modulus was designated as a factor it could be assigned three levels: level 1 200 GPa, level 2 210 GPa and level 3 230 GPa.

In order to illustrate the methodology of performing a factorial analysis we will consider the plane stress analysis of a thin beam. The finite element model used in the analysis is shown in figure 9.18. This model has already been validated against analytical results as described in case study C in chapter 10. The model consists of eight quadratic quadrilateral plane stress elements. Let's assume that the objective of our study is to minimise the maximum horizontal stress in the beam.

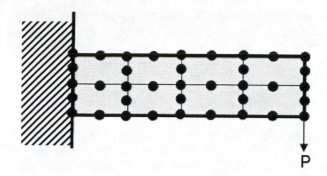

Figure 9.18: Plane stress FE model of a cantilevered beam used for factorial analysis example

The input factors to be used in the analysis, together with the levels used are shown in figure 9.19. It can be seen that four factors have been used in this case: applied load, height of the beam, thickness of the beam and the material Young's modulus.

Factor	Level 1	Level 2	Level 3	Level 4
Applied Load (N)	500	1000	1500	2000
Beam Height (m)	0.3	0.4	0.5	0.6
Beam Thickness (m)	0.03	0.04	0.05	0.06
E, Young's Modulus (GPa)	100	150	200	250

Figure 9.19: Input factors and levels used in the factorial analysis

In order study the effect of the four levels of the four factors sixteen FE models were built and solved as summarised in figure 9.20. This figure illustrates a four level fractional factorial analysis. This matrix is set up in such a way that each factor level appears equally frequently and all combinations of factor levels occur just once in every pair of columns.

Run No	Load	Height	Thickness	E	Stress (MPa)	Transformed Stress
1	1	1	1	1	22.6	-27.08
2	1	2	2	2	9.54	-19.59
3	1	3	3	3	4.88	-13.77
4	1	4	4	4	2.82	-9.00
5	2	1	2	3	17.0	-24.6
6	2	2	1	4	12.7	-22.07
7	2	3	4	1	4.07	-12.19
8	2	4	3	2	3.39	-10.6
9	3	1	3	4	13.6	-22.67
10	3	2	4	3	6.36	-16.07
11	3	3	1	2	8.14	-18.21
12	3	4	2	1	4.24	-12.55
13	4	1	4	2	11.3	-21.06
14	4	2	3	1	7.63	-17.65
15	4	3	2	4	6.1	-15.71
16	4	4	1	3	5.65	-15.04

Figure 9.20: Four level fractional factorial analysis used for the beam problem

The array shown in figure 9.20 is known as an orthogonal array. Experimental design based on such orthogonal arrays was first made popular by the Japanese engineer

G. Taguchi. Taguchi arrays are usually identified with a label such as L4 to indicate an array with four runs. A Taguchi L8 array can be used to determine the effects of up to three factors in four runs. Each run represents a simulation run or experiment and each column represents the factor level setting. For a full factorial design the number of runs required, N, is given by:

$$N = L^m \tag{9.06}$$

Where, L is the number of levels for each factor and m is the number of factors. In this case where we have four factors and four levels a full factorial design would require 256 runs. The benefits of using a L16 array, only requiring 16 runs, in this case are obvious!

The final column in figure 9.20 shows transformed stress. As the purpose of the study is to minimise the maximum horizontal stress in the beam, this value is recorded in the second last column. A minimising function is introduced to transform the recorded stress value into decibels. This has the advantage of allowing factor effects to be added in an unbiased manner. The transformation from measured stress, σ to transformed stress, σT is given by:

$$\sigma_T = -10\log_{10}(\sigma^2) \tag{9.06}$$

The results from the factorial analysis are shown in figure 9.21. In each case an average value for transformed stress is taken from the results shown in figure 9.20. For example for load level 1 the four transformed stress values obtained are averaged to give a transformed stress of 17.36. The mean of the obtained transformed stresses is shown as a dashed line in the figure. In figure 9.21 it should be noted that a less negative number equates to a lower stress.

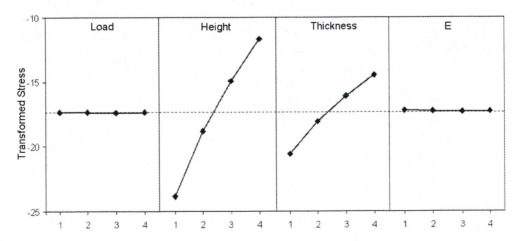

Figure 9.21: Results of the Factorial Analysis

Figure 9.21 clearly shows that the height of the beam has the greatest effect on the measured stress in the horizontal direction in the beam. The thickness of the beam had an equally important but slightly less effect. The applied load and Young's modulus had virtually no effect on the horizontal stress. The graphs in figure 9.21, where a higher value (i.e. less negative) means lower stress, show that in order to minimise horizontal stress the beam height and thickness should be set to their maximum values, i.e. level four in both cases. This result is confirmed by both classical beam theory and by run number four in figure 9.20. Note that the relative importance of each factor is determined not only by the underlying theory of the finite elements but also by that range of values assigned to the factor. We know from simple beam theory that the applied load will definitely have a significant effect on the resultant stress. In this case however, we applied relatively low loads (for such a long and thick beam) over a small range (500-2,000N) and so the applied load did not appear as a significant factor.

The above example, intended as an introduction to the area, illustrates the salient points of how factorial analysis can be used to optimise designs using FEA and how it can be used to reduce the number of simulations required to obtain the required results. If you need to use these techniques in your work then please consult many of the excellent resources available on Taguchi methods and factorial design.

9.8.2 Introducing Uncertainty

In the real world it is impossible to manufacture a component exactly to the specified dimensions as there will always be some margin of error due to the manufacturing process. Similarly, material properties are seldom linear or exact values as we often assume them to be in engineering analysis. When we build an FE model, however, we assume that these values are exact and absolutely precise. Obviously any small uncertainty in the assumed values for model dimensions, material properties or applied loads will have a significant effect on the FE results. In order to take account of this it is sometimes required to introduce uncertainty into a FE model. Probabilistic methods are used so that statistical distributions of values input to the model are part of the FE modelling process.

The idea behind probabilistic methods is that the input parameters of the model are no longer seen as being a discrete value but rather are represented by a statistical distribution. The distribution can take any form that can be mathematically described but is most commonly assigned a normal distribution that is defined by a mean and standard deviation. The values for each input parameter can then be sampled randomly from this distribution and used in the FE model, which will be solved a number of times in order to build up a distribution of the result of interest (e.g. deformation or stress). This methodology allows for confidence limits to be determined and will show up the most likely response of the model.

Let's return to the plane stress beam bending problem as shown in figure 9.18 and described in the previous section. In the previous section, after carrying out the factorial analysis we determined that the most important factor determining the horizontal stress generated in the beam was the beam height. We will now use a probabilistic analysis to introduce uncertainty into the beam height and thus investigate the effects of this uncertainty. The mean height of the beam was assumed to be 0.6m with a normal distribution. Two values for standard deviation were used: 2% and 10% of the mean value, in order to show how the amount of introduced uncertainty (implemented via an increased standard deviation) influences the uncertainty of the FE calculated results. The beam length was set at 5m, the height at 0.6m, the thickness at 0.06m, Young's modulus at 210 GPa and the applied load at 2,000 N.

Most commercial FE solvers will have a probabilistic design module available in which it is possible to specify that a certain model parameter is to be assigned a random normal distribution (also known as a "Gaussian distribution") according to a certain mean and standard deviation. In this case the random distribution was assigned to the applied loading and the model was set to solve 1000 times with a random value for applied load in each case. Due to simplicity of the model used here (only eight elements) this did not take very long.

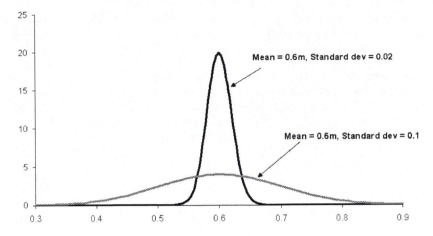

Figure 9.20: Overview of normal distributions of beam height used as an input to the probabilistic analysis

In order to illustrate how the probabilistic design module ran the simulations, figure 9.21 shows a plot of the randomly generated value of beam height versus simulation number for the 2% standard deviation analysis. The Monte Carlo simulation method was used to generate the random values of beam height shown in figure 9.21. There are many different methods available for generating these random values, however Monte Carlo methods are by far the most popular.

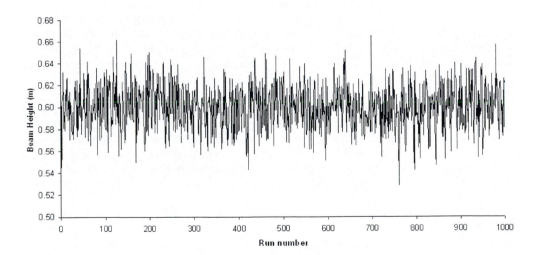

Figure 9.21: Randomly generated beam height versus run number for the 2% standard deviation test.

Figure 9.22 shows a histogram of maximum horizontal stress in the beam when the standard deviation of the beam height was 2%. It can be clearly seen that in this case the stresses are normally distributed with slight positive skew with a mean of 2.8 MPa. From these results we can deduce that with a 2% uncertainty in the beam height then the 95% confidence limits on the maximum stress of the beam are between 2.5 and 3.3 MPa, i.e. the mean value of 2.8 MPa +0.5 MPa, -0.3 MPa.

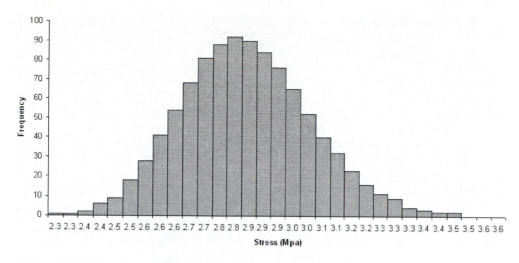

Figure 9.22: Frequency distribution of the maximum horizontal stress in the beam when the uncertainty in beam height is 2%

Figure 9.23 shows a similar plot when the standard deviation of the beam height was 10%. In this case the distribution is clearly skewed with a means stress of 3.1 MPa. In this case the 95% confidence limits were between 1.6 and 6.2 MPa, i.e. the mean value of 3.1 MPa +3.1 MPa, -1.5 MPa.

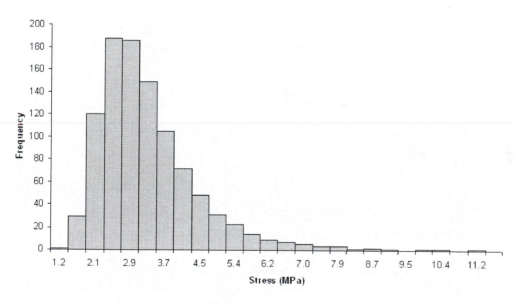

Figure 9.23: Frequency distribution of the maximum horizontal stress in the beam when the uncertainty in beam height is 10%

The above example provides an elementary introduction to the field of probabilistic design using FEA. For further information you should consult your software's documentation to understand how your FE software implements probabilistic design. If you think you may need to use probabilistic design in your work you should obtain a good textbook on the field and brush up on your knowledge of statistics and probability techniques and tools.

9.9. Summary of Chapter 9

After completing this chapter you should:
- Understand the role of post-processing techniques in the FE process.
- Know the procedure used to check your model for validity after solution.
- Understand the difference between various stress and strain results and in particular the difference between averaged and un-averaged stress results.
- Understand each of the methods available for the display of results from a structural finite element analysis.
- Understand the different methods available for verification of FE results and be able to apply them to your model to ensure quality results.
- Be able to apply statistical methods to perform sensitivity tests and probabilistic analysis on your FE model.

10

Case Studies

10.1. Introduction

This final chapter describes nine case studies which attempt to apply the concepts presented in this book. Most of the element types described in chapter 4 are used in one of the case studies. Similarly, many of the material models discussed in chapter 5 are used. In each case the assumptions made when building the finite element models are discussed in detail together with the methodology used to build the models. After the results have been presented a model verification and updating process in undertaken, where relevant. Each case study has been presented in such a way that it should be possible for the reader to replicate the case study using whatever FE software is available and, indeed, this is the intention of this chapter.

10.2. Case Study A: Analysis of a Bridge Structure

A1:   Problem Description

The small railroad bridge shown in figure 10.01(a) is constructed from steel members of I beam cross section, with a cross sectional area of 0.003 m². The loads on one side of the bridge when a train has stopped in the middle of the bridge span are shown in figure 10.02(b). We are required to determine how the structure will deform due to the presence of the train on the bridge and the reaction forces at the supports.

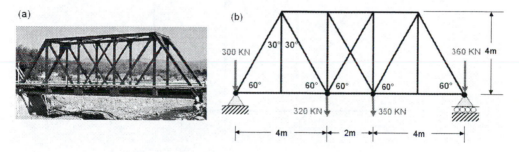

Figure 10.01: Truss bridge details for case study A

A2:   Assumptions

Materials: We are told that the bridge is constructed from steel.
               We will assume a linear elastic material model for steel. This is valid considering the problem, as a bridge structure would not be expected to yield under operating conditions. We will, however, take a note of the assumed yield stress and use this to ensure that our linear elastic model is valid.

Geometry: It should be obvious from the problem description that a 2D truss analysis is appropriate in this case. If you don't know why this is the case, then re-read page 54. The structure is symmetrical about its midpoint, the loading however is not symmetrical so the full geometry must be modelled.

Loading: We are told that the train has stopped on the bridge so it is reasonable to assume that the loads are static, thus static point forces can be applied to the appropriate points as shown in figure 10.01(b).

B.C.'s: We can assume that the boundary conditions will not change over time. We know from the problem description that the left hand support is fixed and the right hand support is a frictionless support, thus, appropriate boundary conditions can be applied to the nodes at these locations.

Solution type: As a linear elastic model is being used and loads are non time varying: a simple small deflection linear static structural analysis should be sufficient in this case.

A3: Finite Element Modelling – Truss Model

Element type: 2D truss element
Material Model: Linear elastic model for structural steel
$E = 210 \times 10^9$ Pa, $\nu = 0.27$, $\sigma_y = 355 \times 10^6$ Pa
Geometry: As per figure 10.01.
Mesh: Each line was divided into one truss element. The mesh used is shown in figure 10.02
Boundary conditions: Applied to nodes as shown in figure 10.02
Loads: Applied to nodes as shown in figure 10.02

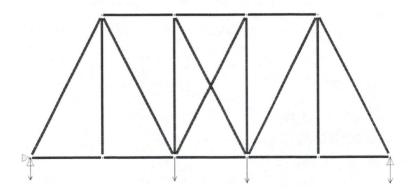

Figure 10.02: Finite Element Model of Truss Bridge using Truss Elements

A4: FE Results

FE predicted reaction forces at left hand support: $F_y = 698$ kN
FE predicted reaction forces at left hand support: $F_x = 145.8 \times 10^{-4}$ N, $F_y = 632$ kN
FE predicted axial stress in member 1: $\sigma_1 = -123.73$ MPa
The FE predicted deformed shape of the structure due to the applied loads is shown in figure 10.03. The maximum deflection in the vertical direction was at the bottom of the structure at the joint slightly to the right of the centre and was 0.00865 m. The maximum deflection in the horizontal direction was at the right hand support and was 0.00319 m. The distribution of axial stress in the model is shown in figure 10.04.

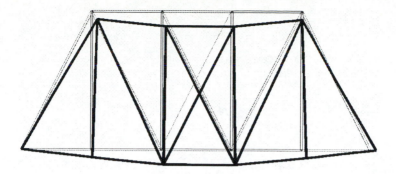

Figure 10.03: FE predicted deformed shape. Deflections are exaggerated and overlaid on the original mesh.

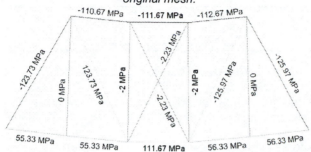

Figure 10.04: FE predicted distribution of Axial Stress in the Finite Element Model

A5:   Results Verification

The deflections predicted by the finite element model would appear to be reasonable. The maximum deflections listed above are in the order of mm which is what one would expect from a bridge with a 10 m span subjected to an operational load. The maximum axial stress in the structure, from figure 10, is 125.97 MPa which is well below the yield stress for the material (355 MPa). This shows that the linear elastic assumption is valid and we can be confident that yielding would not be expected in the structure due to this loading.

In order to further verify the results we can use simple paper based truss analysis calculations. Let's first take moments about the left hand support and assume that the direction of loading (i.e. downwards) is positive:

$$\sum M = 0 = 320{,}000(4) + 350{,}000(6) + 360{,}000(10) - R_{rhs}$$

Where R_{rhs} is the reaction force at the right hand support. Rearranging the above equation and evaluating gives us:

R_{rhs} = -698 kN

Now, let's sum forces in the vertical direction in order to determine the reaction at the left hand support:

$$\sum F_y = 300{,}000 + 320{,}000 + 350{,}000 + 360{,}000 - 698{,}000 - R_{lhs}$$

Evaluating this equation gives us:

R_{lhs} = -632 kN

We can now calculate the force, and hence the stress, in element 1 (the left-most element in the model) as follows:

$$F_1 = \frac{-632,000 + 300,000}{Cos\,30} = -383.36\,kN$$

$$\sigma_1 = \frac{-383,360}{0.003} = -1.27786\,x\,10^8\,Pa = -127.786\,MPa$$

	FEA	Analytical	% Difference
R_{rhs}	632 kN	632 kN	0
R_{lhs}	698 kN	698 kN	0
σ_1	-123.73 MPa	-127.786 MPa	3.17%

Table 10.01: Results Comparison for Case Study A

It is clear from table 10.01 that the FE model shown in figure 10.02 predicts the reaction forces accurately and predicts the stress in element 1 within 4% accuracy.

A6:   Model Update and Improvement

The usual and most obvious method of improving the accuracy of a finite element model is to increase the number of elements in the model and to perform a mesh convergence test. In the case of truss elements, however, this is not always appropriate. Figure 10.05 shows the deformed shape predicted by a re-meshed version of the FE model discussed above, where eight element divisions were used on each line.

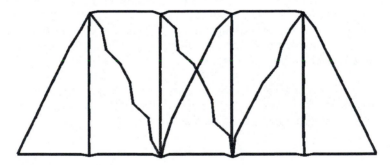

Figure 10.05: Deformed shape predicted by Truss FE model with increased mesh density

It is clear from the figure that this model is unacceptable. In order to understand why, we must return to our original definition of a truss element in chapter 4. Remember that the central assumption of a truss element is that it can only expand or contract and that all joints are assumed to be pin joints. By placing more that one truss element along the structural members of the bridge we have introduced extra pin joints that are not present in the actual structure, hence the incorrect deformed shape shown in figure 10.05. If we remember that the finite element method is only an approximate method in any case, then results within 5% accuracy are very acceptable and thus the original model should be used in this case.

A7:   Conclusion

Given the information available we can conclude that the finite element model predicts the deformation of the bridge structure with acceptable accuracy. We can also conclude that the bridge structure would not be expected to deform significantly due to the loading imparted by the train. We may also conclude that yielding of the bridge structure or failure would not be expected due to this loading.

10.3.   Case Study B: Analysis of a Bicycle Frame

B1:   Problem Description

The idealised bicycle frame shown in figure 10.06(a) is constructed from aluminium tubes. The main part of the frame is made from tubes of outer diameter 20mm and wall thickness 2mm, while the fork structure is made from smaller tubes of outside diameter 10mm and wall thickness 1mm. The geometrical details of the frame and the idealised loading due to an adult rider sitting on the bicycle are shown in figure 10.06(b). We are required:

(ii)   To determine the extent of the deformation of the structure due to the presence of the rider.

(iii)   To determine the von-Mises stress distribution in the structure due to the presence of the rider.

(iv)   To increase the load at the saddle point until yielding of the bicycle structure is observed.

(v)   To investigate if using AZ31 magnesium alloy is a viable alternative to aluminium for manufacturing the bicycle frame.

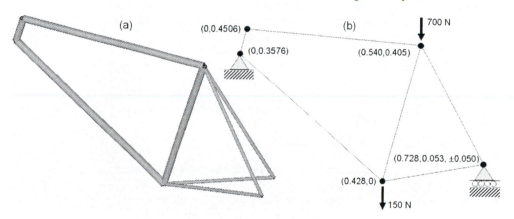

Figure 10.06: Bicycle Frame Details for Case Study B

B2:   Assumptions

Materials: We are told that the bicycle frame is constructed from aluminium. We will assume a linear elastic material model for aluminium. This is valid considering the problem, as a bicycle would not be expected to deform significantly or yield under operating conditions. We will, however, take a note of the assumed yield stress and use this to ensure that our linear elastic model is valid and to determine the load that causes yielding of the structure. For the final part of the problem we will use a linear elastic material model for AZ31 and will also note the yield stress for this material.

Geometry: Given the three dimensional nature of the structure it should be clear that a 3D analysis is required. The bicycle frame consists of a number of long/thin beams that would be expected to bend due to the applied loads so a truss analysis would not be appropriate. A 3D beam analysis is the most suitable geometric model here. A 3D solid model would be overkill and would not estimate bending stresses correctly. The structure and loading are symmetrical about the mid-plane, however, given the simple nature of the geometry it would probably take more time to build a symmetric model than the full model, so in this case the full geometry will be modelled.

Loading: We are not given much information about the loading apart from being told that it is due to a rider sitting on the bicycle. Let's assume that the rider sits down on the bicycle slowly and that they are not moving. Thus we can assume that the loads are static and static point forces can be applied to the appropriate points as shown in figure 10.06(b).

B.C.'s: Given the above assumptions on loading, we can further assume that the boundary conditions will not change over time. We know from the problem description that the left hand support is fixed (to model the reaction from the ground, through the front wheel, through the front fork) and the right hand support is a frictionless support (to model the reaction from the ground through the rear wheel axel) thus, appropriate boundary conditions can be applied to the nodes at these locations.

Solution type: As a linear elastic model is being used and loads are non time varying: a simple small deflection linear static structural analysis should be sufficient in this case.

Yielding: We will assume that yielding has taken place once the von-Mises stress exceeds the uniaxial yield stress for the material. By changing the applied load and checking the resultant von-Mises stress we can determine the applied saddle load that causes the material to yield.

B3:   Finite Element Modelling – Beam Model

Element type: 3D beam element (a pipe element would also be appropriate in this case!)
Material Model: Linear elastic isotropic model for aluminium:
                     $E = 70 \times 10^9$ Pa, $\nu = 0.33$, $\sigma_y = 35 \times 10^6$ Pa
Geometry: As per figure 10.06.
Mesh: Initially each line was broken up into 20 elements. The short line at the handlebar location was broken into 3 elements. The initial mesh used is shown in figure 10.07. Note that in figure 10.07 the beam elements are shown with their cross sectional shape enabled, normally beam elements will just appear as a line.
Boundary conditions: Applied to nodes as shown in figure 10.07. Front support is constrained in all DOF. Rear support is constrained in the Y and Z directions.
Loads: Applied to nodes at saddle and peddle locations as shown in figure 10.07.

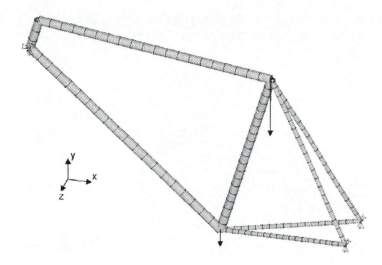

Figure 10.07: Finite Element Model of Bicycle Frame Using Beam Elements

B4:   Results and Discussion

Figure 10.08(a) shows the deformed shape of structure due to the applied loading. It is clear from the figure that both the saddle and pedal points are displacing downwards and forwards, while the rear wheel attachment point is moving forwards. This deformation seems reasonable considering the loads and boundary conditions applied. In order to quantify the deformation, figure 10.08(b) shows a nodal distribution of displacement vector sum (i.e. u+v+w for each node). It can be seen here that the maximum displacement is 0.395 mm and occurs at the saddle point and at the front of the top bar. Again, this level of displacement and its locations seem reasonable based on the applied loads. One would expect the displacement of a bicycle frame when the rider sits on it not to be noticeable to the rider and a deformation of 0.4 mm would certainly not be noticed by the rider.

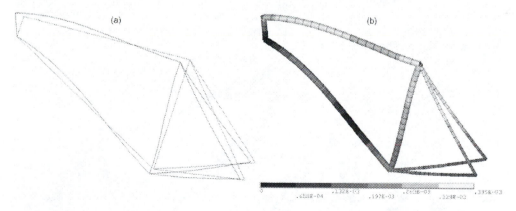

Figure 10.08: FE predicted deformed shape. (a) Deflections are exaggerated and overlaid on the original mesh (b) Displacement Overview, note max displacement is 0.3mm

Figure 10.09(a) shows an element plot of the distribution of von-Mises stress in the beam model. This figure is rather difficult to read, as is normally the case for such beam models, so in order to aid visualisation the stress distribution has been converted into a bar graph and mapped onto the beam model, as shown in figure

10.09(b). It can be seen from the figure that the maximum stress is 0.19 x 10^8 Pa = 19 MPa and is located around the front of the bicycle frame.

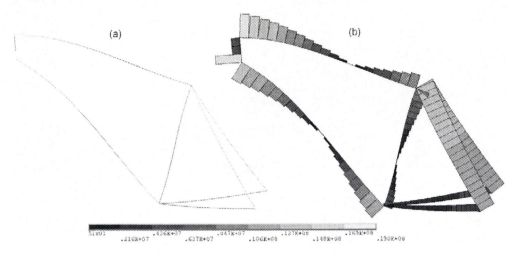

Figure 10.09: (a) FE predicted distribution of von-Mises stress (b) Magnitude of von-Mises stress of each element overlaid on deformed mesh.

The results around the front of the frame and indeed at all of the joints give some cause for concern as stress is changing too rapidly from high values to low values across the joints. This is an indication that a finer mesh is required in these locations and this will be discussed in the next section.

B5:   Model Update and Improvement

There is clearly a problem with the results shown in figure 10.09. In particular, at the top left hand corner of the frame, where the handlebars would normally sit, there is a discontinuity in stress between the two adjoining bars. This is a prime candidate for increasing mesh density (i.e. reducing element size) to investigate if a smoother transition can be obtained. Figure 10.10(a) shows a finer mesh, which uses an element edge length of 3mm throughout. The results for von-Mises stress distribution with this mesh are shown in figure 10.10(b).

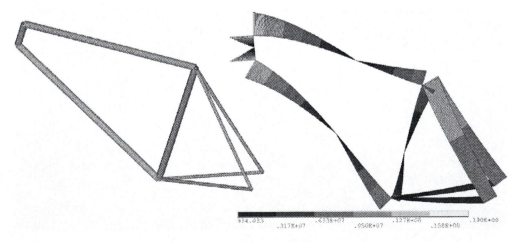

Figure 10.10: (a) Fine Mesh with element edge length of 3mm (b) Magnitude of von-Mises stress of each element overlaid on deformed fine mesh.

It is clear that increasing the mesh density has given a better picture of what is happening in the bicycle frame. The short beam at the front of the bicycle is obviously experiencing bending and the bending stress was not being predicted correctly by the original model.

While the value of increasing the mesh density in regions of rapidly changing stress is readily apparent from the above, having very small elements in main part of the beams (i.e. away from the joints) is of little value since stress is not changing rapidly in these regions. With this in mind a third mesh was generated, in which the element divisions along the lines were "biased" towards the joints. This mesh, as shown in figure 10.11, has small elements where they are needed and has a gradual increase in element size as one moves away from the joints. The results obtained from this mesh are almost identical to those shown in figure 10.10 for the very fine mesh, but in this case many less elements are required.

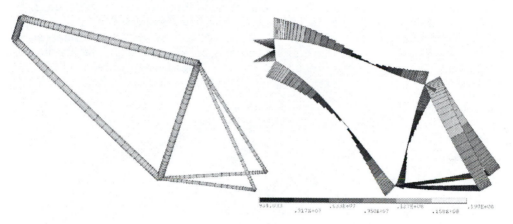

*Figure 10.11: (a) Optimal mesh with gradual transitions between small and large elements
(b) von-Mises stress distribution predicted by optimal mesh.*

B6:   Yielding Analysis

In order to determine the applied saddle load which causes yielding of the bicycle frame a number of analyses were run using the optimal mesh. In each case the applied load at the saddle point was increased, from the initial value of 700N and the resultant maximum von-Mises stress was compared to the yield stress for the material (35 MPa). Once the von-Mises stress exceeded 35 MPa then yielding was assumed to have occurred. The results of this analysis are shown in figure 10.12.

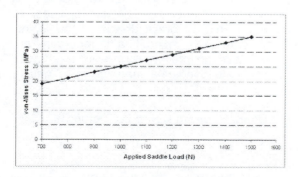

*Figure 10.12: Applied saddle load vs. resultant von-Mises stress for the Aluminium bicycle
frame FE model.*

It is clear from the figure that, in this case, the saddle load required to cause yielding of the structure is 1500 N. Since this loading is equivalent to two and half average adults sitting on the saddle, we can safely say that yielding of the bicycle structure due to a high saddle load should not occur during normal bicycle operation.

B7: Magnesium Alloy Analysis

In order to investigate if AZ31 Magnesium alloy was a suitable candidate material for the bicycle frame. The material properties for the optimal mesh model were changed to correspond to a linear elastic model for AZ31: E = 45 GPa, ν =0.35, σ_y = 160 MPa. The resulting deformation and von-Mises stress results were then checked accordingly.

Figure 10.13(a) shows the deformation of the magnesium alloy frame due to the original loading shown in figure 10.06. It can be seen that the maximum deflection is 0.615mm. While this is double the value experienced by the aluminium frame it is still not significant and most likely would not be noticed by the rider under static conditions. Figure 10.13(b) shows the distribution of von-Mises stress in the magnesium alloy frame due to the original loading. The maximum stress in this case is 19 MPa

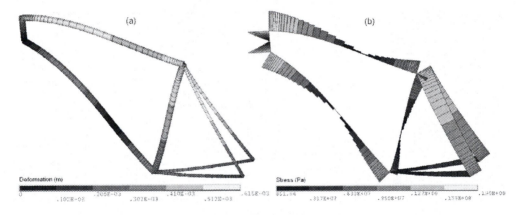

Figure 10.13: (a) Displacement distribution and (b) von-Mises stress distribution for the AZ31 bicycle frame

B8: Conclusion

The case study detailed above illustrates several key points that should be considered in any finite element analysis. The use of linear elastic models and determining the point at which they break down (i.e. the point at which yielding begins) is clearly demonstrated. The importance of generating an appropriate mesh and adapting the mesh based on preliminary results is shown. The use of such models to identify or eliminate candidate materials in a design exercise is also demonstrated.

10.4. Case Study C: Beam Bending Using 2D Elements

C1: Problem Description

In this study we will consider bending of the steel beam shown in figure 10.14 below. The beam is subjected to a tip load of 2,000 N at point A which causes the beam to bend. We are required to determine the vertical deflection of the point A on the beam and are required to find the stress in the x-direction at point B.

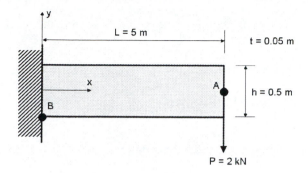

Figure 10.14: Cantilever Beam Problem

In this study we will compare the use of 2D beam elements, three node constant strain triangles (CST), four node quadrilaterals (Q4), six node triangles (LST) and eight node quadrilaterals (Q8) to model the above plane stress cantilevered beam problem and we will compare the accuracy of the results produced by each element type against an analytical solution.

C2:   Assumptions

Materials: We are told that the beam is made from steel.
                We will initially assume a linear elastic material model for steel. We will, however, take a note of the assumed yield stress and use this to ensure that our linear elastic model is valid. Let's assume that E = 210 GPa, G = 80 GPa and ν = 0.27.

Geometry: In this study we will examine two types of model geometry. We will initially assume that a beam analysis is valid and subsequently that a plane stress analysis is valid. The plane stress assumption is valid considering that the beam thickness is an order of magnitude smaller than the other beam dimensions.

Loading: We are not told very much about the nature of the loading so let's initially assume that the loads are static, thus a static point force can be applied to point A as shown in figure 10.14.

B.C.'s: We can assume that the boundary conditions will not change over time. We know from the problem description that the beam is built-in at its left hand edge so we will constrain all DOF at this edge.

Solution type: As a linear elastic model is being used and loads are non time varying: a simple small deflection linear static structural analysis should be sufficient in this case. If yielding is found to occur due to the loads applied then a further large deflection static analysis may subsequently be required.

C3:   Analytical Solution

From standard beam theory we know that the vertical deflection of point A is given by:

$$v_A = \frac{PL^3}{3EI} + \frac{6}{5}\frac{PL}{AG} = 7.679 \times 10^{-4} \, m = 0.768 \, mm$$

Given that the maximum displacement is less that 1mm, the linear elastic assumption made above would seem valid.

Similarly the stress in the x-direction at point B is given by:

$$\sigma = \frac{My}{I} = \frac{PL(-0.25)}{I} = 4.8 \times 10^6 \, Pa = 4.8 \, MPa$$

We will compare these results to those obtained for each of the finite element models described below.

C4:   Finite Element Modelling

(a)   Beam Model

A beam model consisting of one beam element, with two nodes was created in order to compare with the analytical model and the subsequent 2D plane stress models. The model is shown in figure 10.15(a). The left hand node was constrained in all DOF and a vertical load of 2,000 N was applied in a downward direction at the right hand node. The beam element was assigned cross sectional properties according to figure 10.14.

(b)   CST Plane Stress Model

A 2D plane stress model from three node triangular (CST) elements was constructed as shown in figure 10.15(b). The nodes on the left hand edge of the model were constrained in all directions and a vertical load of 2,000N was applied in a downward direction at the appropriate point on the right hand edge.

(c)   Four Node Quadrilateral Plane Stress Model

A 2D plane stress model consisting of four node quadrilateral (Q4) elements was generated as shown in figure 10.15(c). Again, the nodes on the left hand edge of the model were constrained in all directions and a vertical load of 2,000N was applied in a downward direction at the appropriate point on the right hand edge.

(d)   Six Node Triangle Plane Stress Model

A 2D plane stress model from six node triangular (LST) elements was constructed as shown in figure 10.15(d). These elements have a quadratic shape function with mid-side nodes which allows for curved element edges and potentially more accurate results. The nodes on the left hand edge of the model were constrained in all directions and a vertical load of 2,000N was applied in a downward direction at the appropriate point on the right hand edge.

(e)   Eight Node Quadrilateral Plane Stress Model

A 2D plane stress model consisting of eight node quadrilateral (Q4) elements was generated as shown in figure 10.15(e). These elements have a quadratic shape function with mid-side nodes which allows for curved element edges and potentially more accurate results. Again, the nodes on the left hand edge of the model were constrained in all directions and a vertical load of 2,000N was applied in a downward direction at the appropriate point on the right hand edge.

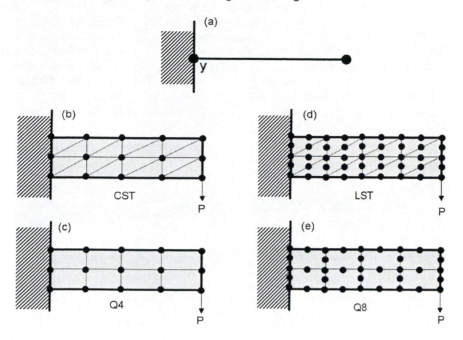

Figure 10.15: Overview of 2D Beam and Plane Stress Elements Used in this Analysis

C5:   Results and Discussion

(a)   Beam Model

The initial beam model consisted of 1 element. This predicted a deflection of 0.616 mm and a stress at point B of 2.4 MPa. Clearly these results were not accurate when compared to the analytical model, so in order to improve the results a convergence test was carried out, whereby the number of (equally spaced) elements in the beam was increased in order to improve accuracy. The results of the beam convergence test are shown in figure 10.16.

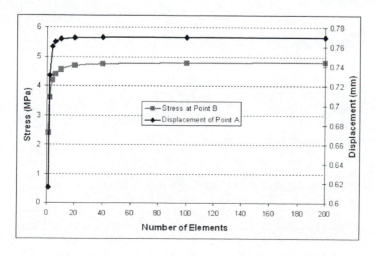

Figure 10.16: Convergence of 2D Beam Element Model

It required approximately 10 evenly spaced beam elements to arrive at the analytical solution for displacement, but required approximately 200 evenly spaced beam elements to approach the analytical solution for stress at point B. This is obviously due to the fact that bending stress is changing very rapidly at the support and thus a fine mesh is required in order to pick up this rapid change in stress. In comparison stress does vary much at the tip end of the beam. In order to take account of this and reduce the number of elements a new beam model was built which had element divisions biased towards the built in edge of the beam. This model had 40 element divisions with the elements at the built-in end of the beam being 15 times smaller than those at the free end of the beam. This model produced the exact analytical solution and is shown in figure 10.17.

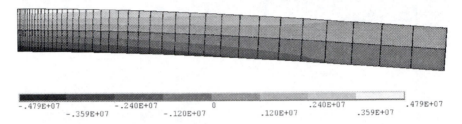

Figure 10.17: Optimal Beam Element Model with Biased Mesh (Element Shape On)

(b)   Three Node Plane Stress Triangles (CST)

The initial mesh used for the CST analysis is as shown in figure 10.15(b). This model, consisting of 16 elements and 15 nodes, predicted a tip deflection of 0.089 mm and a stress at point B of 502160 Pa. Clearly these results are significantly inaccurate which was expected using CST. Figure 10.18 shows a nodal (averaged) contour plot and an element (un-averaged) contour plot of stress in the x-direction for this model. It can clearly be seen from this figure that the CST elements are much too stiff to properly model the bending behaviour of the beam and a much finer mesh is required.

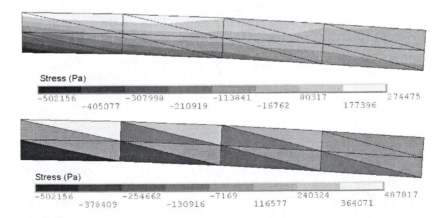

Figure 10.18: Predicted distribution of stress in the x-direction from initial mesh of CST. Un-averaged solution (top) and averaged nodal solution (bottom)

Following the same procedure as before, a mesh convergence test was carried out by increasing the mesh density using equally sized elements throughout. At a mesh size of 256 elements a displacement of 0.682 mm and a stress of 4.4 MPa was predicted. Increasing the mesh density further didn't produce an appreciable improvement in results.

This exercise clearly demonstrates the limitations of the CST element. This element type is highly inaccurate and its use should be avoided in all circumstances. This element has survived in some FE software since the infancy of FEA and it has outlived its usefulness! There are now many more accurate 2D elements available, as we shall see below.

(c)   Four Node Plane Stress Quadrilaterals (Q4)

The initial mesh used in this case is shown in figure 10.15(c). This model which consisted of 8 elements and 15 nodes, predicted a tip deflection of 0.74 mm and a stress at point B of 4.16 MPa. It should be noted here, that even with just eight elements the Q4 element is already more accurate than a CST model with 256 elements!

Repeating our established procedure: a mesh convergence test was carried out on the Q4 model by increasing the mesh density using equally sized elements throughout. Using 200 equally sized elements a predicted tip displacement of 0.769 mm and a stress at point B of 4.79 MPa was obtained, both of which were 98% accurate when compared to the analytical result. The stress distribution in this converged mesh is shown in figure 10.19.

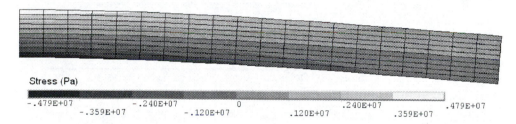

Stress (Pa)

-.479E+07		-.240E+07		0		.240E+07		.479E+07
	-.359E+07		-.120E+07		.120E+07		.359E+07	

Figure 10.19: Predicted distribution of stress in the x-direction from the converged mesh of Q4 elements.

(d)   Six Node Quadratic Plane Stress Triangles (LST)

The initial mesh used in this case is shown in figure 10.15(d). This model which consisted of 16 elements and 90 nodes, predicted a tip deflection of 0.756 mm and a stress at point B of 4.45 MPa. These values are an improvement over the initial mesh of Q4 elements, but in this case it must be taken into account that double the amount of elements have been used.

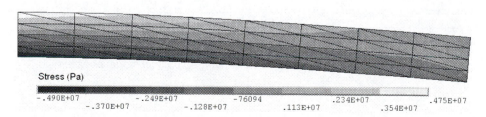

Stress (Pa)

-.490E+07		-.249E+07		-76094		.234E+07		.475E+07
	-.370E+07		-.128E+07		.113E+07		.354E+07	

Figure 10.20: Predicted distribution of stress in the x-direction from the converged mesh of LST elements.

Repeating our mesh convergence procedure on the LST model by increasing the mesh density using equally sized elements throughout showed that by using 64 equally sized elements a predicted tip displacement of 0.765 mm and a stress at point B of 4.75 MPa was obtained, both of which were 99% and 98% accurate respectively when compared to the analytical result. The stress distribution in this converged mesh is shown in figure 10.20.

(e)   Eight Node Quadratic Plane Stress Quadrilaterals (Q8)

The initial mesh used in this case is shown in figure 10.15(e). This model which consisted of 8 elements and 37 nodes, predicted a tip deflection of 0.765 mm and a stress at point B of 4.88 MPa. It should be noted here, that even with just eight elements the Q8 element is already predicting deflection with 99% accuracy and stress with 98% when compared to the analytical results. Due to this level of accuracy further investigation into convergence was not carried out. Figure 10.21 shows the distribution of stress in the x-direction obtained with this model.

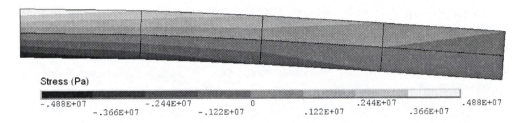

Stress (Pa)

| -.488E+07 | -.244E+07 | 0 | .244E+07 | .488E+07 |
| -.366E+07 | -.122E+07 | .122E+07 | .366E+07 | |

Figure 10.21: Predicted distribution of stress in the x-direction from the converged mesh of Q8 elements.

Model	No of Elements	Deflection (mm)	Stress (MPa)	% Accuracy
Analytical	-	0.768	4.8	-
Beam	4	0.76	4.2	99% / 87%
CST	16	0.089	0.5	11% / 10%
Q4	8	0.74	4.16	96% / 87%
LST	16	0.756	4.45	98% / 93%
Q8	8	0.765	4.88	99% / 98%

Figure 10.22: Summary of results from the initial model of each type.

Model	No of Elements	Deflection (mm)	Stress (MPa)	% Accuracy
Analytical	-	0.768	4.8	-
Beam	40	0.769	4.79	99.8%
CST	256	0.682	4.45	88% / 92%
Q4	200	0.769	4.79	99.8%
LST	64	0.765	4.75	99% / 98%
Q8	8	0.765	4.88	99% / 98%

Figure 10.23: Summary of results from the converged/optimal model of each type.

D6:   Conclusions

Figure 10.22 shows a summary of the results obtained from the initial mesh (as shown in figure 10.15 for each element type). It is clear from this that the CST elements are highly inaccurate and should be avoided at all costs. If deflection is all that is required then the beam element is the obvious choice for this type of analysis as it provides excellent accuracy with very few elements. If, however, prediction of stress is also required then the Q8 elements would appear to be the ideal candidate as they predicted both deflection and stress to a high level of accuracy for very few elements. This result is further reinforced by figure 10.23 which shows the results from the converged model of each element type.

10.5   Case Study D: Plane Strain Analysis of Long Cylinder

D1:   Problem Description

In this study we will be examining the case of a long hollow aluminium cylinder subjected to compressive loading as shown in figure 10.24. The cylinder is loaded via two solid cylinders that compress the hollow cylinder at its top and bottom extremities. A 2kN load is applied to both solid cylinders and these cylinders are made from a much stiffer material than aluminium. The hollow cylinder has an internal radius of 10mm and a wall thickness of 2mm and is 2m long.

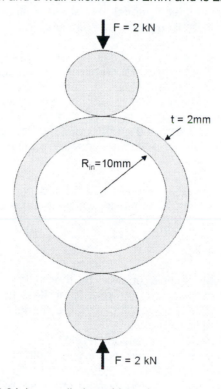

Figure 10.24: Long cylinder subjected to compressive loading

In an experimental study of this problem it was found that the hollow cylinder reduced in height by an average of 0.00807 mm due to this applied loading and the right and left extremities of the hollow cylinder moved out by an average of 0.00705 mm. Strain gauges were placed on the cylinder and it was found that the maximum tensile hoop stress due to the applied loading was 21.8 MPa. We are required to build a finite element model of this problem and validate our model's results against the experimental results. We will initially ignore the solid cylinders and simply assume that they transfer the load via a point to the hollow cylinder. Later on we may need to further consider this assumption.

D2:   Assumptions

Materials: We are told that the cylinder is made from aluminium and the
                  experimental results for deflection and stress would indicate that the
                  cylinder has not yielded due to the applied load. Hence we will
                  assume a linear elastic material model for aluminium. Let's assume
                  that E = 70 GPa and ν = 0.33.

Geometry: We are told that the cylinder is very long and has a continuous load
                  along its length. The cross sectional dimensions of the cylinder are an
                  order of magnitude smaller than its length. Each of these suggests the
                  use of a plane strain model. In addition both the structure and the
                  applied loads are symmetrical about the vertical and horizontal
                  centrelines through the cross section. This means that a ¼ symmetry
                  plane strain model is suitable.

Loading: We are not told very much about the load or how it is applied. In the
                  absence of any further information and given our assumption of a
                  linear elastic material model above, we shall assume that the loading
                  is static, thus enabling a linear static analysis.

B.C.'s: As specified the problem does not contain any boundary conditions.
                  This means we have to creatively come up with some as it is
                  impossible to solve a FEA problem without boundary conditions.
                  Luckily the structure and loading is symmetric and we have already
                  decided to take advantage of this symmetry. Hence, a vertical
                  constraint will be placed on the horizontal cut plane of the ¼ cylinder
                  cross section and a horizontal constraint will be placed on the vertical
                  cut plane.

Solution type: As a linear elastic model is being used and static loads have been
                  assumed, a simple small deflection linear static structural analysis
                  should be sufficient in this case.

D3:   Finite Element Modelling

A 2D plane strain model of one quarter of the cylinder cross section using
quadrilateral elements was generated in order to analyse this problem. The model
used is shown in figure 10.25. The specifics of the FE model are:

Element type: 2D plane strain quadrilaterals
Material Model: Linear elastic isotropic model for aluminium:
                    $E = 70 \times 10^9$ Pa, ν = 0.33, $\sigma_y = 35 \times 10^6$ Pa
Geometry: One quarter of the cylinder cross section was modelled as per
                    the supplied dimensions, as shown in figure 10.25.
Mesh: A global element edge length was applied to the model so that
                    the ¼ cylinder was divided into equal squares. The size of
                    these squares were reduced until convergence was achieved.
B.C.'s: Applied to nodes to take account of symmetry as shown in
                    figure 10.25. The left hand support is constrained in the
                    horizontal direction and the right hand support is constrained in
                    the vertical direction.
Loads: Initially applied as a point force to the topmost node as shown
                    in figure 10.25. We will subsequently investigate different

methods of load application in order to remove the singularity introduced by the point force.

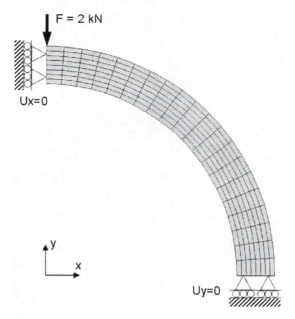

Figure 10.25: ¼ Symmetry plane strain model of long cylinder in compression

D4:   Results and Discussion

The initial model had an element edge length of 0.25mm, effectively dividing the thickness of the cylinder into eight elements. The deformed shape of the cylinder predicted by the analysis is shown in figure 10.26 - in this case the deformed shape has been overlaid on the original shape of the mesh. It can be clearly seen that as expected the cylinder is being compressed vertically which is resulting in lateral expansion due to the applied loading.

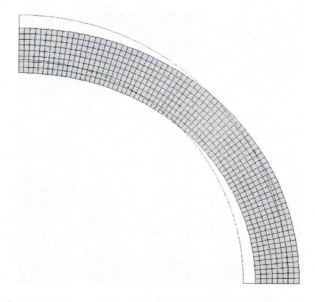

Figure 10.26: Deformed shape of the initial finite element model due to applied loading

The distribution of hoop stress in the ¼ cylinder due to the applied loading is shown in figure 10.27 and the corresponding distribution of radial stress is shown in figure 10.28.

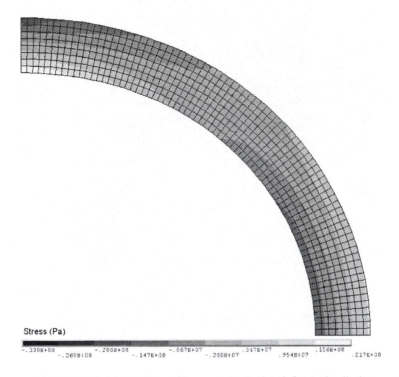

Stress (Pa)

Figure 10.27: Distribution of hoop stress in the deformed cylinder

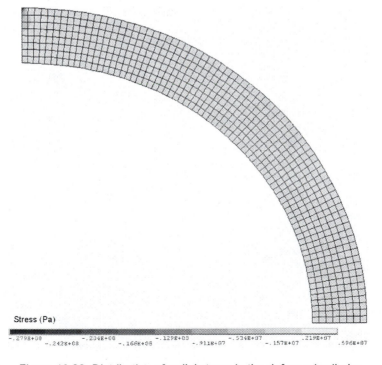

Stress (Pa)

Figure 10.28: Distribution of radial stress in the deformed cylinder

From the distribution of hoop stress it can be seen that the locations of maximum stress are in the expected locations: the top of the cylinder at the inner surface and the outside of the cylinder at the outer surface, with corresponding compressive results on the opposite surfaces. The distribution of radial stress shows a sharp stress concentration at the point of load application. All of the other models showed similar trends.

In order to investigate the effect of applying the load as a point force on the topmost node an alternative model was built whereby the solid cylinders shown in figure 10.24 were modelled via a ¼ solid cylinder placed above the original model. 2D Surface to surface contact elements were used in order to transfer the load from the solid cylinder to the large cylinder. Frictionless contact was specified between the two contacting bodies. The solid cylinder was restrained from moving in the horizontal direction and a point load was applied to the central node on the top surface. The results for hoop stress from this model are shown in figure 10.29 and the results for radial stress are shown in figure 10.30.

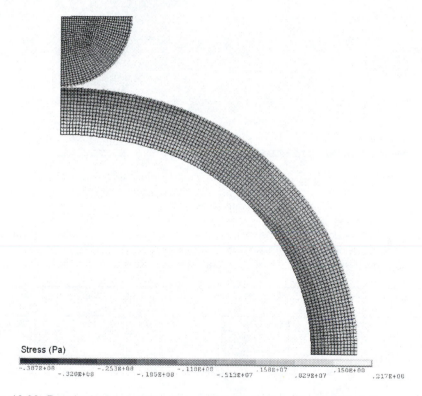

Stress (Pa)

-.387E+08   -.253E+08   -.118E+08   .158E+07   .156E+08
    -.320E+08   -.185E+08   -.513E+07   .829E+07   .217E+08

Figure 10.29: Results for hoop stress from an alternative model using 2D surface to surface frictionless contact

Comparing figures 10.29 and 10.27 shows that there is not much difference in hoop stress between the two models. The maximum value for hoop stress is exactly the same, while the minimum value, for compressive hoop stress, is slightly higher in the model with contact. Similarly, comparing figures 10.30 and 10.28 shows that although the magnitude of compressive radial stress in the hollow cylinder is slightly higher for the contact model, the distribution of stress is much more regular, indicating that load is being transferred in a more realistic manner. Incidentally, it should be noted that in this figure the maximum compressive stress occurs in the solid cylinder at the point of load application. The maximum tensile radial stress, which makes up the majority of the plot, is lower in the model with contact than the previous model.

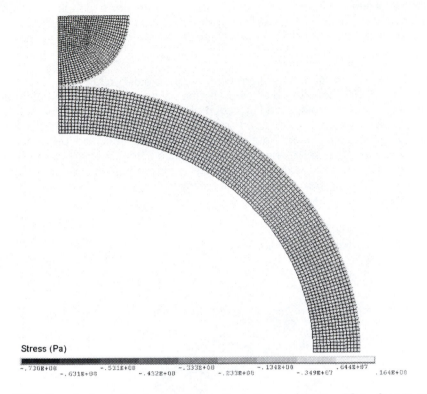

Stress (Pa)

Figure 10.30: Results for radial stress from an alternative model using 2D surface to surface frictionless contact

In order to obtain a converged mesh the problem geometry was divided into equal sided squares of decreasing area until convergence was achieved. The result of the convergence process for displacement and stress are shown in the tables below.

Model	Element edge length	Deflection (mm) Uy / Ux	Target (mm) Uy / Ux	% Accuracy
1	0.25 mm	0.8e-5 / 0.705e-5	0.807e-5 / 0.705e-5	99.1% / 100%
2	0.15 mm	0.802e-5 / 0.705e-5	0.807e-5 / 0.705e-5	99.3% / 100%
3	0.05 mm	0.805e-5 / 0.705e-5	0.807e-5 / 0.705e-5	99.7% / 100%
4	Contact	0.807e-5 / 0.705e-5	0.807e-5 / 0.705e-5	100% / 100%

Figure 10.31: Summary of displacement results from each model

Model	Element edge length	Max. Radial Stress (MPa)	Max Hoop Stress (MPa)	% Accuracy of Hoop Stress
1	0.25 mm	-27.9	21.7	100%
2	0.15 mm	-46.8	21.7	100%
3	0.05 mm	-138	21.7	100%
4	Contact	-73.8	21.7	100%

Figure 10.32: Summary of stress results from each model

The results for radial stress are interesting as in several of the cases they are in excess of the yield stress for the material which suggests that localised yielding may be occurring at the point of load application. Given that the predicted deformations are very small it is highly unlikely that significant plastic deformation is taking place, rather a partial through thickness yielding at the point of load application. In order to investigate this further a model with plastic deformation capable material model should be used. Try this yourself with a bilinear elastic material model for aluminium and see what the results tell you.

D5:   Conclusions

This case study illustrates an appropriate use of plane strain elements and demonstrates how symmetry can be used to create boundary conditions in problems where they don't exist naturally. This study also illustrates the concepts of using increasingly advanced models, by introducing contact in this case, to investigate how increasing the model complexity changes the derived results. Finally, this case study is the first to introduce non-linearity as we introduced contact elements, thus making the analysis a small deflection non-linear static analysis.

10.6.   Case Study E: Axisymmetric Analysis of a Pressure Vessel

E1:   Problem Description

A cylindrical pressure vessel with flat end caps, shown in figure 10.31 is used to contain a fluid under a pressure of 10,000 Pa. We are required to determine the principal stress distributions in the pressure vessel wall due to the internal pressure and to validate these results against analytical theory. Subsequently we are required to determine the magnitude of internal pressure that will cause the pressure vessel wall to yield.

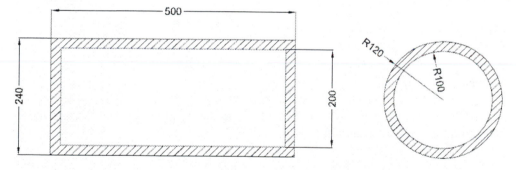

Figure 10.31: Pressure vessel details, all dimensions are in mm.

The relevant dimensions of the pressure vessel, as shown above, are: length = 500mm, diameter = 240mm, wall thickness = 20mm. Initially we can assume that the pressure vessel is made from a mild steel with the following properties: E = 207 GPa, $v = 0.27$ and $\sigma_Y = 240$ MPa.

E2:   Analytical Theory

There are standard theories available for the behaviour of thin and thick walled cylinders subjected to internal pressure. These equations can be found in any text book on mechanics of solids or in any reference book.

For a cylindrical thin walled pressure vessel, it is assumed that there is no stress through the wall thickness, so the stress in the radial direction $\sigma_r = 0$.

The longitudinal stress in a cylindrical thin walled pressure vessel of mid thickness radius r and wall thickness t, subjected to internal pressure, P, is given by:

$$\sigma_l = \frac{Pr}{2t}$$

Similarly, the circumferential or "hoop" stress is given by:

$$\sigma_h = \frac{Pr}{t}$$

As a rough guide, a pressure vessel is considered "thin walled" if its radius is larger than five times it's wall thickness, which applies here. For the sake of completeness, however, we will also consider equations describing the behaviour of thick walled vessels in our analysis. In this case, the longitudinal stress is given by:

$$\sigma_l = \frac{Pr_{in}^2}{r_{out}^2 - r_{in}^2}$$

Similarly the hoop stress for a thick walled vessel is given by:

$$\sigma_h = \frac{Pr_{in}^2 \left(r_{out}^2 + r_{in}^2 \right)}{r^2 \left(r_{out}^2 - r_{in}^2 \right)}$$

Where r, is the point through the wall thickness at which you wish stress to be calculated, i.e. σ_h is maximum at $r = r_{in}$

The radial stress is given by:

$$\sigma_r = \frac{-Pr_{in}^2 \left(r_{out}^2 - r^2 \right)}{r^2 \left(r_{out}^2 - r_{in}^2 \right)}$$

Where, again, r is the point through the wall thickness at which you wish stress to be calculated, i.e. σ_r is maximum at $r = r_{in}$

Thick walled pressure vessel theory also provides a methodology for calculating the displacement of the inner and outer vessel walls:

$$u_r = \frac{(1+v)}{E} \left[\frac{Pr_{in}^2}{\left(r_{out}^2 - r_{in}^2 \right)} \right] r + \frac{(1+v)}{E} \left[\frac{Pr_{in}^2 r_{out}^2}{\left(r_{out}^2 - r_{in}^2 \right)} \right] \frac{1}{r}$$

Using the above equations we can calculate the following:

Quantity	Thin Walled Theory	Thick Walled Theory
Longitudinal stress, σ_l	27,500 Pa	22,727 Pa
Hoop stress, σ_h	55,000 Pa	55,455 Pa
Radial stress, σ_r	-	-10,000 Pa
Deflection of inner wall	-	2.81 x 10^{-6} m
Deflection of outer wall	-	2.64 x 10^{-6} m

E2: FE model assumptions

Materials: We are told that the pressure vessel is made from steel and the analytical results for stress and deflection clearly show that the pressure vessel has not yielded due to the applied load. Hence we will assume a linear elastic material model for mild steel. Later on we are required to determine the applied pressure load that causes yielding. The linear elastic model is suitable here also as it will break down at the yield point and we will not be loading beyond yield.

Geometry: This is obviously an axisymmetric problem as the pressure vessel is cylindrical. We can model ½ of the cross section and ensure that the model is placed so that the axis of symmetry aligns with the global Y-axis to ensure that the axisymmetric assumption is properly implemented.

Loading: We don't know how fast the pressure load is applied or if it is constant etc. In the absence of any further information and given our assumption of a linear elastic material model above, we shall assume that the loading is static, thus enabling a linear static analysis.

B.C.'s: As specified the problem does not naturally contain any boundary conditions, however the fact that we are using ½ of the cross section for the axisymmetric analysis means that we can impose symmetric boundary conditions on the cut edge which will coincide with the axis of symmetry.

Solution type: As a linear elastic model is being used and static loads have been assumed, and we are not going to load the model beyond the yield stress, a simple small deflection linear static structural analysis should be sufficient in this case.

E3: Finite Element Modelling

A 2D axisymmetric model of a slice through the pressure vessel cross section was generated in order to analyse this problem. The initial models used are shown in figure 10.32. The specifics of the FE models are:

Element type: 2D axisymmetric quadratic quadrilaterals

Material Model: Linear elastic isotropic model for mild steel:
$E = 210 \times 10^9$ Pa, $v = 0.27$, $\sigma_y = 240 \times 10^6$ Pa

Geometry: One half of a longitudinal slice through the cylinder cross section was modelled according to the given dimensions as shown in figure 10.32(a). Initially the end caps (horizontal portions at the top and bottom on the vessel) were not modelled, just the vessel wall. Later on the full geometry was modelled as shown in figure 10.32(b)

Mesh: A global element edge length was applied to the model so that the geometry was divided into equal squares. The size of these squares were reduced until convergence was achieved.

B.C.'s: In the first model B.C.'s were applied to the nodes at the top and bottom of the vertical wall in order to constrain these surfaces in the vertical direction. This allowed the vessel to expand without changing its length. In the model with the end caps the B.C.'s were applied to nodes to take account of symmetry as shown in figure 10.31. The left hand edge of the model coincided with the global Y-axis for the problem and this was set by the solver as the axis of symmetry for the axisymmetric assumption.

Loads: Initially applied as a uniform pressure to all internal walls of the geometry as shown in figure 10.32.

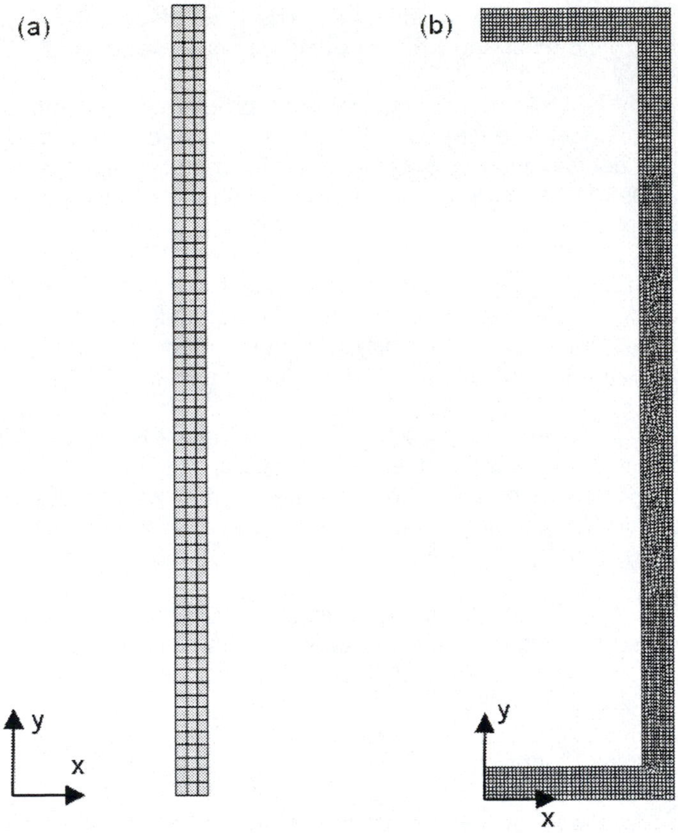

(a)   (b)

Figure 10.32: Initial axisymmetric finite element models of a cylindrical pressure vessel.
Without end caps (a) and with end caps (b)

E4:   FE Results

The results from the simplified model (without end caps) for hoop stress and radial stress are shown in figure 10.33. Since the top and bottom surface of the rectangle were constrained only in the horizontal direction the wall was free to expand and there was no longitudinal stress induced in the wall. Figure 10.33 shows a close up of the bottom of the rectangle which defined the geometry. A mesh convergence process was run and it was found that the results converged with eight element divisions through the vessel wall thickness.

The value of hoop stress predicted by this model on the internal wall surface was 55,455 Pa which is exactly that predicted by the analytical theory. The maximum radial stress also occurred at the internal wall surface and was -9,990 Pa, which is 99.9% of the analytically predicted value of -10,000 Pa. The FE predicted displacements in this case were 2.41 x 10^{-6}m for the outer wall and 2.61 x 10^{-6}m for the inner wall. These values are approximately 90% of those predicted by the analytical theory.

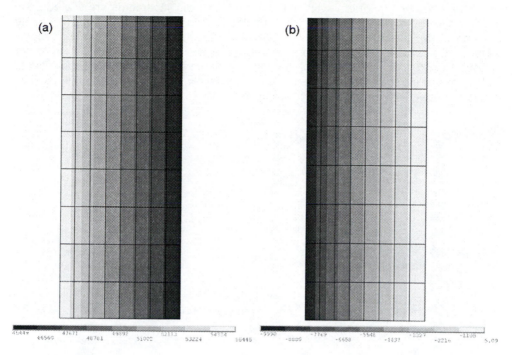

Figure 10.33: Stress results from the model without end caps: (a) Hoop Stress and (b) Radial Stress

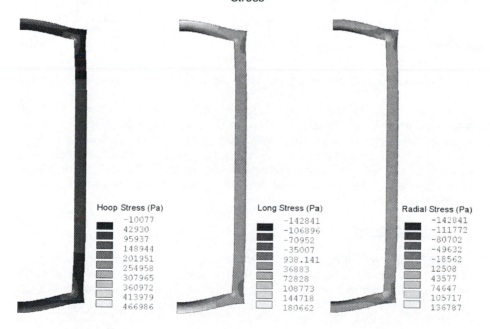

Hoop Stress (Pa)	Long Stress (Pa)	Radial Stress (Pa)
-10077	-142841	-142841
42930	-106896	-111772
95937	-70952	-80702
148944	-35007	-49632
201951	938.141	-18562
254958	36883	12508
307965	72828	43577
360972	108773	74647
413979	144718	105717
466986	180662	136787

Figure 10.34: Initial stress results from the model with end caps

Based on the above results we can conclude that the simplified model without end caps is accurately modelling the expansion behaviour of the cylinder. We can now confidently solve the FE model of the full problem with end caps using a similar mesh density. The initial results from the full model are shown in figure 10.34. Clearly there is a problem due to a stress concentration being introduced due to the sharp corner where the end caps meet the walls of the vessel. The displacements are scaled up in the figure, however the displacement results in this case were similar to those obtained in the previous model, thus indicating that the results are valid. In reality however it would be impossible to manufacture a pressure vessel with such sharp corners. In order to take account of this the corners in the above model were filleted in order to reduce the stress concentration and to allow for more realistic results. The results from the filleted model are shown in figure 10.35.

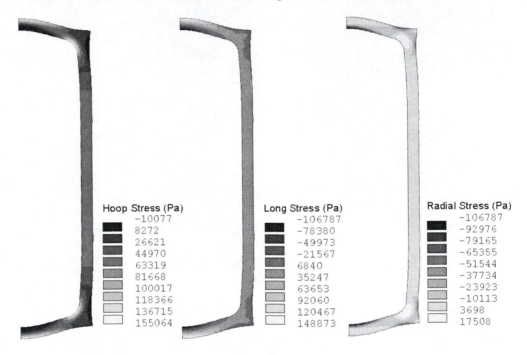

Hoop Stress (Pa)	Long Stress (Pa)	Radial Stress (Pa)
-10077	-106787	-106787
8272	-78380	-92976
26621	-49973	-79165
44970	-21567	-65355
63319	6840	-51544
81668	35247	-37734
100017	63653	-23923
118366	92060	-10113
136715	120467	3698
155064	148873	17508

Figure 10.35: Stress results from the model with end caps and fillets.

It is clear from figure 10.35 that the introduction of the fillet has significantly reduced the stresses, in particular the hoop stress. In order to check the results a group of elements in the centre of the vessel wall, away from the end caps were examined and the results for all three stresses were recorded. This yielded a value of 55,548 Pa for hoop stress, 23,289 Pa for axial stress and -9,990 Pa for radial stress. These values are all very close to the analytical predictions and thus we can conclude that the model is behaving reasonably accurately.

E5:   Yielding Analysis

The final FE model with end caps and fillets was used for the yielding analysis. The applied load was increased in steps and the distribution of von-Mises stress checked after each run until the material yield stress of 240 MPa was recorded. It was found that an internal pressure load of 14.6 MPa was required in order to initiate yielding in the pressure vessel wall. The distribution of von-Mises stress in the pressure vessel wall due to an applied internal pressure of 14.6 MPa is shown in figure 10.36. It can be seen that as expected the yield stress has been found at the filleted regions. It should be noted, however, that the yield stress simultaneously occurs at the outer surface of the centre of the end caps due the large amount of bending that takes place here. This result has serious implications for manufacture of the pressure vessel as any welds or other joints in these regions should be designed with caution.

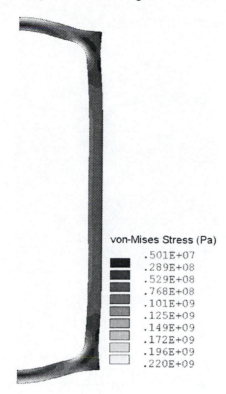

von-Mises Stress (Pa)

	.501E+07
	.289E+08
	.529E+08
	.768E+08
	.101E+09
	.125E+09
	.149E+09
	.172E+09
	.196E+09
	.220E+09

Figure 10.36: Distribution of von-Mises stress in the pressure vessel at the initiation of yielding, due to an applied load of 14.6 MPa.

E6:   Conclusions

This case study illustrates an appropriate use of axisymmetric elements and demonstrates how hollow axisymmetric components may be modelled by correct positioning of the geometry relative to the problem coordinate system. In addition this study also demonstrated how an analytical model may be used to validate a basic FE model which may then be enhanced in order to more closely model the required problem. Finally a yielding analysis was undertaken and a methodology for determining yield using a linear static FEA was outlined.

10.7. Case Study F: Thin Shell Analysis of a Fuselage Panel

F1:   Problem Description

We are required to perform an analysis of a 7075-T6 aluminium alloy inspection panel that makes up part of an aircraft fuselage. The panel is at a location on the aircraft that is relatively un-curved so we may neglect curvature in our analysis. The panel is square (1m x 1m) with a thickness of 5mm and is stiffened via four horizontal and four vertical stiffeners which are equally spaced, as shown in figure 10.37. Both stiffener types are the same thickness, however, the vertical stiffeners are slightly taller than the horizontal stiffeners. The mean external pressure to which the panel is subjected to during a typical flight cycle (take off, cruise, landing) is 5000 Pa. The inspection panel is rigidly fastened to the rest of structure so we may assume that the edges of the panel are clamped. We are required to determine the maximum deflection of the panel due to this pressure and also the stresses at the centre of the panel and at one of the clamped edges. We are also required to validate our FE results against an available analytical theory.

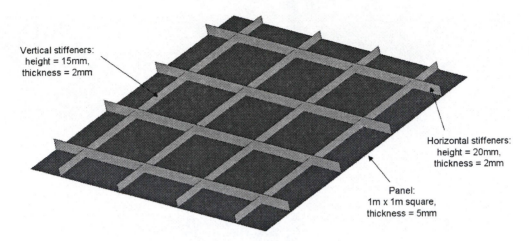

Figure 10.37: Overview of Fuselage Panel Problem

F2:   Assumptions

Materials: We are told that the pressure vessel is made from a 7075-T6 aluminium alloy. We do not know if that applied load is great enough to induce yielding of the structure yet, so we will initially assume a linear elastic model for aluminium, with E = 71.7 GPa, ν = 0.33 and σ_y = 503 MPa. If the yield stress is exceeded during the analysis we may need to use a material model that allows for plastic deformation later on.

Geometry: This is clearly a thin shell problem. The vertical and horizontal dimensions of the panel (1m x 1m) are much larger than its thickness (0.005m) which means that a thin shell assumption is valid. Similarly the stiffeners length (1m) and height (0.015m and 0.020m) are large enough in comparison to their thickness (0.002m) to allow for a thin shell assumption. Aircraft structures are commonly analysed, both analytically and using FEA, using thin shell theory. Hence we will build a finite element model based on areas with no thickness and will mesh these areas with shell elements with an assigned thickness. As the panel is symmetrical in both the horizontal and vertical directions and a uniform pressure is applied to the entire panel face, only ¼ of the panel structure will be modelled.

Loading: We are not given information about how quickly the 5000 Pa pressure load is applied or if it is constant, we are just told that it is the mean value. As such we will assume that the load is a static pressure load. Later on we can check the validity of this assumption by using a time-varying load to simulate the take-off, cruise, landing flight cycle.

B.C.'s: The fact that we are using ¼ of the full model due to symmetry means that we must use appropriate symmetrical boundary conditions on the symmetry edges of the panel. The fact that the panel is effectively clamped on the non-symmetry edges means that we can fix these edges in all DOF.

Solution type: As a linear elastic model is being used and static loads have been assumed, and we are presuming that the loads don't go beyond the yield stress, a simple small deflection linear static structural analysis should be sufficient in this case. Subsequently we will use a non-linear load to simulate the flight cycle and a non-linear static analysis will be used.

F3: Analytical Theory

From plate theory, the maximum deflection of a plate clamped on all edges is given by:

$$w_{max} = 0.00126 P(W)^4 / D$$

Where, P is the applied pressure, W is the width of the plate and D is the flexural rigidity of the plate, which is given by:

$$D = \frac{E' I}{W}$$

Where: $E' = \dfrac{E}{1-v^2}$ for the plate and E' = E for the stiffeners.

These equations all assume that the panel is isotropic. In our case the horizontal stiffeners and vertical stiffeners are slightly different (i.e. they have different heights) so the flexural rigidity D_h and D_v in the horizontal and vertical directions will be different.

In order to determine D_h and D_v we need to first find the distance of the centroid of the plate/stiffener combination from the centroid of the plate. The distance of the centroid from the mid plane of the plate is given by:

$$\bar{z} = \frac{0(Wt_p) + (h_s/2)(4h_s t_s)}{Wt_p + 4h_s t_s}$$

Where, W is the width of the plate, t_p is the thickness of the plate, h_s is the height of the stiffeners, t_s is the thickness of the stiffeners.

The horizontal and vertical flexural rigidities are given by:

$$D_h = \frac{E}{1-v^2}\left[\frac{t_p^3}{12} + t_p \bar{z}^2\right] + E\left[\frac{4t_s h_{Vs}^3}{12W} + \frac{4t_s h_{Vs}}{W}\left(\frac{h_{Vs}}{2} - \bar{z}\right)^2\right]$$

$$D_v = \frac{E}{1-v^2}\left[\frac{t_p^3}{12} + t_p \bar{z}^2\right] + E\left[\frac{4t_s h_{Hs}^3}{12W} + \frac{4t_s h_{Hs}}{W}\left(\frac{h_{Hs}}{2} - \bar{z}\right)^2\right]$$

Where: h_{Hs} is the height of the horizontal stiffeners and h_{Vs} is the height of the vertical stiffeners.

Using the above equations gives the following values in this case:

$$\bar{z}_h = 0.0001757 \qquad\qquad \bar{z}_v = 0.00031$$
$$D_h = 1{,}295.7\,Pa \qquad\qquad D_v = 2336.4\,Pa$$

By using an average value of D we can use the equations governing an isotropic plate, thus we will let D = 1816 Pa.

Thus our equation for maximum plate displacement becomes:

$$w_{max} = 0.00126 P(W)^4 / 1816$$

The stresses at the bottom of the plate and the top of the stiffeners can be determined from:

$$\sigma_{hor} = \frac{E}{1-v^2} \frac{M_{hor}}{D_{hor}}[y] \qquad \text{and} \qquad \sigma_{ver} = \frac{E}{1-v^2} \frac{M_{ver}}{D_{ver}}[y]$$

Where $y = (\dfrac{-t_p}{2} - \bar{z})$ for the bottom of the plate and $y = (h_s - \bar{z})$ for the top of the stiffener.

At the centre of the plate: $M_{hor} = M_{ver} = 0.0231P(W)^2$
At the clamped edges of the plate: $M_{hor} = M_{ver} = -0.0513P(W)^2$

Using all of the above equations gives the following analytical results:

Result	Location	Value
Deflection, W_{max}	Centre of Plate	0.0035 m
σ_{hor}	Bottom of plate at centre	-19.19 MPa
σ_{hor}	Top of stiffener at centre	106.32 MPa
σ_{ver}	Bottom of plate at centre	-10.6 MPa
σ_{ver}	Top of stiffener at centre	58.96 MPa
σ_{hor}	Bottom of plate at edge	42.61 MPa
σ_{hor}	Top of stiffener at edge	-236.1 MPa
σ_{ver}	Bottom of plate at edge	23.64 MPa
σ_{ver}	Top of stiffener at edge	-130.9 MPa

F3:   Finite Element Modelling

A 3D thin shell FE model of one quarter of the fuselage panel was generated in order to solve this problem. The initial model used is shown in figure 10.38. The specifics of the FE models are:

Element type: 3D four node linear shell element
Material Model: Linear elastic isotropic model for 7075-T6 aluminium alloy:
                     E = 71.7 x 10^9 Pa, v = 0.33, σ_y = 503 x 10^6 Pa
Geometry: An area defining one quarter of the mid-plane of the plate was generated. Four additional areas orientated at 90° to this area were then generated to represent the mid-plane of the horizontal and vertical stiffeners. The horizontal stiffeners had a greater height as per the problem description.

Mesh: Initially a reasonably coarse mesh was used as shown in figure 10.38. A mesh convergence process was carried out until a converged mesh was achieved.

B.C.'s: Symmetry boundary conditions were applied to the edges of the plate and stiffeners that lay on the symmetry planes. The other edges of the plate were constrained in all directions in order to simulate the clamping effect.

Loads: Initially applied as a steady uniform pressure to the bottom face of the plate. The pressure loading was subsequently applied as a non-linear load in order to examine a complete flight cycle.

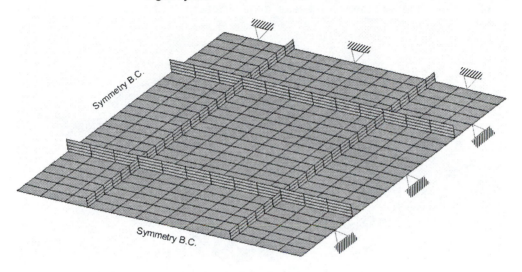

Figure 10.38: Finite element model of ¼ of the fuselage panel

F4:   FE Results

The deformed shape of the ¼ panel is shown overlaid on the original model shape in figure 10.39. As expected, the maximum displacement occurred at the centre of the panel (i.e. bottom left hand edge of the ¼ panel) and was 0.003883 m.

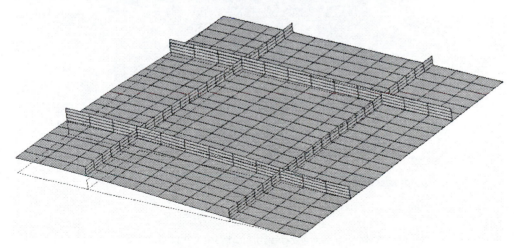

Figure 10.39: Deformed shape of the ¼ fuselage panel

The FE predicted distribution of stress in the stiffened panel for stress in the x-direction (horizontal) is shown in figure 10.40 and stress in the y-direction (vertical) is shown in figure 10.41. It can be seen from figure 10.40 that the maximum tensile stress in the horizontal direction is 72.4 MPa and this is located at the top of the central horizontal stiffener towards the centre of the panel. The maximum compressive stress is 175 MPa and this again occurs at the top of the central horizontal stiffener at a location near the clamped edge of the panel. Similarly, figure 10.41 shows that the maximum tensile stress in the vertical direction occurs at the top of the central vertical stiffener near the centre of the panel and has a value of 53.4 MPa. The maximum compressive stress in this case occurs at the top of the same stiffener near the clamped edge and has a value of 150 MPa. The stress in the panel was much lower than the stresses in the stiffeners in all cases as illustrated by figures 10.42 and 10.43.

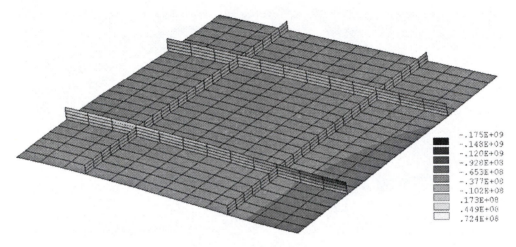

Figure 10.40: Distribution of stress in the horizontal direction in the stiffened panel

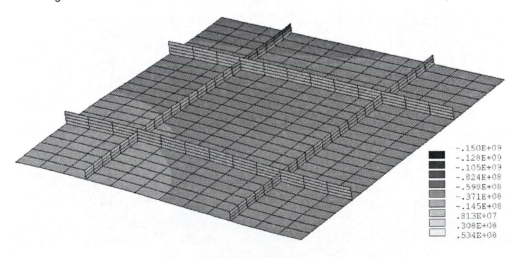

Figure 10.41: Distribution of stress in the vertical direction in the stiffened panel

F5:   Results Comparison

The table in figure 10.44 shows a comparison between the FEA results and the analytical results obtained in section F3. The FEA result for deflection is approximately 8% larger than that predicted by the analytical theory. This is reasonable considering the fact that the FE model takes account of shear deflection which the simple analytical model does not. The results for stress are not so accurate. The results are, however, reasonably close when one considers that the stiffeners are not on the centreline of the plate but we have assumed they were when recording stresses.

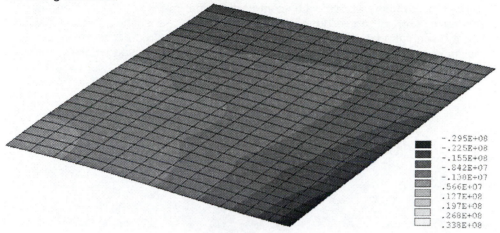

Figure 10.42: Distribution of stress in the horizontal direction in the panel

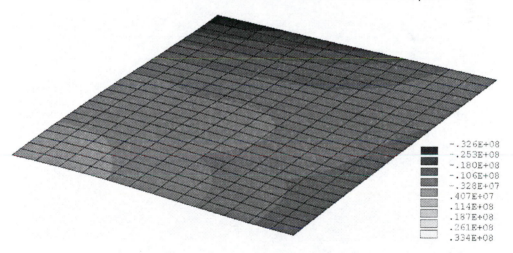

Figure 10.43: Distribution of stress in the vertical direction in the panel

Result	Location	Theory	FEA	% difference
W_{max}	Centre of Plate	0.0035 m	0.0038	8.5%
σ_{hor}	Bottom of plate at centre	-19.19 MPa	-16.86 MPa	13.8%
σ_{hor}	Top of stiffener at centre	106.32 MPa	72.4 MPa	46%
σ_{ver}	Bottom of plate at centre	-10.6 MPa	-16.3 MPa	53%
σ_{ver}	Top of stiffener at centre	58.96 MPa	53.39 MPa	10%
σ_{hor}	Bottom of plate at edge	42.61 MPa	31.8 MPa	33%
σ_{hor}	Top of stiffener at edge	-236.1 MPa	-175 MPa	35%
σ_{ver}	Bottom of plate at edge	23.64 MPa	33.4 MPa	46%
σ_{ver}	Top of stiffener at edge	-130.9 MPa	-150 MPa	15%

Figure 10.44: Comparison of Analytical and FE results for the stiffened plate

F6:   Non-linear Analysis

In the previous section we validated our FE results against analytical results for a line static analysis. In reality a panel on an aircraft will never be subjected to a constant pressure. Pressure will increase after take off as the aircraft climbs to its cruising altitude then remain relatively constant during the majority of the flight before beginning to drop off again during the landing phase. As such, a load that varies during the simulation is more appropriate for this analysis. In order to implement this in the finite element model a non-linear static analysis was used where three load steps were used as shown in figure 10.45.

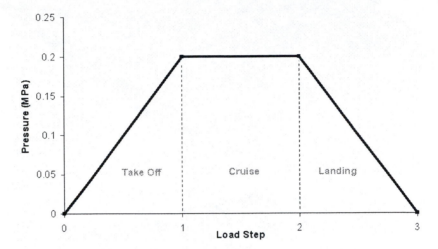

Figure 10.45: Variation of pressure loading on the stiffened panel during the flight cycle.

The results for displacement of the centre of the main sheet (i.e. location of maximum displacement) are shown in figure 10.46. Unsurprisingly we have obtained a response that closely matches the applied load shown in figure 10.45. This is what we should have expected considering that we are still using a static analysis.

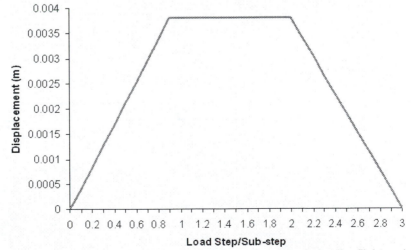

Figure 10.46: Displacement of central node on the plate during the flight cycle.

Had we used a transient analysis and introduced dynamic effects we would most likely have seen a different loading and unloading gradient due to damping effects.

Figure 10.47 shows the development of stress at a number of points in the stiffened sheet. Stress at the centre of the main sheet in both the X and Y directions is shown, together with the stress in the X direction at the central region of the top of the central vertical stiffener (i.e. location of maximum horizontal stress) and the stress in the Y direction at the central region of the top of the central horizontal stiffener. Again these results were as expected.

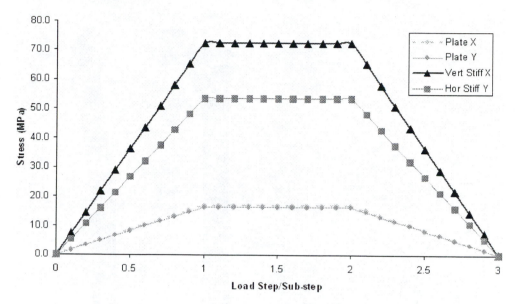

Figure 10.47: Variation of stresses in the stiffened panel during the flight cycle.

The results from the flight cycle analysis are quite limited due to the fact that a large assumption has been made regarding the nature of the load variation and the fact that dynamic effects, such as damping, are not present in the model. The results may still be useful, however, as they can be used as a rough indication of what point in the flight cycle a particular stress was exceeded. This can be useful in the design process to ensure that such a panel does not experience or approach yielding during a flight cycle.

F6:   Conclusions

This case study clearly illustrates the use of shell elements to solve complex 3D problems that can be considered "thin wall" or "thin shell" problems. This study also demonstrates the process of validating shell element models against appropriate analytical models. Finally use of simple non-linear loads with a static analysis is demonstrated and the limitations of such an approach are discussed.

10.8.   Case Study G: 3D Solid Analysis of a Hip Prosthesis

G1:   Problem Description

In this study we will perform an analysis of a human hip replacement prosthesis. This prosthesis is used to replace the head of a femur when the head has become weak due to disease or wear. The femur is the long bone in the upper leg. The head of the femur consists of a neck to which a ball is attached. The ball forms a "ball and socket" joint with the acetabulu, which is essentially a spherical hole in the pelvis. The main part of the femur, below the head, which is known as the diaphysis, is essentially a hollow cylinder. The hollow cylinder consists of cortical bone which is a hard dense material. The interior of the hollow cylinder, known as the intramedullary canal, is filled with cancellous bone which is a spongy soft material. At the bottom of the femur it widens out to form the knee joint. The wider portion at the bottom of the femur is known as the Condylar region.

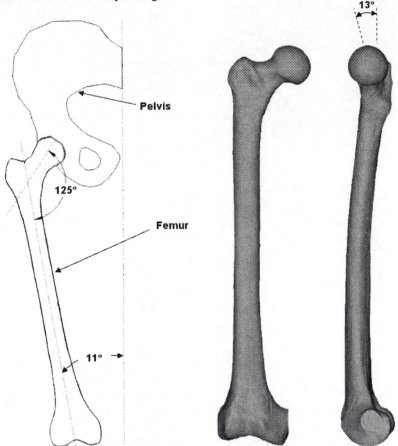

Figure 10.48: Details of a human femur and important angles.

The orientation of the femur inside the human body is characterised by a number of angles as shown in figure 10.48. Obviously each of these angles varies from person to person, however average values are shown in the figure. The femoral head angle defines the angle between the neck and diaphysis and has an average value of 125°. In the frontal plane, the femur is not orientated vertically in the body but rather slightly off vertical. This angle is known as the anterversion angle and has an average value of 11°. From the side the femoral head is also angled slightly forward, as shown on the right of figure 10.48. The average value for this angle is 13°.

During a hip replacement operation the head and neck of the femur are cut off and the inside of the femur is hollowed out to allow insertion of an appropriate hip prosthesis. Figure 10.49 shows an x-ray image of an implanted hip prosthesis. It can be clearly seen that the prosthesis replaces the neck and head of the femur and is inserted down in the intramedullary canal of the femur. The hip prosthesis may be cemented in or may be held in place via an interference fit depending on the type of prosthesis used. The head of the prosthesis is clearly much smaller than the original head of the femur. This is due to the fact that an acetabular cup or "socket liner" is also implanted into the pelvis and the head of the prosthesis is inserted into this new socket. The acetabular cup typically consists of polymeric materials in order to reduce friction. These materials do not show up on an x-ray, hence the void around the prosthesis head in figure 10.49.

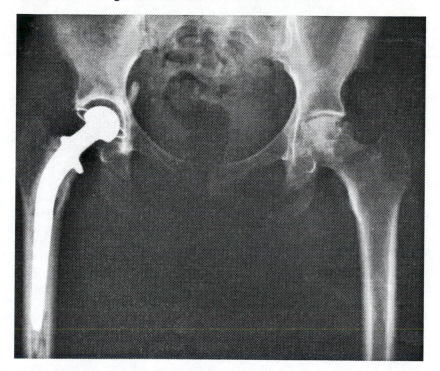

Figure 10.49: X-ray image of an implanted hip prosthesis, courtesy of Wikipedia

It should be clear from the above description that because of the various angles involved and the complex nature of the geometry of the femur and the prosthesis that a full 3D analysis will be required to accurately represent the geometry, hence necessitating the use of 3D solid elements. Having said this it should be noted that many researchers have obtained reasonably accurate results using 2D models!

We will approach this analysis by first building a FE model of the intact femur (i.e. without prosthesis) and then later introduce the prosthesis. We shall make some fairly large assumptions regarding geometry in the absence of further information. The purpose of this study is to demonstrate how to perform such an analysis, if you can understand the process then you can later use a more accurate geometry, obtained, for example, from a CT scan. We are going to assume that the femur has a head diameter of 53mm and a diaphysis length of 0.36m, further dimensions are shown in figure 10.51. We will assume that the prosthesis has a head diameter of 22mm and is constructed from biocompatible 316L steel.

G2:   Assumptions

Materials: Bone is an anisotropic material with different properties in every direction. It is standard practice, however, to assume that cortical bone is a linear elastic orthotropic material and cancellous bone is linear elastic isotropic. The values used for the material constants are shown in the table below. The prosthesis was assigned parameters consistent with a linear elastic model for 316L steel.

Material	Young's Modulus	Shear Modulus	Poisson's Ratio
Cortical Bone	$E_X = 7$ GPa $E_Y = 7$ GPa $E_Z = 11.5$ GPa	$G_{XY} = 2.6$ GPa $G_{YZ} = 3.5$ GPa $G_{ZX} = 3.5$ GPa	$v_{XY} = 0.4$ $v_{ZY} = 0.4$ $v_{ZX} = 0.4$
Cancellous Bone	$E = 0.413$ GPa	N/A	$v = 0.36$
316L Steel	$E = 210$ GPa	N/A	$v = 0.27$

Geometry: A 3D analysis is required due to the complexity of the geometries of both the femur and prosthesis and the fact that they both vary in all three dimensions – with three significant angles as shown in figure 10.48. A full three dimensional solid model of the femur and prosthesis will be required in order to generate a mesh of 3D solid elements.

Loading: The hip joint is obviously subjected to a time dependant loading based on the activity of the patient. In this case we will examine one particular point during the gait cycle: heel strike and hence use a static loading. This is the point during walking when the heel hits the ground and load is transferred onto that foot via the hip. The main loads acting on the femur are due to the body weight, which acts through the pelvis onto the head of the femur, and the force imparted by the abductor muscle on the outside of the upper part of the femur. Typical values for the body weight force and abductor force are 1700N and 1400N respectively. These forces act at particular angles and can be resolved into horizontal and vertical components for convenience, as shown in figure 10.50.

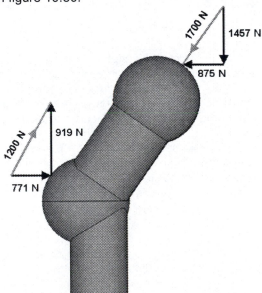

Figure 10.50: Overview of major loads acting on a typical femur

B.C.'s: The femur is constrained from moving down by reaction force from the ground via the knee joint. In order to simulate this the bottom of the femur model will be constrained in all directions. It would perhaps be more accurate to provide a number of spring elements at the bottom of the femur which would simulate the allowable compression of the knee and ankle joints and take account of any other losses due to compression of footwear, other body tissues etc. In this case however given the other large assumptions we are making it is reasonable to constrain the bottom of the femur in all directions.

We will have interfaces between a number of different material types and bodies in this analysis. The cortical bone will interface with the cancellous bone in the interior of the femur. It is reasonable to assume that there is perfect bonding here and hence the elements that make up the cortical and cancellous bone regions can be joined at the interface region. Similarly, the prosthesis will be either cemented in or will be fixed by an interference fit and thus perfect bonding will be assumed here also. Later on a more advanced analysis that uses contact elements may be required to examine effects such as the friction between the implant and the bone and how it affects prosthesis performance.

Solution type: As a linear elastic model is being used and static loads have been assumed, and we are presuming that the loads don't go beyond the yield stress, a simple small deflection linear static structural analysis should be sufficient in this case. This is reasonable considering that we would not expect bones to deflect significantly under normal loading. We also know from experience that bones are brittle and thus do not deform significantly before failure.

G3:   Finite Element Modelling – Intact Femur

A 3D solid finite element of the intact femur was generated in order to solve this problem. The initial model used is shown in figure 10.51. The specifics of the FE models are:

Element type:	3D ten node solid tetrahedral elements
Material Model:	Linear elastic orthotropic model for cortical bone and linear elastic isotropic model for cancellous bone.
Geometry:	A three dimensional geometry was generated from the average values obtained from a stadardised femur. The main geometrical details are shown in figure 10.51. The relevant angles are shown in figure 10.48.
Mesh:	A mesh convergence process was carried out until a converged mesh was achieved.
B.C.'s:	A group of nodes on the lower extremity of the femur model were constrained in all directions.
Loads:	Applied to nodes on the head and greater tronchater of the femur as detailed in figure 10.50

We have obviously made many assumptions in order to arrive at the simplified geometry shown in figure 10.51; however it should be reasonably easy for you to replicate this geometry using your FE software pre-processor. If you find that you are missing a particular dimension then just make an educated guess based on a measurement from figure 10.51. The geometry of the femur has been simplified into two hollow cones which represent the diaphysis of the femur, a hollow wedge which represents the tronchator and neck region and a hollow sphere which represents the femur head.

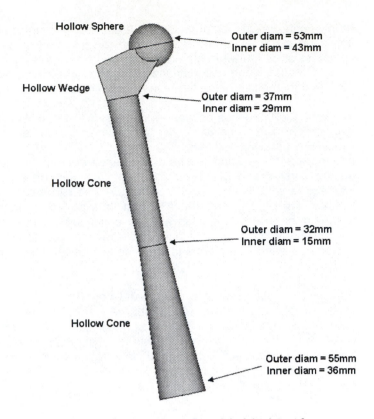

Hollow Sphere

Outer diam = 53mm
Inner diam = 43mm

Hollow Wedge

Outer diam = 37mm
Inner diam = 29mm

Hollow Cone

Outer diam = 32mm
Inner diam = 15mm

Hollow Cone

Outer diam = 55mm
Inner diam = 36mm

Figure 10.51: Details of FE model of the intact femur

G4:   FE Results – Intact Femur

In order to check that the model was behaving properly the abductor muscle load was initially ignored and only the body weight load applied to the femoral head was used. Results for principal strains were examined and the distribution of strain in the axial direction was determined and is shown in figure 10.52. As can be seen from the figure the medial side of the femur diaphysis is in compression while the lateral side of the diaphysis is in tension. This result is as expected for this type of loading and indicates that the model is behaving as expected. The level of strain experienced by the femur is also of the order of magnitude expected.

Figure 10.53(a) shows the corresponding result when the abductor force is included in the finite element model. The level of strain experienced is, however, rather excessive when compared to similar studies available in literature. Upon further examination it was found that an artificial stress concentration has been introduced at the point of application of the abductor force, due to the fact that the load was applied as a point force to a particular node. This is clearly shown in figure 10.53(b) which shows an alternative view of the model. In order to remove the effect of the stress concentration the elements making up the diaphysis were selected and all other elements were removed from the post-processing of results. Figure 10.53(c) shows the distribution of axial strain in the diaphysis only. It can be seen here that the strain levels are more in line with what was expected when compared to similar studies available in literature. It is clear from figure 10.53 that a further stress concentration has been introduced at the junction of the diaphysis and neck region due to the fact that a sharp corner exists here. The model could be further improved by introducing a blend in this region.

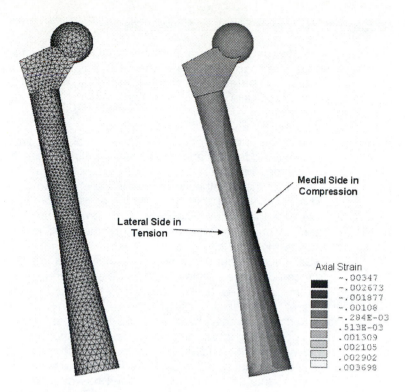

Figure 10.52: Resultant distribution of axial strain in the femur model when only the body weight load is used.

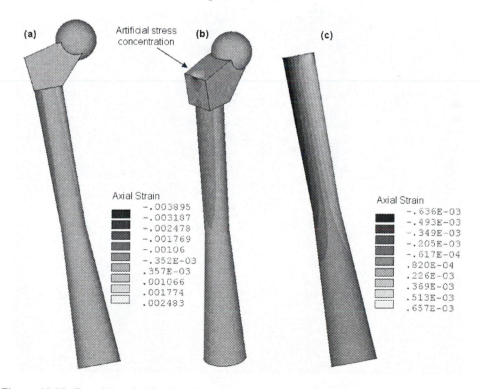

Figure 10.53: Resultant distribution of axial strain in the femur model when abductor load is included.

G5:   Finite Element Modelling – Femur and Prosthesis

The 3D model of the intact femur described above was modified in order to carry out the hip prosthesis analysis. The head and some of the neck region was cut away and the prosthesis solid model was inserted in the hollow femur. The initial model used is shown in figure 10.54. The specifics of the FE models are:

Element type: 3D ten node solid tetrahedral elements
Material Model: Linear elastic orthotropic model for cortical bone, linear elastic isotropic model for cancellous bone and linear elastic isotropic model for 316L steel.
Geometry: Details of the prosthesis geometry are shown in figure 10.54. The 3D model of the intact femur as described above was used for the femur. The head and some of the neck of the femur were cut away and the prosthesis solid model was inserted in the hollow femur model as shown in the figure. It was assumed that the prosthesis exactly filled the cavity inside the cortical bone. Clearly this is not the case in reality, however for the purposes of illustration and to keep the analysis as simple as possible we will ignore this obvious error. The spherical prosthesis head was also ignored as it will be easier to apply loads to the prosthesis neck and the only function of the prosthesis head is to transfer load from the acetabular socket. In reality the prosthesis head is screwed on in any case so this assumption is valid and has been widely used in literature.

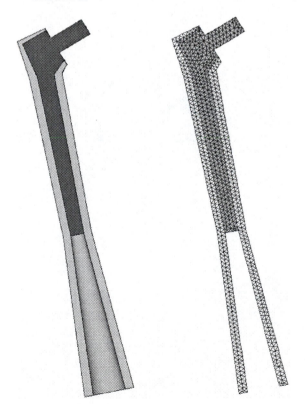

Figure 10.54: Cross section through FE model of cut femur and prosthesis

B.C.'s:

A group of nodes on the lower extremity of the femur model were constrained in all directions. It was assumed that the prosthesis and femur are perfectly bonded thus perfect contact was used between the prosthesis and femur geometries.

Loads:

Applied to nodes on the prosthesis head and on the greater tronchator of the femur as described previously

G6:   FE Results –Femur and Prosthesis

Figure 10.55 shows the distribution of axial strain in a slice through the femur/prosthesis assembly. It can clearly be seen that the distribution of axial strain in the femur has clearly been modified by the presence of the prosthesis. The femur is still behaving as expected with the lateral side in tension and the medial side in compression, with strain concentrated mid way along the length of the diaphysis, however in this case the levels of strain are much smaller than before.

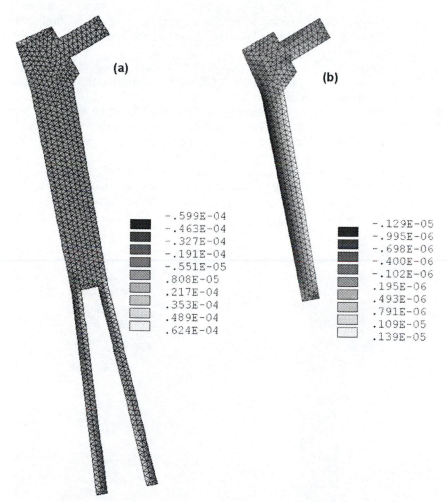

Figure 10.55: Distribution of axial strain in a slice through the combined prosthesis and femur FE model.

This result was expected as the prosthesis is much stiffer than the surrounding bone and the assumptions of 1) perfect bonding and 2) a prosthesis that fills the interior of the cancellous bone perfectly have resulted in a very stiff prosthesis/bone assembly. The fact that the bottom of the prosthesis is a sharp edge and is in perfect contact with the bone at this location has introduced some artificial stress concentrations in this region.

G7:   Conclusions

The objective of this case study was to demonstrate a problem that cannot be solved using 2D elements or shell elements. The femur and implanted femur is such a problem due to the 3D nature of the femur geometry and, in particular, the many important angles involved. This case study also illustrates the use of an orthotropic material model and using dissimilar material models in the one FE model.

There are clearly many problems with the analysis carried out here. The case study has been designed and presented in such a way that it should be easier for you to replicate the work carried out. This case study should not be seen as a model for analysing a realistic hip prosthesis problem, but rather as a first step to understanding some of the concepts involved. In particular, the above analysis has the following problems:

(a)   The geometry of both the femur and prosthesis are highly simplified. In reality geometries obtained from CT scans and/or coordinate measuring machines would be used.

(b)   The assumption of perfect bonding between the prosthesis and femur is rarely valid. In most cases a cement layer is used which may also require modelling and contact elements may be required between the femur and cement and between the cement and prosthesis.

(c)   The loading presented above has been highly simplified. In reality there are many more muscle loads which must be accounted for and many different loading situations (e.g. climbing stairs, standing on one leg, running etc.)

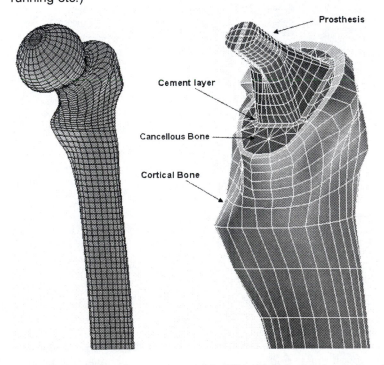

Figure 10.56: Examples of more appropriate FE models for this type of analysis

10.9. Case Study H: Analysis of Forging of a Rectangular Bar

H1:   Problem Description

In this study we will carry out a finite element simulation of forging of a rectangular bar made from annealed copper. Forging is a metal forming process that is used in this case to reduce the height of the bar. The bar is placed between two steel dies after which the upper die moves down to compress the copper bar and hence cause permanent deformation. Figure 10.60 shows a cross sectional overview of the forging operation.

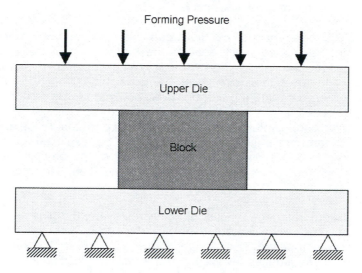

Figure 10.60: Overview of Forging of a Rectangular Bar

In this case we are required to determine the forging pressure required to reduce the height of the bar by 25%. The bar has a square cross section and has a height and width of 0.1m.

H2:   Assumptions

Materials: We are told that the bar is made from annealed copper and we know that the bar will be permanently deformed, so a material model that uses parameters obtained from material tests on annealed copper and allows for plastic deformation is required. We are not given any information on the speed of deformation, thus strain-rate effects can not be included in the analysis. In this case we could use any of the strain rate independent elasto-plastic material models from section 5.5 could be potentially be used, however the simplest model to implement is the bilinear isotropic elasto-plastic model (see section 5.5.2). Experimental tests on the annealed copper material showed that: $E = 110$ GPa, $v = 0.34$, $\sigma_y = 33$ MPa, $\varepsilon_f = 0.6$.

We will assume that the dies are made from tool steel and that they do not deform significantly during the forging process, thus a linear elastic material model for steel may be used.

Geometry: We are told that the geometry to be deformed is a bar with a square rectangular cross section. We are not given any information about the length of the bar, only cross sectional details. As such, it is reasonable to assume that the bar is quite long and that a plane strain analysis is appropriate. This has the added benefit of being directly comparable to plane strain forging analytical theory which is outlined below.

Loading: We are required to determine the required applied pressure load that causes the bar height to be reduced by 25%. This will require performing several analyses in which the pressure load is changed and the resultant deformation of the block is noted. The pressure load is applied to the upper die, which moves to deform the bar, while the lower die remains stationary.

B.C.'s: The lower die does not move during the forging process and provides the resistance which enables the deformation of the block to take place. Accordingly, nodes comprising the entire lower die are constrained in all directions. The upper die moves vertically down in order to deform the bar, however, horizontal motion of the die is not required, accordingly the lower die will be constrained in the horizontal direction, but free to move in the vertical direction. Since the bar has a square cross section symmetry can be exploited to reduce the problem size and hence only one half of the problem needs to be analysed once the correct symmetric boundary conditions are employed.

There is a large amount of contact between the dies and the bar. In this case we will assume that a surface-to-surface contact description is appropriate. This assumption is valid given the relatively large contact surfaces involved and the fact that significant sliding should occur between the bar and dies. We will initially assume that the dies are so well lubricated that friction is zero and subsequently will assume a friction coefficient of 0.05.

Solution type: As we are using an elasto-plastic material model and significant deformation of the bar is expected, a large deformation analysis is appropriate. A strain-rate independent material model is being used which implies a static large displacement analysis. The analysis is further complicated by the existence of large contact surfaces between the dies and bar. This means that the analysis will be highly non-linear but still static. The solution will be split up into sub-steps in order to ensure a stable solution, as described in sections 8.3.2 and 8.9.2.

H3:   Analytical Theory

We can use plane strain forging theory to validate our finite element results as detailed below.

If we assume that the initial yield stress, Y_0, is equal to the uniaxial yield strength for the material, σ_y, then:

$$Y_0 = \sigma_y = 33\,MPa$$

The resultant yield stress due to any deformation, Y_1, is given by:

$$Y_1 = Y_0 + E_{Tan}\varepsilon$$

Where strain, ε, is obtained from the height before, h_0, and after, h_1, deformation:

$$\varepsilon = \ln\left(\frac{h_0}{h_1}\right)$$

And, the tangent modulus, E_{Tan}, as before, is given by:

$$E_{Tan} = \frac{UTS - \sigma_y}{\varepsilon_f - \varepsilon_{proof}}$$

The pressure required to cause the reduction in height of a block of width, b, from h_0 to h_1 is given by:

$$P = Y_1 e^{\left(\frac{\mu b}{h_1}\right)}$$

Where, μ is the coefficient of friction between the forging dies and the bar.

Using the above equations we can calculate the following:

Result	Value		
Y_0	33 MPa		
h_0	0.1		
h_1	0.075		
ε	0.287		
E_{Tan}	167 MPa		
Y_1	81.04 MPa		
μ	0		0.05
b	0.1		0.1
P	81.04 MPa		86.62 MPa

Thus, the analytical model predicts that a pressure of 98.986 MPa is required in order to reduce the bar height by 25%.

H3:   Finite Element Modelling

A 2D plane strain FE model of one half of a cross section through the forging problem was generated in order to examine this problem. The initial model used is shown in figure 10.61. The specifics of the FE models are:

Element type: 2D eight node quadratic plane strain elements
Material Models: Linear elastic isotropic model for tool steel for the dies:
                     $E = 210 \times 10^9$ Pa, $v = 0.27$
                     Bilinear isotropic elasto-plastic model for copper for the bar:
                     $E = 110 \times 10^9$ Pa, $v = 0.34$, $\sigma_y = 33 \times 10^6$ Pa, $E_{Tan} = 167 \times 10^6$ Pa
Geometry: An area defining one half of the cross section through the bar/die assembly was generated.
Mesh: Due to the fact that the bar would be deforming significantly during the FE simulation, a very fine mesh was used for the

rectangle representing the bar. The dies were not expected to deform significantly and their main purpose in the model is to transfer loads to the bar, thus a reasonably coarse mesh was used for the dies. The coarseness of the die mesh was constrained by the contact algorithm used which required a minimum number of nodes on the die contact surfaces in order that the algorithm would provide reliable results.

B.C.'s: The dies were essentially treated as rigid bodies as we are not particularly interested in what is happening within the die geometries.

The area defining the upper die was constrained from moving in the horizontal direction, but free to move in the vertical direction. The lower die was constrained in all directions. Symmetry boundary conditions were applied to the cut edges of the dies and bar that lay on the symmetry plane.

2D surface to surface contact elements were defined between the upper die and the upper surface of the bar and between the lower die and the lower surface of the bar. Initially a coefficient of friction of zero was used and this was later increased to 0.05. Symmetric contact was used in both cases, with each surface acting as both a target and a contact surface.

Loads: A pressure load was applied to the upper surface of the upper die.

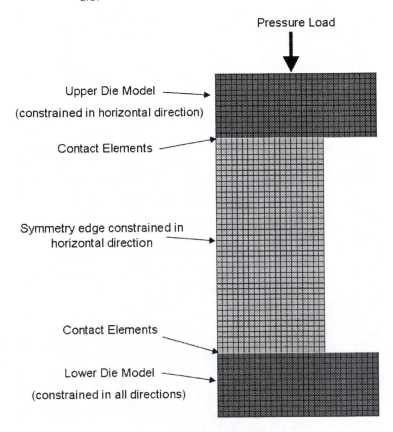

Figure 10.61: The half symmetry plane strain model used to analyse the forging problem

H4:   FE Results – Zero Friction

A pressure load of 100 MPa was applied to the upper die and this resulted in a reduction in height of the bar of approximately 33%. As the solution was set up so that a number of sub-steps were used it was possible to investigate the deformation at each sub-step and hence find the load at which a 25% reduction in height was achieved. In this case the load needed to cause a 25% reduction in height was 77 MPa. The resultant deformation of the FE model is shown in figure 10.62. It can be seen from the figure that, as no friction is present, the block has deformed uniformly across its height. The distribution of plastic strain in the block is shown in figure 10.62 and it can be seen that

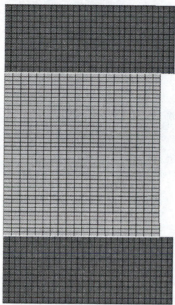

Figure 10.62: Deformed shape of the FE model in the frictionless forming case

H5:   FE Results – Friction = 0.05

In this case a pressure load of approximately 88Mpa was required in order to cause a 25% reduction in the height of the block. The resultant deformation of the FE model is shown in figure 10.63. It can be clearly seen that the introduction of friction into the FEA has caused "barrelling" of the bar, as would be expected. Friction at the contact surfaces with the dies is preventing the nodes on the bar at these regions from moving in the horizontal direction. The nodes in the centre of the bar cross section are not constrained and can move relatively freely in the horizontal direction. The combined result of this is the barrelling effect shown in figure 10.63. Figure 10.64 shows the barrelling effect in more detail by expanding the model across its symmetry plane, thus showing the entire bar cross section. The left hand side of figure 10.64 shows the original un-deformed bar while the right hand side shows the bar after forming.

Figure 10.65 shows a symmetry expanded view of the distribution of von-Mises strain in the deformed bar. The distribution is as expected and corresponds well to theoretical models and to experimental observations. Note that the maximum von-Mises strain is 0.486 which is well below the elongation at break (i.e. failure strain) which is 0.6.

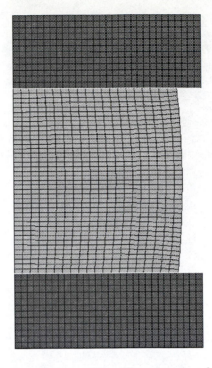

Figure 10.63: Deformed shape of the FE model when friction is included

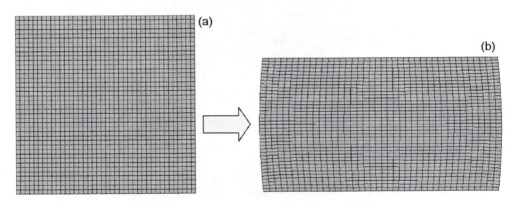

Figure 10.64: Expanded FE model before (a) and after (b) the forging process

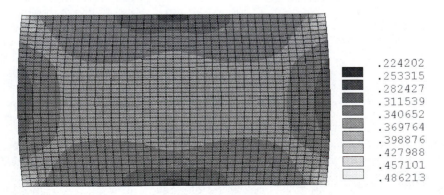

.224202
.253315
.282427
.311539
.340652
.369764
.398876
.427988
.457101
.486213

Figure 10.65: Distribution of von-Mises strain in the expanded FE model before after forging

H6:   Conclusions

When we compare the analytical and the finite element predictions of the pressure required to reduce the block height, as shown in the table below, it is immediately clear that there are some small differences.

	Analytical	FEA	% Difference
Friction = 0	81.04 MPa	77 MPa	4.9 %
Friction = 0.15	86.62 MPa	88 MPa	1.5 %

It can be seen from the table that the results for forging with friction are much closer than those obtained for frictionless forming. Both models have, however, predicted results which are within 5% of the analytically predicted results which is satisfactory.

This case study illustrates and applies a number of the advanced concepts presented earlier in the book, including: non-linear analysis, load steps/sub-steps, use of contact elements and non-linear materials.

10.10. Case Study I: Analysis of Reinforced Concrete Strip Foundation

I1:   Problem Description

Strip foundations are used as foundations for most walls built from brick or concrete block. They essentially consist of a strip of concrete placed in a trench dug into the ground, sometimes with steel reinforcement, which is used as a base for the wall during building. The purpose of the foundation is to redistribute the loads from the walls into the surrounding ground which is normally soil in such a manner that settlement is limited and failure of the underlying soil is avoided. Figure 10.66 shows an cross section through a typical reinforced concrete strip foundation used to support a 200mm thick wall.

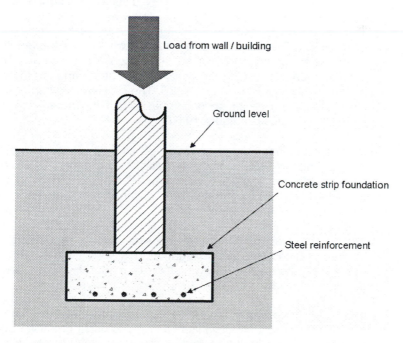

Figure 10.66: Overview of a reinforced concrete strip foundation

Wall loading will normally cause the concrete strip to bend with tension on its lower surface and compression of the upper surface. Concrete is very strong in compression but relatively weak in tension, so steel reinforcing rods (known as "rebar") are sometimes placed in the concrete near the bottom of the strip in order to help resist bending.

We are required to perform a finite element analysis to determine the wall loading required to induce failure in a 3m long reinforced strip foundation used to support a 200mm thick block wall. The foundation is 600mm wide and 300mm deep and is made from normal strength concrete. Four steel reinforcing rods (rebar) running along the length of the strip are spaced equally across the width and are approximately 50mm from the bottom of the foundation. The steel rods have a diameter of 12mm. For the failure analysis we will assume that the foundation is simply supported.

Subsequently we will introduce a model for the subsoil underneath the foundation to investigate settlement of the foundation due loads transferred from the walls.

I2:   Assumptions

Part 1: Foundation Failure

Materials: The foundation is made from reinforced concrete so we must use an appropriate concrete material model and a model for the steel reinforcement. A simple linear elastic isotropic material model will suffice for the steel since the concrete will fail long before the yield stress of the steel is approached.

A concrete material model which predicts both crushing (due to excessive compression) and cracking (due to excessive tension) is required. Concrete material models are discussed in section 5.7.

Geometry: The foundation is basically a rectangular cross section beam of length 3m, height 0.5m and width 0.5. The wall which runs the length of the beam is placed on the top surface at the centre of the beam. The steel reinforcement runs along the length of the beam and is placed at the bottom, as illustrated in figure 10.66. The problem is symmetric across the width of the beam so only one half of the beam needs to be modelled, as shown in figure 10.67(b).

The wall does not need to be modelled as the wall simply transfers its weight, and the weight of the building, to the central 200mm of the strip foundation. Thus the wall can be represented by applying a pressure load to the central portion of the strip foundation model, as shown in figure 10.67(b).

Loading: We are required to determine the wall loading required to induce failure in the concrete strip foundation thus we will use a pressure load applied to the centre of the strip to model the wall loading as described above.

B.C.'s: In order to investigate failure we will assume that the foundation is simply supported. In reality this is unlikely to occur directly, however, it may occur indirectly due to settlement of subsoil at the central portion of the foundation as shown in figure 10.68.

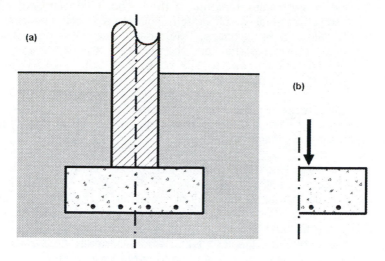

Figure 10.67: Symmetry in strip foundation problem

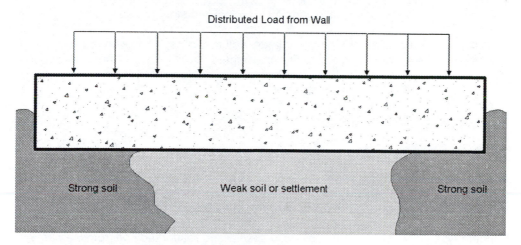

Figure 10.68: Illustration of how a weak pocket of soil or settlement can produce a simply supported concrete beam

Solution type: The concrete material model is a non-linear model however the analysis is otherwise relatively straightforward. Concrete is a brittle material which will not deform significantly before failure so a small deflection analysis is required. Hence, a small deflection non-linear static analysis should suffice.

Part 2: Foundation Settlement

In order to investigate settlement of the foundation due to compression or failure of the subsoil underneath the foundation we will need to introduce a finite element model of the subsoil and subsequently define contact elements between the foundation model and the soil model. We will re-use the finite element model built in part 1 and build the subsoil model underneath it.

Materials: Soil is a complex material to model as it consists not just of the granular material (which we term soil) but also contains large amounts of liquid and some gases. This means that soil is generally a highly compressible material and behaves highly non-linearly. We will assume that the soil in this case is a clay type soil and assign appropriate parameters.

The Drucker-Prager material model is commonly used in finite element codes to model granular materials such as soil. This material model is discussed in section 5.7.

Geometry: The same geometry used for the concrete strip foundation in Part 1 will be re-used here.
In order to allow for distribution of loads in the soil and avoid misinterpretation of localised effects a volume of soil six times as deep as the foundation strip and 12 times as wide will be used. We will ignore that fact the foundation would normally be buried and there would be soil at the side and on top of a portion of the strip.

Loading: We are required to determine the settlement of the foundation due to the average wall loading for a typical two story house.

B.C.'s: Since we are using a large volume of soil we may fix the outer surfaces of the soil in appropriate directions as there should be enough soil volume to allow for any settlement without being effected by the boundary conditions.

I3: <u>Finite Element Modelling</u>

Part 1: Foundation Failure

A 3D solid model representing one half of width of the strip generated in order to solve this problem. A 3D solid model was necessary in order to implement specialised concrete elements which allow for specification of rebar details.

Element type: 3D eight node solid specialised reinforced concrete elements
Material Models: Linear elastic isotropic model for steel for the rebar:
$E = 210 \times 10^9$ Pa, $\nu = 0.27$
Concrete material model with damage:
$E = 25 \times 10^9$ Pa, $\nu = 0.17$,
Uniaxial Compressive strength (crushing): $\sigma_{fc} = 2.5 \times 10^6$ Pa
Uniaxial Tensile strength (cracking): $\sigma_{ft} = 30 \times 10^6$ Pa
Shear transfer coefficient (open crack) = 0.3
Shear transfer coefficient (closed crack) = 0.9

The shear transfer coefficients describe the ability of an open or closed crack to transfer loads. These coefficients can vary from 0 to 1, with 0 indicating a very smooth crack and 1 indicating a rough crack surface.

The material model here is known as a "smeared" reinforcement model as it distributes the effect of the rebar throughout the element based on a volume fraction which describes the relative cross sectional are of rebar to concrete.

In this case the volume fraction used was 0.0006818. The rebar can be angled in any direction: in this case it was set parallel to the length of the strip.

Geometry: A volume defining on half of the width of the strip was generated.

Mesh: A regular brick mesh was used for the analysis. As the wall lies on the central 0.2m of the 0.6m wide strip, the width of the model volume was divided into three elements, which allowed for easy application of loads. Appropriate element divisions were then applied to the other dimensions of the model in order to ensure a regular cube mesh was generated. The model used is shown in figure 10.69

B.C.'s: In order to model the simply supported strip, nodes at the bottom surface of the strip near the ends were constrained in the vertical direction, as shown in figure 10.69

Loads: A pressure load was applied to the upper surface of the top central elements as shown in figure 10.69. The load was subsequently varied in order to determine the load required to induce failure in the foundation.

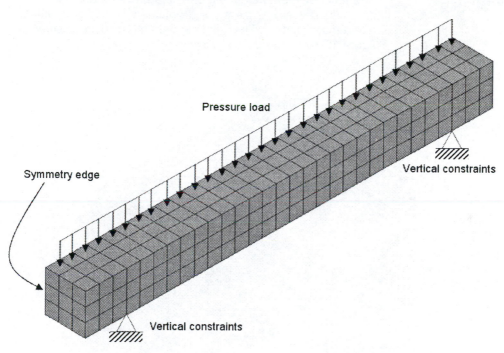

Figure 10.69: Finite element model used to investigate foundation failure

Part 2: Foundation Settlement

A 3D solid model representing one half of width of the foundation strip was used, as in the previous analysis. A 3D solid model of the subsoil was generated underneath the foundation.

Element type: 3D eight node solid specialised reinforced concrete elements for the strip and 3D eight node homogenous solid elements for the soil.

Material Models: Linear elastic isotropic model for steel for the rebar:
$E = 210 \times 10^9$ Pa, $\nu = 0.27$
Concrete material model with damage: (as before)
$E = 25 \times 10^9$ Pa, $\nu = 0.17$,
Uniaxial Compressive strength (crushing): $\sigma_{fc} = 2.5 \times 10^6$ Pa
Uniaxial Tensile strength (cracking): $\sigma_{ft} = 30 \times 10^6$ Pa
Shear transfer coefficient (open crack) = 0.3
Shear transfer coefficient (closed crack) = 0.9
Drucker-Prager model for clay type soil:
$E = 13 \times 10^6$ Pa, $\nu = 0.45$,
Cohesion value = 10,840 N/m^2
Angle of internal friction = 10°
Dilatancy angle = 0°

Geometry: A volume defining on half of the width of the strip was generated as before.
A second volume defining the soil was generated underneath the strip. This volume had a depth of 3m, a width of 3m and a length equal to the strip.

Mesh: The finite element model used for the previous analysis of the strip was reused; hence the soil volume was meshed with the same mesh density. The finite element model used is shown in figure 10.70.

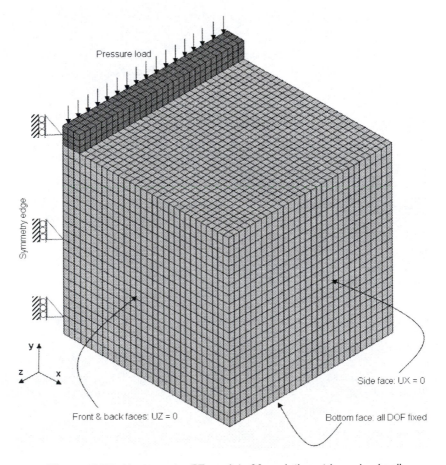

Figure 10.70: ½ symmetry FE model of foundation strip and subsoil

B.C.'s: Symmetrical boundary conditions were placed on the symmetry cut plane of the strip and soil volumes as indicated in figure 10.70. The base of the soil volume was constrained in all DOF, while the other faces were constrained as shown in figure 10.70.

Loads: A pressure load was applied to the upper surface of the top central elements of the strip as before.

I4:   Results

Part 1: Foundation Failure

Figure 10.71 shows the deformed shape of the strip foundation just before failure when simply supported as shown in figure 10.69. The deformation in figure 10.71 has been exaggerated for clarity. At this point the maximum displacement at the centre of the strip was 0.00056m, i.e. approximately half a millimetre. The applied pressure load at this point was 159,000 Pa. It was found that, once the applied pressure was increased above this point, the strip began to fail catastrophically. Cracks immediately formed through the thickness of the central region of the strip.

Figure 10.72 shows a cracking/crushing plot for the strip when the applied pressure is increased to 160,000 Pa. This plot shows where cracks have formed in the model by generating a circle in the plane of the crack at any element that has cracked. Cracks can be set to be displayed at either integration points (as is the case here) or at the element centroid. The first, second and third crack in each element are indicated by a particular colour to allow visualisation of the first and subsequent direction of cracking. It is clear from figure 10.72 that the strip foundation would be completely destroyed at this point as it has effectively cracked completely through its thickness. Luckily the chances of the strip foundation experiencing this type of loading are relatively low as a different foundation design would be used for simple supports and the wall loading for a typical two story house is of the order of 50 KN/m.

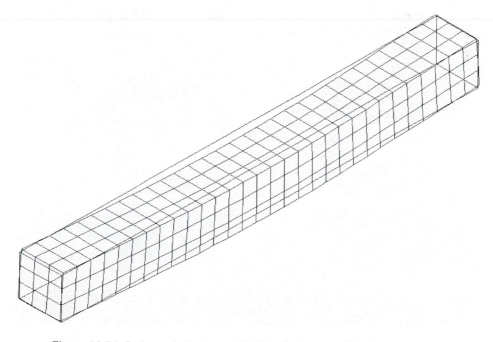

Figure 10.71: Deformed shape on strip foundation model just before failure.

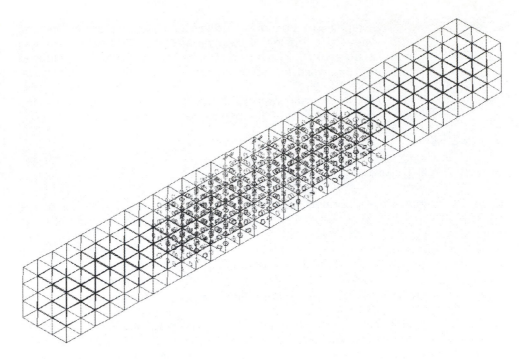

Figure 10.72: Cracking plot for the strip foundation model with an applied load of 160 kPa.

Part 2: Settlement

Figure 10.73 shows the predicted settlement of the foundation due to the applied loading. Figure 10.73(a) shows the settlement for the combined strip and soil model and figure 10.73(b) shows the same plot with the strip removed for clarity. It can be seen that the maximum soil settlement occurs, as expected, in the region directly under the strip foundation and is of the order of 0.9 mm. This level of settlement is acceptable given the dimensions of the foundation and would not cause any serious problems were it to occur in practice.

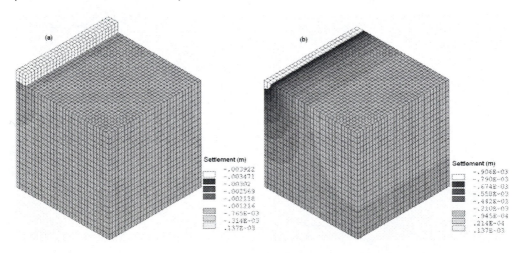

Figure 10.73: Plots of settlement of (a) the combined strip/soil model and (b) the soil FE model.

Figure 10.74(a) shows a vector plot of nodal displacement in the soil model. This plot is useful in order to visualise how the soil is settling due to the applied wall and foundation loads. It should be noted that an artificial effect has been introduced because of the boundary conditions used, whereby the soil is moving upwards at the boundary condition furthest from the foundation strip. It is likely that this effect can be eliminated by using a wider volume for the soil FE model.

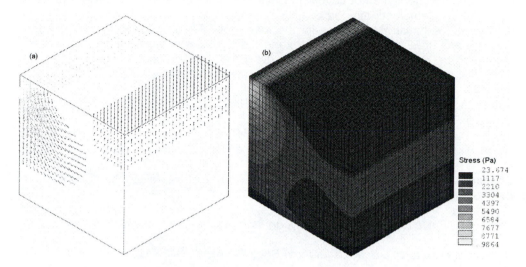

Figure 10.74: (a) Vector plot showing displacement of soil due to wall loading (b) distribution of hydrostatic stress in the soil

Figure 10.74(b) shows a plot of hydrostatic pressure in the soil FE model. This plot clearly illustrates the "bulb of pressure" phenomena which is experienced under all such strip foundations and shows why removing soil or excavating in this region can have catastrophic results!

I5:   Conclusions

The failure analysis of the concrete strip foundation shows that a wall loading far in excess of that expected from an average two story house is required to induce failure even under the highly un-favorable simply supported loading scenario. The reinforced concrete strip is modeled using elements which use a smeared reinforcement approach. This doesn't accurately model the real situation where the steel reinforcement is biased towards the bottom of the strip in order to more effectively resist bending. The model could be improved by using a homogenous concrete element for the strip and using beam elements to model the reinforcement. In such a case the beam elements would have to be generated carefully so that they share nodes with the solid concrete elements. In such a model it is likely that a greater resistance to bending could be realised.

The settlement analysis showed that there was no significant settlement of the strip foundation or subsoil due to the applied load. Analysis of the results showed that the model was behaving as expected by exhibiting the "bulb of pressure". This model could be improved by widening the soil volume in order to eliminate localized effects.

Bibliography

There are many excellent books available covering the theory of finite element analysis. Most of these books go into much more detail on the formulation of problems using various types of finite elements than this text. Also the formulation of more complex elements not covered in this text can be found in some of the texts listed below. Remember that the objective of this book was to be a practical guide to FEA not an exhaustive reference for the theory of FEA. This book is also aimed at novice users of FEA. Once you understand the concepts in this book you will benefit greatly from consulting any of the texts listed below. Those listed below, in alphabetical order, are the texts which I personally find the most useful.

Adams, V. and Askenazi, A., *Building Better Products with Finite Element Analysis*, OnWord Press, 1999

Bathe, K.J., *Finite Element Procedures,* Prentice Hall, 2nd edition, 1995

Chandrupatla, T.R. and Belegundu, A.D., *Introduction to Finite Elements in Engineering*, Prentice Hall, 3rd edition, 2002

Fagan, M., *Finite Element Analysis: Theory and Practice*, Prentice Hall, 1996

Hellen, T., *How to Use Beam, Plate and Shell Elements*, NAFEMS, 2007

Zienkiewicz, O.C and Taylor, R.L., *The Finite Element Method*, Butterworth-Heinemann, 5th edition, 2000

Index